Shock Wave Science and Technology Reference Library

The new Springer collection, Shock Wave Science and Technology Reference Library, conceived in the style of the famous Handbuch der Physik, has as its principal motivation to assemble authoritative, state-of-the-art, archival reference articles by leading scientists and engineers in the field of shock wave research and its applications. A numbered and bounded collection, this reference library will consist of specifically commissioned volumes with internationally renowned experts as editors and contributing authors. Each volume consists of a small collection of extensive, topical and independent surveys and reviews. Typical articles start at an elementary level that is accessible to non-specialists and beginners. The main part of the articles deals with the most recent advances in the field with focus on experiment, instrumentation, theory, and modeling. Finally, prospects and opportunities for new developments are examined. Last but not least, the authors offer expert advice and cautions that are valuable for both the novice and the well-seasoned specialist.

Shock Wave Science and Technology Reference Library

Collection Editors

Hans Grönig

Hans Grönig is Professor emeritus at the Shock Wave Laboratory of RWTH Aachen University, Germany. He obtained his Dr. rer. nat. degree in Mechanical Engineering and then worked as postdoctoral fellow at GALCIT, Pasadena, for one year. For more than 50 years he has been engaged in many aspects of mainly experimental shock wave research including hypersonics, gaseous and dust detonations. For about 10 years he was Editor-in-Chief of the journal Shock Waves.

Yasuyuki Horie

Professor Yasuyuki (Yuki) Horie is internationally recognized for his contributions in high-pressure shock compression of solids and energetic materials modeling. He is a co-chief editor of the Springer series on Shock Wave and High Pressure Phenomena and the Shock Wave Science and Technology Reference Library, and a Liaison editor of the journal Shock Waves. He is a Fellow of the American Physical Society, and Secretary of the International Institute of Shock Wave Research. His current interests include fundamental understanding of (a) the impact sensitivity of energetic solids and its relation to microstructure attributes such as particle size distribution and interface morphology, and (b) heterogeneous and nonequilibrium effects in shock compression of solids at the mesoscale.

Kazuyoshi Takayama

Professor Kazuyoshi Takayama obtained his doctoral degree from Tohoku University in 1970 and was then appointed lecturer at the Institute of High Speed Mechanics, Tohoku University, promoted to associate professor in 1975 and to professor in 1986. He was appointed director of the Shock Wave Research Center at the Institute of High Speed Mechanics in 1988. The Institute of High Speed Mechanics was restructured as the Institute of Fluid Science in 1989. He retired in 2004 and became emeritus professor of Tohoku University. In 1990 he launched Shock Waves, an international journal, taking on the role of managing editor and in 2002 became editor-in-chief. He was elected president of the Japan Society for Aeronautical and Space Sciences for one year in 2000 and was chairman of the Japanese Society of Shock Wave Research in 2000. He was appointed president of the International Shock Wave Institute in 2005. His research interests range from fundamental shock wave studies to the interdisciplinary application of shock wave research.

M. E. H. van Dongen (Ed.)

Shock Wave Science and Technology Reference Library, Vol. 1

Multiphase Flows I

With 188 Figures, 6 in Color and 11 Tables

 Springer

Marinus E.H. van Dongen
Eindhoven University of Technology
Department of Applied Physics
Eindhoven, The Netherlands
Email: m.e.h.v.dongen@tue.nl

Marinus E.H. van Dongen

Professor Marinus (Rini) van Dongen is a physicist who has been active in research and education in fluid dynamics, physical gas dynamics, physical transport phenomena, waves in porous media, waves with phase transition, nucleation and condensation in real gases and in bio-fluid dynamics. He is a member of the J.M. Burgerscentrum, Research School for Fluid Mechanics. He has been affiliated with Eindhoven University of Technology and part-time with Twente University, Department of Mechanical Engineering.

Library of Congress Control Number: 2006928273

ISBN-10 3-540-35845-5 Springer Berlin Heidelberg New York
ISBN-13 978-3-540-35845-9 Springer Berlin Heidelberg New York

Springer is a part of Springer Science+Business Media.

springer.com

Springer Berlin Heidelberg 2007

Typesetting by the Authors and SPi using a Springer LATEX macro package
Cover design: eStudio Calamar Steinen

Printed on acid-free paper   SPIN: 11420712   54/3100/SPi – 5 4 3 2 1 0

Preface

Shock waves in multi-phase media refer to a rich variety of phenomena of interest to physicists, chemists, mechanical engineers, fluid dynamicists and aeronautical engineers. Today, new applications in biomedical engineering are being explored. For the description of shocks in multi-phase media, elements from thermodynamics and thermal physics, mechanics and fluid mechanics have to be combined with elements from applied mathematics. To study shocks in multi-phase media use of dedicated and sophisticated experimental facilities is required while these experimental studies provide new insights into the state of matter, ranging from micro-bubbles in distilled water to yield stresses and plastic deformation in porous materials.

This book contains chapters on shocks and expansion waves in complex liquids with bubble clouds and single bubbles, shock waves in superfluid helium, shock waves related to phase transition and shocks interacting with solid foams, textiles and porous and granular materials. Different authors will focus on biomedical applications. Some of these are widely used in medical practice, such as extracorporeal shock wave lithotripsy (ESWL), applied successfully to the non-invasive disintegration of kidney stones.

The book opens with a section on shock waves in bubbly liquids (van Wijngaarden). The reader will be directly confronted with the peculiar wave properties of such a bubbly liquid, its different wave speeds and its strong dispersion. It will be shown that shocks in bubbly liquids are formed as a result of a competition of non-linear steepening effects and of dispersion, so that such shocks are essentially different from single phase shocks. The structure of these shocks in bubbly liquids will be addressed in physical and mathematical detail.

Free bubbles, subjected to shock wave loading, start to collapse, thereby loosing their symmetry (Tomita). Liquid microjets are formed with speeds of hundreds of meters per second, colliding on the bubble wall and giving rise to high values of pressures and tensile stresses. When such a bubble collapses near a solid wall, the microjet may cause cracks or pits, depending on geometry, wave amplitude and wave form. Collapsing bubbles may also generate secondary shocks, interfering with other bubbles attached to a solid wall and thereby causing wall damage. A peculiar phenomenon, also associated with bubble collapse is sonoluminescence, the generation of a high power light pulse of extremely short duration.

Strong pulse-shaped shocks in water, reflecting from a free interface, will lead to strong rarefaction waves and possibly to strong cavitation effects (Kedrinskii). Such experiments have led to the characterisation of even pure water as a multi-phase medium consisting of a small volume fraction of micro-bubbles, with diameters ranging from the nanoscale to the microscale and with concentrations on the order of $10^5 - 10^6$ cm^{-3}. This has direct consequences for the dynamic strength of a liquid subjected to strong tensile stress waves. The initial stages of disintegration of liquids and of solids are discussed and compared. An extensive survey is given of recent developments in the application of shock waves in medical practice (Kedrinskii, Tomita), with a discussion of the physical principles involved.

Shocks in cryogenic liquids, in particular shocks in superfluid HeII, have peculiar properties (Murakami) that can be understood on the basis of a two-fluid model. In this model, HeII consists of a superfluid component with zero viscosity and a normal component with relative concentrations depending on temperature. The concepts of the two-fluid model and their consequences for wave propagation in general and for shock waves in particular are outlined. It is shown that two different types of shock waves exist: a normal compression shock (with peculiar properties) and an entropy shock or a thermal shock wave. Experimental facilities are described that enable one to observe such shocks and to study their characteristics.

Waves in wet vapours and in wet carrier gas–vapour mixtures are characterised by the interplay of different relaxation processes, associated with transfer of momentum, heat, and mass between the droplet cloud and gas/vapour (Guha). The latter process refers to phase transition, which has important energetic consequences due to the relatively large value of the latent heat. The relaxation processes strongly affect the structure of shock waves. Different types of partly and fully dispersed shocks are distinguished. Their properties and structure are described in detail and the overall jump conditions are specified.

When dry steam or a humid carrier gas is subjected to a fast isentropic expansion, it is possible that the gas/vapour is brought to a highly supersaturated state. This leads to strong production of the smallest stable droplets possible, a process called homogeneous nucleation. The droplets grow until a new phase equilibrium is attained. Again, it is the latent heat that leads to important consequences for the flow field (Delale, Schnerr and van Dongen). Such a situation is met in convergent–divergent nozzles, or Laval nozzles, in unsteady expansion waves and in steady sonic or supersonic flows around a sharp edge, the so-called Prandtl–Meyer corner flows. In steady flow, the amount of latent heat that can be "absorbed" by the flow is limited. This leads to thermal choking and to the formation of condensation-induced shocks and, in slender nozzles, to condensation-induced oscillations. Results of asymptotic analysis and of numerical modelling are compared with experimental observations.

Shock compression in a gas/vapour does not easily lead to the formation of a liquid. This is because shock compression is accompanied by a temperature increase (shocks in HeII form a possible exception), which brings the gas/vapour further away from vapour–liquid equilibrium. Still, if the specific heat is sufficiently large, the temperature increase may be significantly reduced, such that anomalous gas dynamic behaviour occurs with the liquefaction shock as the most spectacular example (Meier). The theoretical background is explained and interesting observations are described for liquefaction shocks, wave splitting and of vortical instabilities, with a discussion of open questions that still remain.

The last chapter starts with a description of shock waves interacting with perforated rigid sheets, with textile materials, with flexible foams, for different shock strengths and for different inclinations (Skews). From carefully designed experiments a clear physical picture is obtained of the nature and the importance of the different processes involved. Special attention is given to peak pressure amplification that is observed for porous slabs adjacent to a solid wall, subjected to shock wave compression.

When describing wave propagation in porous media, two different compression wave modes have to be distinguished (Smeulders and van Dongen). This is predicted by linear theory, which yields important information on dispersion and damping, impedances and amplitude ratios and the importance of boundary conditions. Different non-linear theoretical models will be discussed. Results are compared with shock wave reflection experiments with rigid and flexible porous materials.

While the first examples of shock interactions with porous materials deal with shock waves of moderate strengths, the last contribution to this book (Golub and Mirova) covers compaction, plastic deformation, and destruction of granules caused by the interaction of a porous material with very strong shocks (1–10 GPa). Attention is given to modelling of such interactions, thereby reducing as much as possible its full physical complexity.

When studying shock waves in multi-phase media, one meets phenomena on largely different length and time scales, but strongly coupled. Thermal effects, entangled quantised vortices, cavitation and collapse, periodic formation of shocks, wave splitting, anomalous thermodynamic behaviour, interfering relaxation processes and the constructive interference of small discrete spherical shocks to global shocks are encountered. This book necessarily only captures a limited part of these and related phenomena. Nevertheless, the contributions to this book will serve as an overview and as an introduction to the fascinating world of discontinuous transport phenomena.

M.E.H. van Dongen

Contents

Part III Shock Waves Interacting with Solid Foams, Textiles, Porous and Granular Media

List of Contributors

Can F. Delale
Faculty of Aeronautics and
Astronautics, Istanbul Technical
University, 34469 Maslak
Istanbul, Turkey
and
TÜBİTAK Feza Gürsey Institute
P.O. Box 6
81220 Çengelköy
Istanbul, Turkey
delale@itu.edu.tr

Marinus E.H. van Dongen
Department of Applied Physics
Eindhoven University of Technology
P.O. Box 513
5600 MB, Eindhoven
The Netherlands
m.e.h.v.Dongen@tue.nl

Victor Golub
Institute for High Energy Densities
Associated Institute for High
Temperatures, Russian Academy of
Science, Izhorskaja 13/19
Moscow 125412, Russia
golub@ihed.ras.ru

Abhijit Guha
Aerospace Engineering Department
University of Bristol
Bristol BS8 1TR, UK

Valery K. Kedrinskii
Lavrentyev Institute of
Hydrodynamics SB RAS
Lavrentyev prosrect 15
630090 Novosibirsk
Russian Federation
kedr@hydro.nsc.ru

Gerd E.A. Meier
Am Menzelberg 6
D-37077 Göttingen
Germany
geameier@web.de

Olga Mirova
Institute for High Energy
Densities, Associated
Institute for High Temperatures
Russian Academy of Science
Izhorskaja 13/19
Moscow 125412
Russia

Masahide Murakami
Graduate School of Systems
and Information Engineering
University of Tsukuba
Tsukuba, 305-8573, Japan
murakami@kz.tsukuba.ac.jp

Günter H. Schnerr
Lehrstuhl für Fluidmechanik –
Fachgebiet Gasdynamik
Technische Universität München
Boltzmannstrasse 15
D-85747, Garching
Germany
schnerr@flm.mw.tu-muenchen.de

Beric W. Skews
School of Mechanical,
Industrial and Aeronautical
Engineering, University of the
Witwatersrand
PO WITS, 2050
Johannesburg
South Africa
bskews@mech.wits.ac.za

David Smeulders
GeoTechnology Department
Delft University of Technology

P.O. Box 5028
2600GA Delft
The Netherlands
d.m.j.smeulders@tudelft.nl

Yukio Tomita
Faculty of Education, Hokkaido
University of Education
P.O. Box 040-8567
1-2 Hachiman-cho, Hakodate
Hokkaido, Japan
tomita@cc.hokkyodai.ac.jp

Leen van Wijngaarden
J.M. Burgers Centre for Fluid
Dynamics, Physics of Fluids Group
University of Twente Enschede
The Netherlands

Part I

Shock Waves in Complex Liquids

1

Shock Waves in Bubbly Liquids

Leen van Wijngaarden

1.1 Introduction

The structure of shock waves in bubbly liquids is governed by the behavior of the bubbles. Therefore this chapter starts with a survey of bubble dynamics. This includes the Rayleigh–Plesset equation for bubble oscillations, and the related Minnaert frequency for volume oscillations. A striking property of bubbly liquids is the low sound velocity even at small gas concentration. This is also discussed in Sect. 1.2, together with a survey of linear acoustic waves. Whereas in single phase media a shock wave is formed as a balance between the tendency of a compressive wave to nonlinear steepening on one hand and viscous dissipation on the other, the mechanism opposing steepening is in bubbly liquids dispersion rather than viscous dissipation. This leads as in other areas to the Korteweg–deVries (KdV) equation. Another mechanism of importance is relaxation. These various subjects are treated in Sects. 1.3 and 1.4 before in Sects. 1.4 and 1.5 shock waves are dealt with, strong shocks in Sects. 1.5 and 1.6 and moderately strong in Sect. 1.5/Sect. 1.6.

Since moderately strong waves obey the KdV equation there is an opportunity to put the so-called inverse scattering theory to a test. According to this an expansion wave evolves, as opposed to a compressive wave, in a finite number of solitons. In Sect. 1.6 experiments with an expansion wave are described and comparison with the corresponding theory is made.

1.2 Elements of Bubble Dynamics

Central in dealing with shock waves in bubbly flows is the response of a bubble to a pressure perturbation in the surrounding liquid. This is described by the Rayleigh–Plesset equation

$$\rho_l R \ddot{R} + \frac{3}{2}\rho_l \left(\dot{R}\right)^2 = p_g - p_\infty - \frac{4\mu_l}{R}\dot{R}. \qquad (1.1)$$

The radius of a bubble, supposed to be spherical, is indicated with R, the dot on R indicating time derivative, the liquid density with ρ_l, the liquid viscosity with μ_l, the gas pressure in the bubble with p_g and the pressure in the incompressible liquid far away from the bubble with p_∞. With cavitation bubbles vapor pressure and surface tension are of importance as well. For our subject they have minor importance. If we exclude diffusion of gas from outside into the bubble or vice versa, the mass of gas inside the bubble remains constant

$$R^3 \rho_g = \text{constant.} \tag{1.2}$$

To fully describe the bubble behavior, we need the energy equation both for the surrounding liquid as well as for the gas inside the bubble. Because of the huge difference in heat capacity between liquid and gas we can consider the liquid to be constant in temperature. The energy equation for a perfect gas is

$$\rho_g C_p \frac{D_g T_g}{D_g t} - \frac{D_g p_g}{D_g t} = K \nabla^2 T_g. \tag{1.3}$$

In this equation T_g is the temperature of the gas in the bubble, K and C_p the heat conduction and specific heat, respectively, whereas $D_g / D_g t$ is the material derivative.

The heat diffusion coefficient for the gas inside a bubble, χ, is defined by

$$\chi = \frac{K}{\rho_g C_p}. \tag{1.4}$$

In air χ is about $2.10^{-5} \, \text{m}^2 \, \text{s}^{-1}$. With a gas temperature variation of angular frequency ω the outside temperature penetrates to a depth of order $(\chi/\omega)^{1/2}$. This permits in a few limiting cases to avoid the need to use the energy equation as such. To identify such cases we compare $(\chi/\omega)^{1/2}$ with the bubble radius R and the wavelength l_g of the pressure wave in the gas in the bubble. When l_g exceeds R and the latter is small with respect to $(\chi/\omega)^{1/2}$, we may consider the gas as isothermal. When the first assumption still holds but R is large with respect to $(\chi/\omega)^{1/2}$, the gas content is adiabatic. This remains the case, at increasing frequency, when the wavelength gets smaller than R. Finally, however, because $(\chi/\omega)^{1/2}$ shrinks at increasing frequency as $(\omega)^{-1/2}$ and l_g as $(\omega)^{-1}$ pressure changes are isothermal again. A summary is provided in Table 1.1.

An important quantity is the natural frequency for volume oscillations of bubbles, the so-called Minnaert [1] frequency. With a polytropic index κ (in air κ is one for isothermal and 1.4 for adiabatic behavior), this is

$$f_b = \frac{\omega_b}{2\pi} = \frac{1}{2\pi} \left(\frac{3\kappa p_g}{\rho_l R^2} \right)^{1/2}. \tag{1.5}$$

This frequency can be deduced from (1.1) when small excursions from equilibrium are assumed. For air bubbles of a few mm radius $f_b R \sim 3 \, \text{KHz} \, \text{mm}$

Table 1.1. Thermodynamic behavior of oscillating gas bubbles

Frequency range	Lengths	Thermodynamic behavior
Low	$1 < \frac{\chi}{\omega R^2} < \frac{l_\mathrm{g}}{R}$	Isothermal
Moderately high	$\frac{\chi}{\omega R^2} < 1 < \frac{l_\mathrm{g}}{R}$	Adiabatic
High	$\frac{\chi}{\omega R^2} < \frac{l_\mathrm{g}}{R} < 1$	Adiabatic
Very high	$\frac{l_\mathrm{g}}{R} \ll \frac{\chi}{\omega R^2} \ll 1$	Isothermal

gives a good and quick estimate. For frequencies of this order of magnitude the wavelength l_g is in air several cm, much larger than the bubble radius. This allows us to consider the pressure inside the bubble as uniform, which as first observed by Nigmatulin and Kabeev [2], permits a simplification of (1.3). With a perfect gas with $\gamma = C_p/C_v$ we have as constitutive equation

$$\frac{\gamma}{\gamma - 1} p_\mathrm{g} = C_p \rho_\mathrm{g} T_\mathrm{g}. \tag{1.6}$$

The continuity equation for the gas, moving with velocity $\boldsymbol{v}$, inside the bubble is

$$\frac{D_\mathrm{g} \rho_\mathrm{g}}{D_\mathrm{g} t} + \rho_\mathrm{g} \nabla . \boldsymbol{v} = 0. \tag{1.7}$$

Using (1.6) and (1.7) and introducing the radial coordinate r inside the bubble we obtain from (1.3)

$$r^2 \frac{D_\mathrm{g} p_\mathrm{g}}{D_\mathrm{g} t} - \gamma p_\mathrm{g} \frac{\partial v}{\partial r} = (\gamma - 1) \frac{\partial}{\partial r} \left(r^2 \frac{\partial T}{\partial r} \right). \tag{1.8}$$

This is still general. Now we take the pressure uniform inside the bubble, $D_\mathrm{g} p_\mathrm{g}/D_\mathrm{g} t$ becomes $\partial p_\mathrm{g}/\partial t$. We can integrate the terms of (1.8) over r and obtain, with $(v)_{r=R} = \partial R/\partial t$,

$$\frac{\partial p_\mathrm{g}}{\partial t} = \frac{3}{R} \left[(\gamma - 1) K \left(\frac{\partial T}{\partial R} \right)_{r=R} - \gamma p_\mathrm{g} \frac{\partial R}{\partial t} \right]. \tag{1.9}$$

The relation (1.9) will turn out to be useful when we come to shock waves.

Now consider a mixture of liquid and bubbles, with concentration by volume α. At frequencies well below ω_b we can neglect the inertia of the fluid pushed away by an expanding bubble or rushing towards where a bubble collapses and take the pressure in the liquid and in the gas to be the same locally. We have a homogeneous fluid with density

$$\rho = \alpha \rho_\mathrm{g} + (1 - \alpha) \rho_\mathrm{l}. \tag{1.10}$$

If we further assume that there is no relative velocity between bubbles and liquid, we have the condition that the mass of gas in a unit mass of the mixture must remain constant

$$\frac{\alpha \rho_g}{\alpha \rho_g + (1 - \alpha) \rho_l} = \text{constant.} \tag{1.11}$$

With the help of equations of state for gas (1.6) and liquid we can calculate the speed of sound using (1.10) and the expression for the compressibility

$$\frac{1}{c^2} = \frac{\mathrm{d}\rho}{\mathrm{d}p} = (1 - \alpha) \frac{\mathrm{d}\rho_l}{\mathrm{d}p} + (\rho_g - \rho_l) \frac{\mathrm{d}\alpha}{\mathrm{d}p} + \alpha \frac{\mathrm{d}\rho_g}{\mathrm{d}p}. \tag{1.12}$$

In the case of adiabatic behavior of the gas this gives

$$\frac{1}{c_{\mathrm{ad}}^2} = \frac{\mathrm{d}\rho}{\mathrm{d}p} = \frac{(1 - \alpha)^2}{c_l^2} + \frac{\alpha^2}{c_g^2} + \frac{\rho_l \alpha (1 - \alpha)}{\gamma p}. \tag{1.13}$$

In Fig. 1.1 are drawn curves of c_{ad} as described by (1.13) for different values of the ambient pressure.

A striking property of bubbly suspensions becomes immediately clear, the extremely low velocity of sound, even at a small concentration by volume of 10^{-2} of only $100\,\mathrm{m\,s^{-1}}$ at atmospheric conditions. This is small with respect to both the velocity of sound in air and in water. At a concentration of 50% it sinks to only $20\,\mathrm{m\,s^{-1}}$. For use in our discussion of shock waves we shall often simplify (1.13) to

$$c_{\mathrm{ad}}^2 = \frac{\gamma p_g}{\rho_l \alpha (1 - \alpha)} \tag{1.14}$$

which approximates (1.13) very well when α is not close to either zero or unity. For isothermal behavior we have similarly

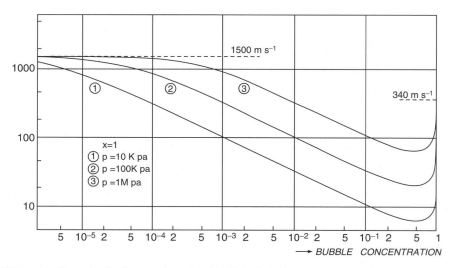

Fig. 1.1. Sound velocity as given by (1.13) in bubbly mixture as a function of gas concentration by volume. The three curves are for different ambient pressures

$$(c_{\text{iso}})^2 = \frac{p_{\text{g}}}{\rho_1 \alpha (1 - \alpha)}. \tag{1.14a}$$

When the frequency to which bubbles are subjected in the suspension is no longer far below the resonance frequency (1.5) the pressures in liquid and in gas are no longer the same. For small amplitude acoustic waves the presence of the bubbles was accounted for by considering these as point scatterers by Foldy [3] who applied multiple scattering theory to obtain the dispersion equation for linear acoustic waves. van Wijngaarden [4] proposed for waves of arbitrary amplitude the following model:

Similar to (1.7) we can for the bubbly mixture as a whole write the conservation of mass as

$$\frac{D\rho}{Dt} + \rho \nabla.\boldsymbol{u} = 0, \tag{1.15}$$

and the momentum equation

$$\frac{D\rho\boldsymbol{u}}{Dt} + \nabla.\rho\boldsymbol{u}\boldsymbol{u} + \nabla p = 0. \tag{1.16}$$

Just as ρ is the local averaged density so is $\boldsymbol{u}$ the local averaged velocity and p the local averaged pressure. In (1.15) and (1.16) the material derivative is

$$\frac{D}{Dt} = \frac{\partial}{\partial t} + \boldsymbol{u}.\nabla. \tag{1.17}$$

The idea in van Wijngaarden [4] now is to assume the Rayleigh–Plesset equation (1.1) to connect this averaged pressure p with the local gas pressure in the bubbles giving

$$\rho_1 R \frac{D^2 R}{Dt^2} + \frac{3}{2}\rho_1 \left(\frac{DR}{Dt}\right)^2 = p_{\text{g}} - p - \frac{4\mu_1}{R}\frac{DR}{Dt}. \tag{1.18}$$

The local volume concentration α is connected with the local bubble radius R and the number density of bubbles n through

$$\alpha = \frac{4}{3}\pi n R^3. \tag{1.19}$$

For the shock waves to be discussed in the following, we shall ignore breaking up of bubbles. Then the number density obeys

$$\frac{\partial n}{\partial t} + \nabla.n\boldsymbol{v} = 0. \tag{1.20}$$

The equations (1.15)–(1.19), especially (1.18) were formulated by van Wijngaarden [4] without specifying the range of validity. Caflisch et al. [5] have shown that these equations are correct for small α. They also showed that for a wave entering a quiescent medium, the convective velocities could be neglected.

We consider small amplitude acoustic waves. By linearising (1.1)–(1.3), and (1.16)–(1.20), and assuming a monochromatic wave of the form $\exp(ikx - \omega t)$ one obtains as dispersion equation

$$\frac{k^2}{\omega^2} = \frac{1}{c^2} \frac{\omega_b^2}{\{(\omega_b^2 - \omega^2) - i\delta\omega\omega_b\}}. \tag{1.21}$$

For details see Prosperetti [6]. For $\omega \ll \omega_b$, ω^2/k^2 approaches the right hand side of (1.14) or (1.14a) depending on the changes in the bubble being adiabatic or isothermal. For larger values of ω (1.21) must be used with for c and ω_b the expressions in (1.14), or (1.14a), and (1.5), respectively. It is interesting to note from (1.5) and (1.14) that both in the adiabatic case and in the isothermal case the quantity ω_b^2/c^2 is the same, approximately $4\pi nR$, and further that for both cases

$$c^2\alpha^2 = \text{constant.} \tag{1.22}$$

The coefficient δ in (1.21) represents the wave attenuation. It consists of a part due to viscosity, coming from the last term on the right hand side of (1.18), a part due to thermal conduction, and a part due to acoustic radiation, arising when in the Rayleigh–Plesset equation (1.1) the liquid compressibility is taken into account also, (Prosperetti [6]). For a general time dependent behavior the thermal attenuation cannot be represented by a coefficient as in (1.21). The full energy equation, taking however the simplification (1.9) into account, must be applied. The relation (1.21) says that linear waves are dispersive. The wave equation, ignoring dissipation, is, van Wijngaarden [4],

$$\frac{\partial^2 p}{\partial t^2} - c_0 \frac{\partial^2 p}{\partial x^2} - \frac{R_0^2}{3\alpha_0 (1 - \alpha_0)} \frac{\partial^4 p}{\partial t^2 \partial x^2} = 0, \tag{1.23}$$

in which the subscript 0 refers to the equilibrium situation. Equation (1.23) holds for waves in both x directions. Often we want to look at waves traveling in one direction only, certainly when we have shock waves in mind. An equation valid for waves traveling to the right can be derived in the following way, provided the dispersion is weak. Let us first neglect dispersion. Then the operator in (1.23) can be split up

$$\left(\frac{\partial}{\partial t} - c_0 \frac{\partial}{\partial x} \right) \left(\frac{\partial}{\partial t} + c_0 \frac{\partial}{\partial x} \right) p = 0. \tag{1.24}$$

For a nondispersive wave going to the right $(\partial/\partial t + c_0\partial/\partial x)p = 0$. For a right going wave with dispersion this quantity is small, of order ε, say, when the dispersion is small. Therefore we may write the first factor in (1.24) as it occurs in the equation with dispersion (1.23) as $-2c_0\partial/\partial x + O(\varepsilon)$. Then, correct to $O(\varepsilon)$, (1.23) becomes

$$-2c_0 \frac{\partial}{\partial x} \left(\frac{\partial}{\partial t} + c_0 \frac{\partial}{\partial x} \right) p - \frac{R_0^2 c_0^2}{3\alpha_0 (1 - \alpha_0)} \frac{\partial^3 p}{\partial x^3} = 0.$$

Integration with respect to x finally gives

$$\frac{\partial p}{\partial t} + c_0 \frac{\partial p}{\partial x} + \frac{R_0^2 c_0}{6\alpha_0 (1 - \alpha_0)} \frac{\partial^2 p}{\partial x^2} = 0. \tag{1.25}$$

This has the same form as the linearized KdV equation for water waves (see e.g., Whitham [7]), a similarity which will prove to be useful in the following.

1.3 Nonlinear Compressive Waves

Since shock waves arise in general from steepening of compressive waves, we start by looking at such a wave. For this purpose we ignore temporarily the effects of bubbles other than their contribution to compressibility. In an x, t plane a right going characteristic has the direction

$$\frac{\mathrm{d}x}{\mathrm{d}t} = u + c = c_0 + u + (c - c_0). \tag{1.26}$$

Along the characteristic the sum of the velocity u and σ is constant where

$$\sigma = \int_{\alpha_0}^{\alpha} \frac{c\mathrm{d}\rho}{\rho}. \tag{1.27}$$

In a general motion (1.26) represents a characteristic (see any textbook on gasdynamics e.g., Liepmann and Roshko [8]); the special feature of a simple wave is that all right going characteristics are straight. The wave runs into an undisturbed medium and then $u = \sigma$, because along the left characteristics $u - \sigma = 0$. It follows from this and (1.26) that in a compressive wave a more compressed part travels faster than a less compressed part and that as a result the wave profile becomes steeper and steeper until at some point the slope is infinite. Making use of (1.10) and (1.14a) it follows that, approximating α $(1 - \alpha)$ by α, for the isothermal case

$$\sigma = c_{\mathrm{iso},0}\alpha_0 \ln \frac{\alpha_0}{\alpha} \quad \text{whereas} \quad c_{\mathrm{iso}} - c_{\mathrm{iso},0} = c_{\mathrm{iso},0} \left(\frac{\alpha_0 - \alpha}{\alpha} \right).$$

Hence a certain value of α travels in a simple wave with a velocity

$$\frac{\mathrm{d}x}{\mathrm{d}t} = c_{\mathrm{iso},0} + c_{\mathrm{iso},0} \left(\frac{\alpha_0 - \alpha}{\alpha} + \alpha_0 \ln \frac{\alpha_0}{\alpha} \right). \tag{1.28}$$

For other values of the polytropic index similar relations can be derived. In particular we have for the adiabatic case

$$\frac{\mathrm{d}x}{\mathrm{d}t} = c_{\mathrm{ad},0} + c_{\mathrm{ad},0} \frac{\alpha_0 - \alpha}{\alpha_0} \left(\frac{\gamma + 1}{2} + \alpha_0 \right). \tag{1.29}$$

Both in (1.28) and in (1.29) the second term in brackets is small in dilute suspensions. This term corresponds to the excess velocity u in (1.26) and can hence be neglected when the gas concentration is small.

At some point a shock wave forms in a single phase gas. This happens also in a bubbly fluid. However the structure is quite different. The Hugoniot relations are the same. Before the discussion of the structure we give them here. Consider a steady shock traveling with velocity U into an undisturbed bubbly suspension with pressure p_0 and void fraction α_0. The corresponding quantities at the other side are p_1 and α_1. We will restrict to low values of the void fraction and therefore the density may be taken as $\rho_l(1 - \alpha)$, see (1.10). Let the shock travel from right to left along the x-axis. Then in a frame moving with velocity U to the left the flow is steady. Conservation of mass requires

$$\rho_l(1 - \alpha_0)U = \rho_l(1 - \alpha_1)(U + u_1). \tag{1.30}$$

Next the conservation of momentum must be formulated. Although within the shock all kinds of out-of-equilibrium conditions prevail, equilibrium is reached far upstream and down stream from the shock, resulting in

$$p_0 + \rho_l(1 - \alpha_0)U^2 = p_1 + \rho_l(1 - \alpha_1)(U + u_1)^2, \text{ or using } (1.30)$$
$$p_1 = p_0 - \rho_l(1 - \alpha_0)Uu_1. \tag{1.31}$$

If behind the shock also the temperature is in equilibrium with the surrounding liquid we have from (1.11)

$$\frac{p_0\alpha_0}{1 - \alpha_0} = \frac{p_1\alpha_1}{1 - \alpha_1}. \tag{1.32}$$

In the opposite case in which just behind the shock no heat exchange between bubbles and liquid is supposed to have taken place, this must be replaced by, see (1.11),

$$\frac{p_0^{1/\gamma}\alpha_0}{1 - \alpha_0} = \frac{p_1^{1/\gamma}\alpha_1}{1 - \alpha_1}. \tag{1.33}$$

From (1.30)–(1.33) we obtain the following important expressions for the shock propagation velocity U for isothermal and adiabatic changes in the gas, respectively,

$$\frac{U^2}{c_{iso,0}^2} = \frac{p_1}{p_0}, \tag{1.34}$$

$$\frac{U^2}{c_{iso,0}^2} = \frac{p_1/p_0 - 1}{1 - (p_0/p_1)^{1/\gamma}}. \tag{1.35}$$

The above Hugoniot relations and the shock speed relation (1.34) were first given by the pioneers Campbell and Pitcher [9] in this subject. They found reasonable agreement with experiments but did not investigate the inner structure of the shocks.

1.4 Mechanisms Opposing Steepening of Compressive Waves

1.4.1 Viscous Stresses

Apart, of course, from shear stresses viscosity gives the bubbly liquid a bulk viscosity, first noted by Taylor [10] and most easily demonstrated by leaving out in (1.1) the inertia terms

$$p_{\mathrm{g}} - p = \frac{4\mu_1}{R}\dot{R}. \tag{1.36}$$

With help of (1.10), (1.15), and (1.19) and neglecting changes in the number density n (they are of order α as follows from 1.20) this can be written as

$$p_{\mathrm{g}} - p = \frac{4\mu_1}{3\alpha}\nabla.\boldsymbol{u}. \tag{1.37}$$

In gases or fluids in nonequilibrium a difference between the mechanically defined pressure, through the trace of the stress tensor, and the thermodynamic pressure, the one in the constitutive relation, is expressed as the right hand side of (1.37) and the quantity preceding the divergence is called bulk viscosity.

Additional dissipation stems from the relative motion between bubbles and liquid. To see this we introduce the equation of motion of a single bubble in the suspension. Until now we did not need this since we assumed that bubbles move with the liquid. However, when the mixture as a whole is accelerated and the same pressure gradient acts on the continuous phase and on the bubbles, the bubbles acquire a velocity $\boldsymbol{v}$ different from the liquid. For one dimensional motion this relative motion is governed by

$$\frac{\mathrm{d}}{\mathrm{d}t}m\left(v-u\right) + f\left(v-u\right) = \rho_1 W\frac{\mathrm{D}u}{\mathrm{D}t}. \tag{1.38}$$

In this equation W is the volume of a bubble and m the added mass, van Wijngaarden [11]

$$m = 1/2\rho_1 W(1+2.78\alpha). \tag{1.39}$$

The first term on the left hand side of (1.38) is the rate of change of the bubble impulse, the second is the viscous friction. For a clean liquid, devoid of surfactants

$$f = 12\pi\mu_1 R. \tag{1.40}$$

The work done by this frictional force adds to the dissipation but this, see van Wijngaarden [12], is not an important effect.

1.4.2 Dispersion

We have seen that bubble oscillations make linear waves dispersive, see (1.21) and (1.25). This mechanism may ease the steepening of a compressive wave.

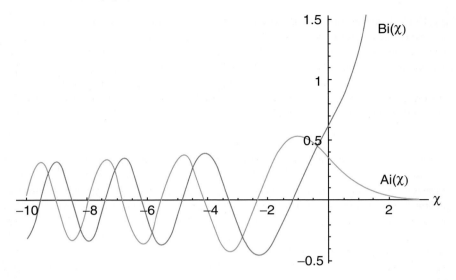

Fig. 1.2. How dispersion spreads, Ai(x), an original pressure step. The curve Ai(x) shown is a solution of (1.25)

Consider a linear wave obeying (1.25). Let at time $t = 0$ the wave profile be a step of strength $\Delta p/p_0$. Then (see e.g., Whitham [7], Sect. 13.6) the pressure evolves according to

$$\frac{p - p_0}{p_0} = \frac{\Delta p}{p_0} \int_z^\infty \mathrm{Ai}(s)\mathrm{d}s, \quad \text{where } z = \frac{x - c_0 t}{\left\{ \dfrac{R_0^2 c_0 t}{6\alpha_0(1-\alpha_0)} \right\}^{1/3}}. \tag{1.41}$$

Here Ai is the Airy function as drawn in Fig. 1.2, which shows the broadening of the pressure profile.

1.4.3 Relaxation

In the theory of shock waves in a single phase gas a well known relaxation process has to do with the nontranslational degrees of freedom of the molecules. These require much more collision times compared with the translational degrees of freedom in order to be in thermal equilibrium with the surroundings. An excellent survey on the implications of this for the propagation of shock waves is given by Lighthill [13]. From these we mention some which are of interest for our purpose. If the rotational and vibrational degrees of freedom do not interact at all with the surroundings, if they are "frozen," the speed of sound has the highest value, the frozen speed c_f, say. Each time a particular degree of freedom starts to take part in the process , a corresponding relaxation time has to elapse for that, the speed of sound decreases. At zero frequency all the internal degrees of freedom are in equilibrium and the speed of sound has the lowest value which we indicate with c_e. Consider now waves traveling at a speed near, but not quite at, c_e. It can be shown that the influence of the nonequilibrium degrees of freedom is expressed as a diffusion with effective diffusion coefficient

$$\tau(c_f^2 - c_e^2), \tag{1.42}$$

when there is only one internal degree of freedom with relaxation time τ. Of course, these molecular degrees of freedom do not play a role in a bubbly suspension but we can think of the bubbles as molecules. One degree of freedom is the relative velocity of a bubble with respect to the liquid and the other the heat exchange with the surrounding liquid. We shall first, by analogy with the molecular example, inspect what the relative motion, described in Sect. 1.4.1 of this chapter, brings in this connection. Equations (1.38–1.40) show that there exists a relaxation time τ_v to accommodate the velocity of a bubble to that of the liquid

$$\tau_v = \frac{R^2}{18\nu_1}. \tag{1.43}$$

With a bubble of radius $R = 1\,\text{mm}$ and $\nu_1 = \mu_1/\rho_1 = 10^{-6}\,\text{m}^2\,\text{s}^{-1}$, the value for water, $\tau_v = 0.06\,\text{s}$. This means that for motions with a time scale smaller than $0.06\,\text{s}$, or frequencies above about 16 Hz bubbles feel no frictional force and move according to (1.38) but with $f = 0$. In this limit the mixture stiffens because a pressure rise is only partly used to compress bubbles. For the rest bubbles are just accelerated out of an elementary volume. The resulting speed of sound is then, van Wijngaarden [12], Noordzij and van Wijngaarden [14] for adiabatic and isothermal motion, respectively,

$$c_{*ad}^2 = \frac{\gamma p\,(1 + 2\alpha)}{\rho_1\alpha\,(1 - \alpha)}, \qquad c_{*\,iso}^2 = \frac{p\,(1 + 2\alpha)}{\rho_1\alpha\,(1 - \alpha)}. \tag{1.44}$$

Comparing these values with those in (1.14) and (1.14a) shows that for both kinds of thermodynamic behavior the sound velocity is higher compared with the case in which bubbles move with the liquid, which corresponds with the equilibrium speed of sound c_e. The relations in (1.44) are just as those in (1.14) and (1.14a) valid far below the natural bubble frequency. Otherwise they have to be replaced by expressions as in (1.21) in which c and ω_b have the proper value. Noordzij and van Wijngaarden [14] show that, when other effects as heat conduction and dispersion are left out, isothermal pressure waves obey

$$\frac{\partial}{\partial t}\left(c_{*iso}^2\frac{\partial^2 p}{\partial x^2} - \frac{\partial^2 p}{\partial t^2}\right) + \tau_v^{-1}\left(c_{iso}^2\frac{\partial^2 p}{\partial x^2} - \frac{\partial^2 p}{\partial t^2}\right) = 0. \tag{1.45}$$

We now deal with (1.45) in the same way as we did with (1.23) to obtain (1.25). First we choose a wave traveling from left to right at fairly low frequency and take

$$\frac{\partial}{\partial t} + c_{\text{iso}} \frac{\partial}{\partial x} \sim 0.$$

Next we write (1.45) as

$$-c_{\text{iso}} \frac{\partial}{\partial x} \left(c_{*\text{iso}}^2 \frac{\partial^2 p}{\partial x^2} - c_{\text{iso}}^2 \frac{\partial^2 p}{\partial x^2} \right) + 2\tau_v^{-1} c_{\text{iso}} \frac{\partial}{\partial x} \left(c_{\text{iso}} \frac{\partial p}{\partial x} + \frac{\partial p}{\partial t} \right) = 0.$$

Integrating out the derivative with respect to x finally gives

$$\frac{\partial p}{\partial t} + c_{\text{iso}} \frac{\partial p}{dx} - \frac{1}{2} \tau_v \left(c_{*\text{iso}}^2 - c_{\text{iso}}^2 \right) \frac{\partial^2 p}{\partial x^2} = 0. \tag{1.46}$$

This is exactly as with the molecular relaxation discussed above. We see that the higher order terms, the first two terms on the left hand side of (1.45) act on the wave traveling near the low frequency equilibrium speed as diffusion with coefficient, compare (1.42),

$$\frac{1}{2} \tau_v \left(c_{*\text{iso}}^2 - c_{\text{iso}}^2 \right). \tag{1.47}$$

In a linear wave a signal eventually dies out by this diffusion but in a nonlinear wave it has, as we shall see, a quite different effect.

Next we discuss heat exchange between bubbles and liquid. Here we have met already the extreme sound velocities, viz. c_{ad} and c_{iso} as defined in (1.14) and (1.14a). However the relaxation process involved in the increasing accommodation of the bubble temperature to the liquid temperature is not so easy to describe as with the relative motion. This is due to the complicated interplay between the compression and expansion of the bubbles on one hand and the energy budget on the other. The latter is described by (1.3) and the boundary relation (1.9). It is hard for example to define a characteristic relaxation time associated with the heat exchange. If such a time could be found, let us call it $\tilde{\tau}$, the diffusion coefficient similar to (1.47) would be

$$\frac{1}{2} \tilde{\tau} \left(c_{\text{ad}}^2 - c_{\text{iso}}^2 \right). \tag{1.48}$$

1.5 Strong Shock Waves

We will now try to describe shock waves in bubbly suspensions, using the concepts and results of previous sections. Pioneers were Campbell and Pitcher [9] who carried out experiments and gave the expression (1.34) for the propagation velocity of the shock waves. They also compared their results with this relation. They carried out the experiments in a vertical shock tube partly filled with an aqueous solution of glycerol in which small air bubbles are suspended. The glycerol is added to keep bubbles small, in their case with radius of the order of 10^{-4} m. Above this bubbly mixture there is an air filled space closed with a seal. This space is evacuated and subsequently the seal is punctured admitting atmospheric air to enter the tube. The shock wave in air thus created hits the surface of the mixture and now a shock is formed in this mixture. The pressure is recorded at several places along the wall of the tube with help of transducers. All the subsequent investigators used a similar device. In Fig. 1.3 a picture is given of the device used by Noordzij and van Wijngaarden [14].

Campbell and Pitcher [9] experimented with pressure ratios of moderate and large strength. Although the theory regarding processes summarized in Sect. 1.4.3 apply to weak disturbances, the expressions for the speed of the shocks, (1.34) and (1.35) are valid for strong shocks as well. Campbell and Pitcher [9] found good agreement with the isothermal relation (1.34). They did not report data on the shock structure, or the shock thickness. Let us inspect what mechanism would in this case resist the steepening. As a first candidate we consider the bulk viscosity expressed in (1.37). We consider the shock as steady. Just as (1.31) holds for the downstream condition indicated with the subscript 1 it holds for any arbitrary location in the shock

$$p = p_0 - \rho_1(1 - \alpha_0)Uu. \tag{1.49}$$

The corresponding relation for mass conservation, derived from (1.30), is

$$u = \frac{U(\alpha - \alpha_0)}{1 - \alpha}. \tag{1.50}$$

With help of (1.50) and assuming isothermal behavior (the finding of Campbell and Pitcher [9]) we obtain from (1.37) and (1.49), van Wijngaarden [15],

$$(\alpha_0 - \alpha)(\alpha - \alpha_1) = -\frac{4}{3}\frac{\nu_1}{U}\frac{d\alpha}{dx}. \tag{1.51}$$

Strictly speaking we could not apply this to the experiments of Campbell and Pitcher [9] because their shocks are strong but for an order of magnitude analysis it suffices. Solution of (1.51) gives

$$\alpha = \frac{\alpha_0 + \alpha_1}{2} - \frac{\alpha_0 - \alpha_1}{2}\tanh\frac{3U\left(1 - \frac{p_0}{p_1}\right)x}{8\nu_1}. \tag{1.52}$$

This has the same form as the famous Taylor shock, see Lighthill [13], in single phase gases. It follows from (1.52) that the thickness of the shock is of the order of

$$d_v \sim \frac{\nu_1}{U\alpha_0(1 - p_0/p_1)}. \tag{1.53}$$

A typical set of experimental values of Campbell and Pitcher [9] is : $R = 10^{-4}$ m, $\nu_1 = 1.2 \times 10^{-5}$ m^2 s^{-1}, $U = 41$ m s^{-1}, $\alpha_0 = 5 \times 10^{-2}$, $p_0/p_1 = 1/6$. Introducing this into (1.53) gives for the right hand side about 10^{-5}m, which is smaller than the bubble radius. Just as in a gas the shock thickness

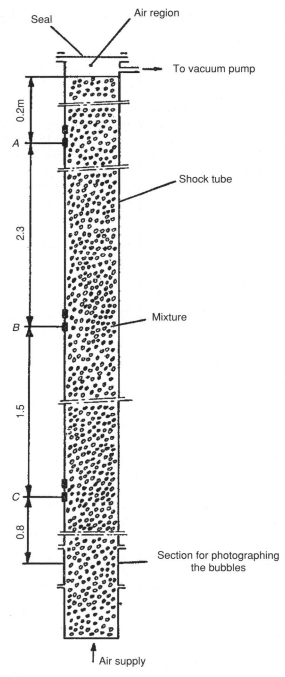

Fig. 1.3. Apparatus for measurements on shock waves, from Noordzij and van Wijngaarden [14]. The (*filled square*) indicate the location of pressure transducers

exceeds the mean free path, here the shock should at least cover a few bubble diameters. Since bubbles of radius 10^{-4} m were used this rules out viscosity as the dominant quantity determining the thickness of the shock. We have to look at the mechanisms of dispersion and relaxation. We know how to describe these only for waves of small and moderate strength and hence we will in the following restrict to that type of shocks. Before doing that we make some further remarks on strong shocks. Although no theory exists for strong shocks, except of course the Hugoniot relations, qualitatively they show the same structure as the weaker ones, as shown e.g. in the experiments of Tan and Bankoff [16]. They too used an aqueous solution of glycerol in water and small bubbles, containing argon or nitrogen. So, as long as bubbles remain intact and do not get broken up, much what is said in the following on weak shocks, applies qualitatively to strong shocks as well.

1.6 Shock Waves of Moderate and Weak Strength

We saw that due to convection and compression a compressive wave steepens, (1.28) and (1.29), and that viscous dissipation alone is not the opposing mechanism. For weak perturbations, in particular for weak shocks we may, see Lighthill [13], add the terms describing these processes in the evolution equations. For example, taking α small and assuming isothermal behavior, summation of the nonlinear growth following from (1.28) and the effects of dispersion and dissipation displayed in (1.25) and (1.37), respectively, leads to the evolution equation for $\tilde{p} = (p_{\mathrm{g}} - p_0)/p_0$, for a wave traveling to the right,

$$\frac{\partial \tilde{p}}{\partial t} + c_{\mathrm{iso},0}\frac{\partial \tilde{p}}{\partial x} + c_{\mathrm{iso},0}\tilde{p}\frac{\partial \tilde{p}}{\partial x} + \frac{R_0^2 c_{\mathrm{iso},0}}{6\alpha_0}\frac{\partial^3 \tilde{p}}{\partial x^3} - \frac{2}{3}\frac{\nu_1}{\alpha_0}\frac{\partial^2 \tilde{p}}{\partial x^2} = 0. \qquad (1.54)$$

The first two terms on the left hand side represent propagation with the sound velocity $c_{\mathrm{iso},0}$ in the undisturbed medium in which the running coordinate is x. The third term represents nonlinear steepening in isothermal conditions. The two other terms represent dispersion and dissipation (by bulk viscosity) respectively. This is the so-called Burgers–KdV equation, known from gravity wave theory and other fields, see e.g., Whitham [7]. In formulating the Hugoniot relations we considered a shock going to the left. To treat that case we first formulate the counterpart of (1.54) for waves traveling to the left. In the wave operator in (1.24) we choose now, in x^*, t coordinates $\frac{\partial}{\partial t} - c_{\mathrm{iso},0}\frac{\partial}{\partial x^*} \sim 0$, giving instead of (1.54)

$$\frac{\partial \tilde{p}}{\partial t} - c_{\mathrm{iso},0}\frac{\partial \tilde{p}}{\partial x^*} - c_{\mathrm{iso},0}\tilde{p}\frac{\partial \tilde{p}}{\partial x^*} - \frac{R_0^2 c_{\mathrm{iso},0}}{6\alpha_0}\frac{\partial^3 \tilde{p}}{\partial x^{*3}} - \frac{2}{3}\frac{\nu_1}{\alpha_0}\frac{\partial^2 \tilde{p}}{\partial x^{*2}} = 0.$$

Next we look for steady solutions moving to the left with velocity U. Introducing $\tilde{p} = \tilde{p}(x^* + Ut) = \tilde{p}(x)$ whence $\partial/\partial t = U\mathrm{d}/\mathrm{d}x$ and integrating with respect to x, we obtain

$$(c_{\text{iso},0} - U)\tilde{p} + \frac{1}{2}c_{\text{iso},0}\tilde{p}^2 + \frac{1}{6}\frac{R_0^2 c_{\text{iso},0}}{\alpha_0}\frac{\mathrm{d}^2\tilde{p}}{\mathrm{d}x^2} + \frac{2}{3}\frac{\nu_1}{\alpha_0}\frac{\mathrm{d}\tilde{p}}{\mathrm{d}x} = 0. \qquad (1.55)$$

We now use the assumption that the shock is weak to write

$$c_{\text{iso},0} - U = -\frac{1}{2}c_{\text{iso},0}\left(\frac{U^2 - c_{\text{iso},0}^2}{c_{\text{iso},0}^2}\right) = -\frac{1}{2}c_{\text{iso},0}\tilde{p}_1, \qquad (1.56)$$

where (1.34) has been used. Introducing this into (1.55) gives

$$\tilde{p}\left(\tilde{p} - \tilde{p}_1\right) + \frac{1}{3}\frac{R_0^2}{\alpha_0}\frac{\mathrm{d}^2\tilde{p}}{\mathrm{d}x^2} + \frac{4}{3}\frac{\nu_1}{U\alpha_0}\frac{\mathrm{d}\tilde{p}}{\mathrm{d}x} = 0. \qquad (1.57)$$

In the last term U appears instead of $c_{\text{iso},0}$, which is correct in the applied approximation, in order to preserve the similarity with (1.51).

It is instructive to derive this equation also from the Hugoniot relations in combination with the extended Rayleigh–Plesset equation (1.18). We start by writing the conservation of momentum (1.49) with help of (1.50) as

$$p + \rho_1 U^2(\alpha - \alpha_0) = p_0. \qquad (1.58)$$

Next we relate the pressure with the gas pressure through (1.18). In Sect. 1.5 we only took the viscous term into account which led to (1.51). Now we take also the first inertia term neglecting the second order term, with $(\mathrm{D}R/\mathrm{D}t)^2$. Hence, from (1.55) and (1.18), writing $U\mathrm{d}R/\mathrm{d}x$ for $\mathrm{D}R/\mathrm{D}t$, we have

$$p_0 = p_{\text{g}} - \rho_1 R_0 U^2 \frac{\mathrm{d}^2 R}{\mathrm{d}x^2} - \frac{4\mu_1 U}{R_0}\frac{\mathrm{d}R}{\mathrm{d}x} + \rho_1 U^2(\alpha - \alpha_0). \qquad (1.59)$$

From (1.19) it follows that in the present approximation $\frac{\mathrm{d}R}{\mathrm{d}x} = \frac{R_0}{3\alpha_0}\frac{\mathrm{d}\alpha}{\mathrm{d}x}$. For isothermal behavior $\tilde{p}$ can be expressed in terms of α by

$$\tilde{p} = (p_{\text{g}} - p_0)/p_0 = \frac{\alpha - \alpha_0}{\alpha}. \qquad (1.60)$$

Further we have, from (1.34), $U^2 = p_0/\rho_1\alpha_1$. Using these expressions (1.59) leads to

$$(\alpha - \alpha_0)(\alpha - \alpha_1) - \frac{R_0^2}{3}\frac{\mathrm{d}^2\alpha}{\mathrm{d}x^2} - \frac{\nu_1}{3U}\frac{\mathrm{d}\alpha}{\mathrm{d}x} = 0. \qquad (1.61)$$

This equation can with help of (1.60) be cast in terms of $\tilde{p}$, resulting again in (1.57).

If adiabatic conditions prevail, the equations are slightly different. We recall that in the absence of dissipation and dispersion a finite disturbance travels with the excess velocity $(c - c_0) + \sigma$, see Sect. 1.3. With adiabatic behavior

$$c - c_{\text{ad},0} = c_{\text{ad},0}\tilde{p}\frac{\gamma+1}{2\gamma}, \quad \text{and} \quad \sigma = \alpha_0 c_{\text{ad},0}\tilde{p}.$$

Just as in the isothermal case the contribution by σ can be neglected for small α. Then the counterpart of (1.54) reads

$$\frac{\partial \tilde{p}}{\partial t} + c_{ad,0}\frac{\partial \tilde{p}}{\partial x} + \frac{\gamma+1}{2\gamma}c_{ad,0}\tilde{p}\frac{\partial \tilde{p}}{\partial x} + \frac{R_0^2 c_{ad,0}}{6\alpha_0}\frac{\partial^3 \tilde{p}}{\partial x^3} - \frac{2}{3}\frac{\nu_1}{\alpha_0}\frac{\partial^2 \tilde{p}}{\partial x^2} = 0. \qquad (1.62)$$

Instead of (1.56) we have in the adiabatic case, using (1.35)

$$c_{ad,0} - U = -c_{ad,0}\tilde{p}_1\frac{1+\gamma}{4\gamma}. \qquad (1.63)$$

Using (1.63) we find in the same way that led from (1.54) to (1.57), from (1.62)

$$\frac{\gamma+1}{2\gamma}\tilde{p}(\tilde{p}-\tilde{p}_1) + \frac{R_0^2}{3\alpha_0}\frac{d^2\tilde{p}}{dx^2} + \frac{4}{3}\frac{\nu_1}{U\alpha_0}\frac{d\tilde{p}}{dx} = 0. \qquad (1.64)$$

The nonlinear equations (1.57) and (1.64) cannot be solved analytically but a lot can be learned by investigating the front and the back. On the low pressure side $\tilde{p}$ is close to zero and we may replace the quantity in brackets in (1.57) by $-\tilde{p}_1$ reducing (1.57) to the linear equation

$$\frac{R_0^2}{3\alpha_0}\frac{d^2\tilde{p}}{dx^2} + \frac{4\nu_1}{3\alpha_0 U}\frac{d\tilde{p}}{dx} - \tilde{p}\tilde{p}_1 = 0. \qquad (1.65)$$

The solution that vanishes for x→−∞ is

$$\tilde{p} \sim \exp\left[-\frac{2\nu_1}{UR_0^2}x + \left\{\left(\frac{2\nu_1}{UR_0^2}\right)^2 + \frac{3\alpha_0(p_1-p_0)}{p_0 R_0^2}\right\}^{1/2}x\right]. \qquad (1.66)$$

It follows from (1.65) that with $R_0 = 0$ we regain the front of the tanh in (1.52). In typical experiments the quantity $\left\{\frac{3\alpha_0(p_1-p_0)}{p_0 R_0^2}\right\}^{1/2}$ is however dominant. In an experiment of Noordzij and van Wijngaarden [14], to be discussed later in this section, $R_0 = 1.32\,\mathrm{mm}$, $\alpha_0 = 3.28\%$, $(p_1-p_0)/p_0 = 0.09$ and $\nu_1 = 7.10^{-6}\,\mathrm{m^2\,s^{-1}}$. With these data

$$2\nu_1/UR_0^2 = 0.13 \quad \text{and} \quad \left\{\frac{3\alpha_0(p_1-p_0)}{p_0 R_0^2}\right\}^{1/2} = 71.3.$$

For this kind of bubbles the second of these terms is clearly much larger than the first and the slope at the front is determined by dispersion. We shall therefore assume

$$\frac{2\nu_1}{UR_0^2} \ll \left\{\frac{3\alpha_0(p_1-p_0)}{p_0 R_0^2}\right\}^{1/2}. \qquad (1.67)$$

The characteristic scale d_A at the front is then for isothermal conditions

$$(d_A)_{iso} = R_0\left(\frac{p_0}{3\alpha_0(p_1-p_0)}\right)^{1/2}. \qquad (1.68)$$

Likewise for adiabatic waves, from (1.64)

$$d_{A,\mathrm{ad}} = R_0 \left\{ \frac{\gamma + 1}{6\gamma\alpha_0} \frac{p_0}{p_1 - p_0} \right\}^{1/2}.$$ (1.69)

Next we inspect the back side of the shock where $\tilde{p}$ is close to $\tilde{p}_1$. Now taking the adiabatic case, we obtain from linearization of (1.64) around $\tilde{p}_1$,

$$\frac{R_0^2}{3\alpha_0} \frac{\mathrm{d}^2 \, (\tilde{p}_1 - \tilde{p})}{\mathrm{d}x^2} + \frac{4\nu_1}{3U\alpha_0} \frac{\mathrm{d} \, (\tilde{p}_1 - \tilde{p})}{\mathrm{d}x} + \frac{2\gamma}{\gamma + 1}\tilde{p}_1 \, (\tilde{p}_1 - \tilde{p}) = 0.$$ (1.70)

The solution which vanishes at $x \to \infty$ is

$$\tilde{p}_1 - \tilde{p} \sim \exp\left[-\frac{2\nu_1}{UR_0^2}x + \left\{ \left(\frac{2\nu_1}{UR_0^2}\right)^2 - \frac{6\gamma}{\gamma + 1} \frac{(p_1 - p_0)\,\alpha_0}{p_0 R_0^2} \right\}^{1/2} x \right].$$ (1.71)

This predicts that under circumstances for which (1.67) holds, the pressure has the form of damped waves with wavelength λ, see (1.69)

$$\lambda = 2\pi d_{A,\mathrm{ad}} = 2\pi R_0 \left\{ \frac{\gamma + 1}{6\gamma\alpha_0} \frac{p_0}{p_1 - p_0} \right\}^{1/2}.$$ (1.72)

Similarly, from (1.68) for isothermal conditions

$$\lambda = 2\pi d_{A,\mathrm{iso}} = 2\pi R_0 \left(\frac{p_0}{3\alpha_0 \, (p_1 - p_0)} \right)^{1/2}.$$ (1.73)

These relations show that the thickness of the shock wave is related to bubble radius rather than molecular mean free path as in single phase gases. For a continuum theory to be valid the thickness should cover at least a few bubble diameters for a single experiment. As (1.69) and (1.72) show this is certainly true for small α and weak shocks.

How do these predictions compare with experiments? After the pioneering experiments by Campbell and Pitcher [9] who did not investigate the shock structure, experiments of the shock tube type were made by Noordzij and van Wijngaarden [14], Nakoryakov et al. [17], Beylich and Gülhan [18], Tan and Bankoff [16], Kameda and Matsumoto [19]. These experiments show that in the first meter or so after the shock has been formed the structure agrees with the above analysis: a steep rise at the front and waves at the back side. A typical example is in Fig. 1.4, from Noordzij and van Wijngaarden [14], who called this a type A shock, a nomenclature adopted later also by Watanabe and Prosperetti [20].

So, qualitatively, this A type corresponds with theory. Quantitatively is a different matter as far as the early, [14, 17], experiments are concerned. Let us consider first the thickness and wavelength. When there is agreement it is agreement with the adiabatic rather than the isothermal relation. This can

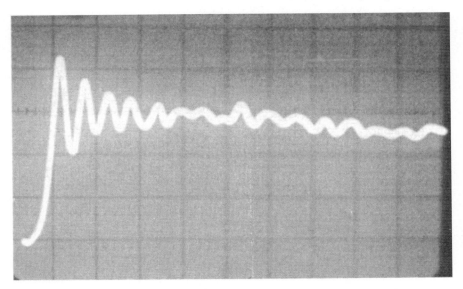

Fig. 1.4. An example of an A shock. Shown is the recording of the pressure in the bubbly suspension. $R_0 = 1.32\,\mathrm{mm}$, $\alpha_0 = 3.28\%$. From Noordzij and van Wijngaarden [14]

be understood; according to (1.69) and (1.14) the time needed for the shock to pass from the lower to the higher pressure side $d_{A,\mathrm{ad}}/c_{\mathrm{ad}}$ is of the order $R_0\left(\frac{\rho_l}{p_1-p_0}\right)^{1/2}$. The corresponding temperature penetration inside the bubble is

$$\left(R_0\chi\left(\frac{\rho_l}{p_1-p_0}\right)^{1/2}\right)^{1/2}. \tag{1.74}$$

With $R_0 = 10^{-3}\,\mathrm{m}$, $\chi = 10^{-5}\,\mathrm{m^2\,s^{-1}}$ this is only a small fraction of R_0. While many of the results in [9,14], agree well with (1.35) and (1.72), others do not. The reason is as pointed out by Nigmatulin [21], that a steady solution of (1.62) has not been reached at a small distance from the entrance in the mixture. Watanabe and Prosperetti [20] set out to compare in a precise way the experimental results with a numerical simulation using the equations proposed by Caflisch et al. [5], which are equivalent with the equations (1.15)–(1.20) given here. Because they were aware of the possible deviations displayed in shocks measured close to the entrance, they compared their results specifically with those obtained by Beylich and Gülhan [18], whose measurements were further from the entrance. They found qualitative agreement, but considerable disagreement in details of the waves at the back of the shock. Figure 1.5 is taken from Watanabe and Prosperetti [20]. It shows experimental results of [18] for an A shock together with results from their, [20], numerical simulation. (The dotted line will be discussed later).

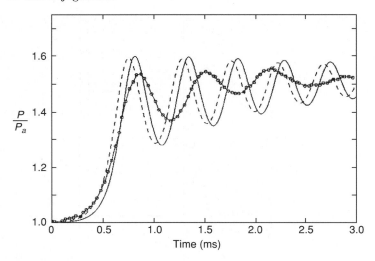

Fig. 1.5. Pressure recording in *A* shock. Line with circles shows experimental results of Beylich and Gülhan [18]. *Solid line* shows numerical simulation of Watanabe and Prosperetti [20], whereas the *dotted line* includes relative motion in these simulations

As we will see, other results of the numerical simulation agreed quite well with experimental observations, so it remained a mystery for a while what the reason of the discrepancy as displayed in Fig. 1.5 is. This mystery was solved by Kameda and Matsumoto [19, 22]. While doing experiments they realized that whereas the theories discussed above assume an homogeneous suspension of bubbles, this is not always the case in experiments. In particular the influence of walls renders in vertical devices the transverse distribution inhomogeneous. So, they did two-dimensional numerical calculations for bubble distributions, random in vertical and parabolic in transverse direction. A result is shown in Fig. 1.6a. There is now good agreement with the experimental results. In addition Kameda and Matsumoto [21] did experiments in which care was taken that the bubble distribution was fairly homogeneous. With that precaution they obtained good agreement with the results , for the wave form and wave shape, of experiments. An example is shown in Fig. 1.6b.

Whereas the type *A* shocks are steady solutions of the Burgers–KdV equation (1.62), in all mentioned experiments they were only observed close to the entrance of the shock tube. Deep down they evolve into other forms. This transition is caused by relaxation, discussed in Sect. 1.4. In the beginning steepening is resisted by dispersion as we have seen (as opposed to molecular diffusion in single phase gases). As time proceeds relaxation starts to play an active role. It is, with diffusion coefficients as in (1.47) and (1.48), not able to resist the strong steepening as in the front of an *A* shock. However it may do so at the back of a shock where the slope is much smaller. What we get to observe then is a shock of the type *B*, in the nomenclature of Noordzij and van Wijngaarden [14] and Watanabe and Prosperetti [20].

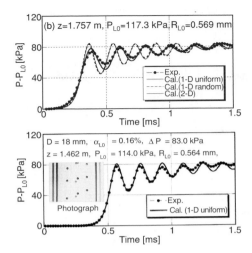

Fig. 1.6. (a) Comparison of pressures in an A-shock obtained in experiment and by computation; (b) idem for uniform bubbly flow. By permission from Kameda and Matsumoto [21]

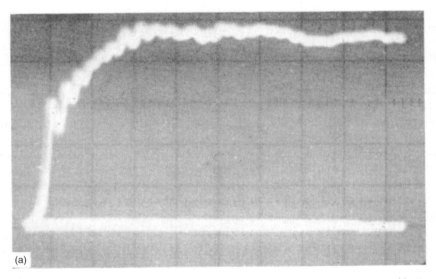

Fig. 1.7. A type B shock wave. $p_1/p_0 = 1.42$, $R_0 = 1.28\,\text{mm}$, $\alpha_0 = 2.19\%$. From Noordzij and van Wijngaarden [14]

In Fig. 1.7 such a shock is shown. In some cases they remain unchanged but in other cases they develop into completely smooth waves, C shocks as shown in Fig. 1.8, where all three types show up. For single phase gases B and

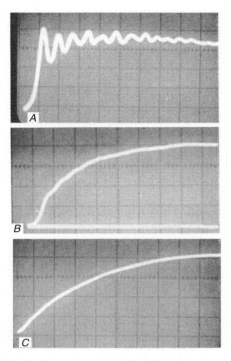

Fig. 1.8. A, B, and C shocks as measured by Noordzij and van Wijngaarden [14]

C shocks are discussed by Lighthill [13], who calls them partly dispersed and fully dispersed, respectively. Noordzij and van Wijngaarden [14], in search for the relevant relaxation process, attributed the observed transition to B and C shocks to relative motion. Later Nigmatulin and colleagues, see Nigmatulin [21], pointed out that thermal relaxation could fully explain the observed phenomena. This became particularly clear through the numerical simulation by Watanabe and Prosperetti [20]. Since this work contributed significantly to the understanding of shocks in bubbly liquids, we will mention some of their results in more detail. In the first place they showed that, when neither of the relaxation processes is allowed for, that is when adiabatic or isothermal behavior is assumed, the transition to B and C waves does not take place. Further, that including relative motion does give some change but not significantly. An example of this is shown in Fig. 1.5, where the dotted line indicates the simulation results including relative motion. Finally that the experimental results are well explained when the energy equations (1.3) and (1.9) are fully taken into account and not when a fixed polytropic exponent is used in the simulation. We will first discuss the C shocks, the fully dispersed ones. In these shocks steepening is so gentle that relaxation alone can provide the resistance necessary for a steady profile. One of the reasons why Noordzij and van Wijngaarden [14] identified relative motion as the relevant relaxation is the agreement with experimentally observed shock thickness. The diffusion coefficient associated with relaxation of relative motion is given in (1.47). If we introduce this quantity instead of the bulk viscosity ν_1/α_0 in (1.53) and use the definitions (1.44) and (1.14a) of c_{*iso} and c_{iso}, respectively, we obtain as an estimate of the shock thickness of a C shock

$$(d_C)_{rel} \sim \frac{1}{2} \frac{\tau_v \left\{(1 + 2\alpha_0) - 1\right\} U}{p_1/p_0 - 1}. \qquad (1.75)$$

Putting in numbers it appears that this is of the order of meters! The C shocks are quite thick indeed, and Noordzij and van Wijngaarden found, see Table 2 of their paper, [14], good agreement with (1.75). However the real relaxation mechanism is of thermal nature. The associated diffusion coefficient is given in (1.48), however with unknown relaxation time $\tilde{\tau}$. Working out the corresponding shock thickness we obtain

$$(d_C)_{th} \sim \frac{1}{2} \frac{\tilde{\tau} U (\gamma - 1)}{p_1/p_0 - 1}. \qquad (1.76)$$

Comparison with (1.75) shows that if the latter agrees with experiments this means that $\tilde{\tau}$ is of the same order as τ_v. It suggests itself to inspect whether the penetration time of heat in a bubble is a candidate for $\tilde{\tau}$. As follows from solution of (1.38), using (1.39) and (1.40), an initial velocity difference between bubbles and liquid relaxes to zero as

$$e^{-t/\tau_v} = e^{-\frac{18\nu_1 t}{R^2}}.$$

On the other hand, if a conducting sphere with initial temperature zero is at its surface exposed to a temperature T_0, the temperature in the center behaves for large times as, [23],

$$T \sim T_0 \left(1 - e^{-\frac{\pi^2 \chi t}{R^2}}\right).$$

This gives a relaxation time

$$\tilde{\tau} = R_0^2/10\chi. \qquad (1.77)$$

For air χ is $2.10^{-5} \, \text{m}^2 \, \text{s}^{-1}$, the experiments in [14] are done in an aqueous solution of glycerin with a viscosity $\nu_1 = 7 \times 10^{-6} \, \text{m}^2 \, \text{s}^{-1}$. Then comparing (1.43) with (1.77) shows that indeed $\tilde{\tau}$ is of the same order of magnitude as τ_v.

Having dealt with the C shock we now turn to the B shocks. Their occurrence is caused by the fact that C shocks can only exist over a limited range of pressure ratios p_1/p_0. To understand this we recall that in a wave equation, such as (1.45), the high frequency terms act as diffusion on a wave propagating near the low frequency limit. The high frequency wavelets at the front of a C shock travel with a speed which cannot be larger than c_{ad}. On the other hand in a steady wave it should travel with the shock, propagating with velocity U. This limits U to $c_{ad,0}$, and using (1.14a) and (1.35) this gives

$$\frac{p_1}{p_0} \leq \gamma. \tag{1.78}$$

This argument has been given by Lighthill [13] and by Zel'dovich and Raizer [24] for gases and by Nigmatulin and Shagapov, see Nigmatulin [21], for bubbly flows. For pressure ratios p_1/p_0 larger than γ a steady C shock cannot exist and a B type of steady shock is the end shape. This has, see the example in Fig. 1.7, a steep front which is essentially an A shock which is followed by a C shock. In the first steep, part the pressure rises from p_0 to an intermediate value p_*, say, and in the smooth part the final value p_1 is reached. The value of p_* follows from the fact that both parts travel at the same velocity U, whence we have, from (1.34) and (1.35)

$$\frac{p_1}{p_0} = \frac{\frac{p_*}{p_0} - 1}{1 - \left(\frac{p_0}{p_*}\right)^{1/\gamma}}. \tag{1.79}$$

What are indicated here as B shocks are called partly dispersed shocks by Lighthill [13]. The transition of B shocks to C shocks is further elucidated by Watanabe and Prosperetti [20] in terms of the propagation of small amplitude wavelets in front and at the back of the shock. For this we turn our attention to the energy. If mass and momentum are conserved what happens to the energy? In a strong hydraulic jump kinetic energy is lost and dissipated in the shock. However here the energy balance is made by radiation of kinetic energy away from the shock, in a way similar to the undular bore in hydraulics, see Benjamin [25]. Consider the A shock. As we have seen, there are waves behind the shock with wavelength λ given in (1.72). These waves have a frequency ω connected with the wave number $2\pi/\lambda$ by the dispersion equation (1.21). The phase velocity ω/k equals the shock velocity U but the energy travels at the group velocity c_{gr} defined as

$$c_{gr} = \frac{\partial \omega}{\partial k}. \tag{1.80}$$

At the high frequencies of the waves at the back of an A shock the group velocity is smaller than the phase velocity $c_{ph} = c_{ad,0}$. Hence in a reference frame moving with the shock, energy is radiated downstream from the shock, instead of being dissipated. How is this in a B and a C shock?. For this we return to the dispersion equation (1.21). Denoting the, approximately, constant quantity ω_b/c with q and neglecting the term with δ, we have from (1.21)

$$\frac{\omega}{k} = \frac{\omega_b}{q} \left(1 - \frac{\omega^2}{\omega_b^2}\right)^{1/2}. \tag{1.81}$$

The resonance frequency ω_b rises from the value in (1.5) with $\kappa = 1$ at low frequencies to the value with $\kappa = \gamma$ at high frequencies, so we can interpret it as a function of frequency. Then we can derive from (1.81) the relation

$$\frac{d}{d\omega}\left(\frac{\omega}{k}\right) = \frac{\omega}{q}\frac{1}{(\omega_b^2 - \omega^2)^{1/2}}\left(\frac{\omega_b}{\omega}\frac{d\omega_b}{d\omega} - 1\right). \tag{1.82}$$

At low frequencies the right hand side of (1.82) is positive. We can write the left hand side as

$$\frac{d}{d\omega}c_{ph} = k^{-1}\left(1 - \frac{c_{ph}}{c_{gr}}\right). \tag{1.83}$$

At low frequency the group velocity is larger than the phase velocity, whereas at high frequencies, where the right hand side of (1.82) is negative, (1.83) shows that c_{gr} is smaller than c_{ph} as we have already seen. Small amplitude waves can travel with the shock when their phase velocity equals the shock velocity U. The energy must be radiated away from the shock in order to satisfy the balance of kinetic energy. This means that low frequency waves can exist near the front. They have a group velocity larger than U and radiate energy into the undisturbed flow. High frequency waves with a group velocity less than U can exist near the backside and radiate energy away in the downstream region. All this holds for shock waves in which the energy is radiated instead of dissipated. The criterion (1.78) for the transition of C to B shocks is valid only for those circumstances. Watanabe and Prosperetti [20] investigated this aspect in some detail and find that (1.78) is valid for $(R/\chi)(p_0/\rho)^{1/2}$ larger than about 10^3. In water at atmospheric conditions this means bubbles with radius larger than a millimeter.

Finally we recommend considerable caution in treating thermodynamic changes of gas in the bubbles during passage of a shock wave with a constant polytropic index. This is useful for certain overall considerations, e.g., for the calculation of the shock speed. Also, the A shock can, as we have seen, taken to be adiabatic. However in dealing with B and C shocks it may lead to erroneous conclusions. This is shown in an instructive way by Watanabe and Prosperetti [20]. They derive from the relation (1.9)

$$\frac{\partial}{\partial t}(p_g R^{3\gamma}) = 3(\gamma - 1)R^{3\gamma-1}K\left(\frac{\partial T}{\partial r}\right)_{r=R}.$$

Clearly changes are adiabatic when there is no heat transfer. We can, following the gas through a B shock, characterize at each instant the state of the gas by an effective polytropic index

$$\kappa_e = -\frac{1}{3}\frac{\ln(p/p_0)}{\ln(R/R_0)}$$

If the gas would behave as one with a fixed polytropic index, κ_e would not change through the shock. Watanabe and Prosperetti show that in fact there is a continuous change both through a B shock (significant) as well as through a C shock (less large).

1.7 Solitons in bubbly flows

In the preceding sections we have shown which mechanisms oppose steepening of compressive waves in a bubbly flow and to what extent these contribute to the formation of steady shock waves. One finding was that nonlinearity, dispersion and dissipation are combined in equations of the KdV–Burgers type, (1.54) and (1.62). In particular the role of this equation in the formation of A shocks was discussed. It was observed that these shocks occur in many of the cited experiments in the first meter or so of the shock tube where a steady state has not been reached. An interesting opportunity to check in quite another way the validity of these equations for perturbations in bubbly flows presents itself when experiments with initial signals other than a step are done. Such experiments were performed and reported by Kutznetsov et al. [26] and by van Wijngaarden and Roelofsen and reported in [27]. Since in the latter the most accurate comparison with experiments was made we shall here discuss their experiment. A shock tube of the same type as drawn in Fig. 1.3 was used. The air region however now consists of two separated parts, a low-pressure region with length h_1 and a high-pressure region with length h_2. At time t_0^* a seal separates these parts. After puncturing of the seal a complicated sequence of compression and expansion waves hits the bubbly suspension located in the tube at $x^* > 0$. Through a special choice of the pressures p^+ and p_0 in the two parts of the upper space and of the lengths h_1 and h_2 the experiment is arranged in such a way that the upper part of the bubbly flow is hit by a shock followed by an expansion wave. Hence the surface of the bubbly suspension is at $t^* = 0$ perturbed by a signal in the form of a rectangular triangle. The initial condition for the pressure perturbation $\tilde{p}$ is, x^* pointing into the mixture

$$\tilde{p} = \frac{p - p_0}{p_0} = \frac{p^+ - p_0}{p_0}\frac{x^*}{l}, 0 < x^* < l \qquad (1.84)$$
$$= 0, x^* \geq l, x^* \leq 0.$$

The initial pressure perturbation has finite extension l. For times $t^* > 0$ this perturbation evolves into the mixture. In the case of a pressure jump this evolution is into an A shock as we have seen. In the case of the expansion wave described by (1.84) the evolution is into a continuous wave train moving rapidly in the negative x^* direction and a series of solitons moving in positive x direction. The number, N, of these solitons can for the initial profile (1.84) be predicted exactly. The way this is done goes as follows. We start with (1.62) with x^* instead of x and we neglect the viscous term since this is unimportant here. Next we define new coordinates x', t' by

$$x' = x^* - c_0 t^*, t' = t^*. \qquad (1.85)$$

Here we have written $c_{ad,0}$ as c_0. Indicating the quantity R_0^2/α_0 with β we then obtain

$$\frac{\partial \tilde{p}}{\partial t'} + \frac{\gamma + 1}{2\gamma}c_0\tilde{p}\frac{\partial \tilde{p}}{\partial x'} + \frac{\beta c_0}{6}\frac{\partial^3 \tilde{p}}{\partial x'^3} = 0. \qquad (1.86)$$

This is cast in the standard form of the KdV equation

$$\frac{\partial V}{\partial t} - 6V\frac{\partial V}{\partial x} + \frac{\partial^3 V}{\partial x^3} = 0 \qquad (1.87)$$

by choosing the dimensionless quantities x, t, and V as

$$x' = \left\{\frac{\beta l\gamma}{3(\gamma+1)}\right\}^{1/3} x, \quad t' = \frac{2l\gamma}{(\gamma+1)c_0}t, \quad \tilde{p} = -\left\{\frac{72\beta\gamma}{l^2(\gamma+1)}\right\}^{1/3} V. \qquad (1.88)$$

Next we consider the Schrödinger equation for the function φ

$$\frac{d^2\varphi}{dx^2} + \{\lambda - V(x,t)\}\varphi = 0. \qquad (1.89)$$

The theory of the KdV equation teaches, see e.g., Whitham [7], that the eigenvalues λ_n of (1.89) are independent of time when V satisfies (1.87). They are all negative, $\lambda_n = -b_n^2$ and ordered such that $b_1 > b_2 > \longrightarrow b_N > 0$, with N finite. Moreover the Sturm–Liouville theory of differential equations says that when the solution of (1.89) corresponding with a certain λ_n has, say, q zeros, then the solution corresponding with the next, λ_{n+1} has one more zero etc. For $\lambda \to -\infty$ the solution of (1.89) is $\cosh x$ which has no zeros for real x. Hence λ_1 has one zero, λ_2 has two zeros and because all λ_n's are negative we catch all eigenvalues if we solve (1.89) with $\lambda = 0$. Thereby the initial value is from (1.84), (1.85) and (1.88)

$$V(x,0) = -\frac{p^+ - p_0}{p_0}\frac{x}{6}, x < \left\{\frac{3l^2(\gamma+1)}{\beta\gamma}\right\}^{2/3}$$

$$= 0, x \le 0, x \ge \left\{\frac{3l^2(\gamma+1)}{\beta\gamma}\right\}^{2/3}. \qquad (1.90)$$

This we introduce into (1.89) and perform yet another transformation

$$x = \left(\frac{6p_0}{p^+ - p_0}\right)^{1/3} y$$

to obtain finally the equations

$$\frac{d^2\varphi}{dy^2} + y\varphi = 0, \text{ in } \quad 0 < y < Q. \qquad (1.91)$$

$$\frac{d^2\varphi}{dy^2} = 0, \quad \text{in} \quad y \le 0 \text{ and in } y \ge Q, \text{ where}$$

$$Q = \left\{\frac{p^+ - p_0}{p_0}\frac{(\gamma+1)l^2}{2\beta\gamma}\right\}^{1/3}. \qquad (1.92)$$

We impose on (1.91) the boundary conditions

$$y = 0 : \frac{d\varphi}{dy} = 0, \varphi = 1. \tag{1.93}$$

In quantum mechanics the wave function φ in (1.89) is solved, starting from the known potential V. Here $V(x,t)$ is in its development in time found with help of the eigenvalues of (1.89) and boundary condition (1.93) with $V(x,t) = V(x,0)$. Therefore this is called the inverse scattering method. The nice circumstance of the choice (1.90) of $V(x,0)$ is that (1.91)–(1.93) can be solved analytically, (1.91) being the Airy equation. We met the Airy functions $Ai(x)$ and $Bi(x)$ before, see Fig. 1.2. The solution of (1.91)–(1.93) is

$$y < 0, \varphi = 1,$$
$$0 \leq y \leq Q, \varphi = F(y) = 1.408 \text{Ai}(-y) + 0.813 Bi(-y), \tag{1.94}$$
$$y > Q, \varphi = \left(\frac{dF}{dy}\right)_{y=Q}(y - Q) + F(Q).$$

The function $F(y)$ is drawn in Fig. 1.9. All we have to do now to predict the number of solitons is to determine for a given value of Q, which follows from the particular choice of the experiment, the number of zeros. The part $y < 0$ has no zeros, the zeros between 0 and Q follow from Fig. 1.9, and the part $y > Q$ contributes one additional zero when at $y = Q$

$$\frac{1}{\varphi}\frac{d\varphi}{dy} < 0.$$

In [27] this is done for a number of experiments. We report one of these. The experimental values are $(p^+ - p^0)/p_0 = 0.52, \alpha = 0.74\%, l = 0.24\text{m}, Q = 6.6$. From Fig. 1.9 it follows that the number of zeros is 4 between $y = 0$ and $y = Q$.

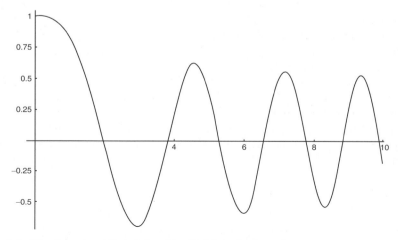

Fig. 1.9. The function $F(y)$ defined in (1.94) in de text. From van Wijngaarden [27]

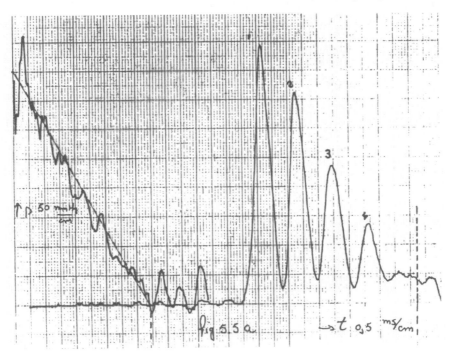

Fig. 1.10. At left the triangular initial pressure perturbation . At the right the pressure signal taken 0.47 m below the interface. It shows the four (predicted) solitons. From van Wijngaarden [27]

Figure 1.10 contains the corresponding pressure recording. The pressure was recorded at a distance of 0.47 m below the interface between air and bubbly liquid. It shows indeed four solitons.

Crighton [28] comments on this "This also amounts to an impressive confirmation of the theory of bubbly liquids. There is a spectacular difference between prediction or observation for a pure liquid or gas phase where the triangular pulse simply lengthens as $t^{1/2}$, remaining triangular of constant area, and with a leading diffusion or relaxation-controlled shock; and the prediction or observation of the response of a liquid with a minute gas bubble concentration where a finite number of rank-ordered solitons may be produced, with no shock-like features under appropriate circumstances."

This praise of the theory of bubbly liquids seems to the present author an appropriate closing of this chapter on shock waves and solitons in bubbly liquids.

References

1. Minnaert, M.: On musical air bubbles and the sound of running streams. Phil. Mag. **16**, 235–245 (1933)

2. Nigmatulin, R.I. and Khabeev, N.S.: Heat exchange between a gas bubble and a liquid. Fluid Dyn. **9**, 890–899 (1974)
3. Foldy, L.J.: The multiple scattering of waves. Phys. Rev. **67**, 107–119 (1945)
4. van Wijngaarden, L. On the equations of motion for mixtures of liquid and gas, bubbles, J. Fluid Mech. **33**, 465–474 (1968)
5. Caflisch, R.E., Miksis, M.J., Papanicolaou, G.C., Ting, L.: Effective equations for wave propagation in bubbly liquids. J. Fluid Mech. **153**, 259–273 (1985)
6. Prosperetti, A.: The thermal behavior of oscillating gas bubbles. J. Fluid Mech. **222**, 587–616 (1991)
7. Whitham, G.B.: Linear and Nonlinear Waves. Mc Graw-Hill (1974)
8. Liepmann, H.W., Roshko, A.: Elements of Gasdynamics. Wiley (1957)
9. Campbell, I.J., Pitcher, A.S.: Shock waves in a liquid containing gas bubbles. Proc. Roy. Soc. Lond. A **243**, 534–545 (1958)
10. Taylor, G.I.: The two coefficients of viscosity for a liquid containing air bubbles. Proc. R. Soc. Lond. A **225**, 473–477 (1953)
11. van Wijngaarden, L.: Hydrodynamic interaction between gas bubbles in liquid. J. Fluid. Mech. **77**, 27–44 (1976)
12. van Wijngaarden, L.: One-dimensional flow of liquids containing small gas bubbles. Ann. Rev. Fluid Mech. **4**, 369–396 (1972)
13. Lighthill, M.J.: Viscosity effects in sound waves of finite amplitude. In: Hussaini, M.J. (ed) Collective works of Sir James Lighthill, vol.1, Oxford University Press (1997)
14. Noordzij, L., van Wijngaarden, L.: Relaxation effects, caused by relative motion, on shock waves in gas-bubble/liquid mixtures. J. Fluid Mech. **66**, 115–143 (1974)
15. van Wijngaarden, L.: On the structure of shock waves in liquid- bubble mixtures. Appl. Sc. Res. **22**, 366–381 (1970)
16. Tan, M.J., Bankoff, S.G.: Strong shock waves propagating through a bubbly mixture. Exp. Fluids **2**, 159–165 (1984)
17. Nakoryakov, V.E., Shreiber, I.R., Gasenko, V.G. Moderate –strength waves in the liquids containing gas bubbles. Fluid Mech. – Sov. Res. **10**, 51–66 (1981)
18. Beylich, A.E., Gülhan, A.: On the structure of nonlinear waves in liquids with gas bubbles. Phys. Fluids **A2**, 1412–1428 (1990)
19. Kameda, M., Matsumoto, Y. Shock waves in a liquid containing small gas bubbles. Phys. Fluids **8**, 322–335 (1996)
20. Watanabe, M., Prosperetti, A.: Shock waves in dilute bubbly liquids. J. Fluid Mech. **274**, 349–381 (1994)
21. Nigmatulin, R.I.: Dynamics of Multiphase Media, vol. 2. Hemisphere (1991)
22. Kameda, M., Matsumoto, Y. Structure of shock waves in a liquid containing gas bubbles. Proceedings of the IUTAM Symposium on waves in Liquid/Gas and Liquid/Vapor Two-Phase Syst, Morioka S. van Wijngaarden L. (eds.) Kluwer, 117–127 (1995)
23. Carslaw, H.S., Jaeger, J.C.: Conduction of Heat in Solids. Oxford University Press (1959)
24. Zel`dovich, Ya. B., Raizer, Yu.P, Physics of Shock Waves and High-Temperature Hydrodynamic Phenomena. Dover (2002)
25. Benjamin,T.B.: Conjugate-flow theory for heterogeneous compressible fluids, with. application to non- uniform suspensions of gas bubbles in liquids. J. Fluid Mech. **54**, 545–563 (1972)

26. Kutznetsov, V.V., Nakoryakov, V.E., Pokusaev, B.G., Shreiber, I.R.: Propagation of perturbations in a gas-liquid mixture. J. Fluid Mech. **85**, 85–96 (1978)
27. van Wijngaarden, L.: Evolving Solitons in Bubbly Flows. Acta Applicandae Mathematicae **39**, 507–516 (1995)
28. Crighton, D.G.: Applications of KdV. Acta Applicandae Mathematicae **39**, 39–67 (1995)

2

Interaction of a Shock Wave with a Single Bubble

Yukio Tomita

2.1 Introduction

Shock wave–bubble interaction is an important phenomenon in hydrodynamics because of its special feature of energy concentration, which leads to the generation of high values of pressure and temperature within a small region of the liquid surrounding a bubble. A spherical shock wave in a liquid, caused by a spark discharge or a microexplosive, is characterized by its short rise-time and pulse duration, due to the low compressibility of the liquid and the corresponding large value of the wave speed. Moreover, there is only little dissipation, since the shocks are relatively weak in terms of shock Mach number. This nature enables to realize highly efficient energy accumulation into a small volume of liquid. However liquid in general is not perfectly pure but inhomogeneous, containing impurities in the form of minute bubbles and particles which can act as cavitation nuclei. Cavitation bubbles existing in liquid give full play to their abilities for applications when they interact with shock waves. The collapse process of a cavitation bubble impacted by a shock wave is influenced by such factors as the initial conditions of bubble size and configuration, bubble number distribution, and those associated with the characteristics of shock waves. A particularly important property of a strong shock wave–bubble interaction is the formation of a liquid jet with high velocity which can induce locally high pressures and shear forces.

Shock wave phenomena generated by bubble collapse have been extensively investigated by many researchers for such situations as a single bubble [1], multibubbles system [2–4] and bubble clusters [5]. Fundamental knowledge and characteristics of underwater shock waves and acoustics are also summarized in the books on cavitation and bubble dynamics [6–8]. A bubble can collapse violently resulting in the generation of high temperature and high pressure even in the case of a low amount of shock wave energy provided that the optimum conditions are achieved [9–12]. Under these conditions physical and chemical reactions may be caused inside the bubble or at the bubble surroundings. Locally high pressures are very useful for microscopic cleaning as a positive effect, but sometimes repeated forces due to cavitation blows produce a negative result in the form of fatigue failure of material, called cavitation erosion. Blake and Gibson [13] reported an excellent review of cavitation bubbles near various kinds of boundaries. To elucidate the mechanism of cavitation erosion, numerous attempts of research works have been made hitherto and summarized by Tomita and Shima [9] and Philipp and Lauterborn [14]. Steinberg [15] wrote a brief review on cavitation bubble collapse near a rigid boundary. On the other hand, high temperatures generated inside a bubble or a cavity play an important role either as a source of hot spot or a cause of luminescence. Highlights of the applications of shock wave–bubble interaction are Extracorporeal Shock Wave Lithotripsy (ESWL), laser surgery e.g., ophthalmology [16] and other medical applications, in which cavitation and related phenomena are inevitably induced. In the early treatment of the ESWL, it was considered that tissue damage primarily resulted from missshot of shock focusing, but later cavitation bubble collapse was also recognized to be an important factor for destructive action to the tissue. Figure 2.1 shows cavitation phenomena including shock wave–bubble interaction in the neighborhood of a model solid sphere [17]. A focused shock wave is visualized on the right, reflecting at the curved surface of the sphere and propagating back to the original shock direction. The interaction of this shock wave with a number of cavitation bubbles results in numerous numbers of small rings as indicated in Fig. 2.1b. All of which are the evidence of shock waves emitting from individual bubble collapses. Impulsive pressures generated by these microscopic shock waves are responsible for either calculi fragmentation or tissue damage when bubbles are positioned on the material surface or very close to it. The high values of pressure and temperature caused by shock wave focusing provide essentially important conditions for effective applications to science and technology, especially in the advanced field of medicine and some results can be found in the book by Srivastava et al. [18]. Shock wave–bubble interaction induced liquid jets may act as microsyringes, delivering a drug to a region of interest [19]. Recently the shock wave–cell interaction has received much attention because nano/microbubbles are capable of enhancing noninvasive cytoplasmic molecular delivery in the presence of ultrasound [20]. Collapse of nano/microshell-capsulated bubbles might generate nanoscale cavitation bubbles, leading to transient permeabilization of the cell membrane. It is also pointed out that interaction of an ultrasound-induced shock wave with a cavitation bubble is primarily responsible for skin permeabilization [21].

In the following sections, current topics of the interaction of a shock wave with a single bubble will be described mainly based on the author's results, including some results of others. In Sect. 2.2, the shock wave–bubble (or cavity) interaction is introduced in the absence of any boundary. It contains the early stage of the phenomena after the impact of a shock wave with a bubble, followed by the bubble surface deformation, wave phenomena at the gas–liquid interface and liquid jet formation. In Sect. 2.3, the shock wave–bubble interaction near various boundaries including solid walls and deformable surfaces is

RS

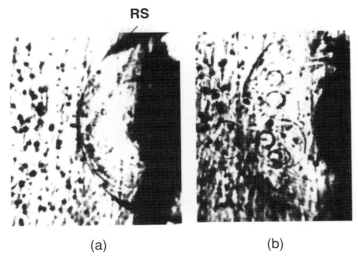

(a)   (b)

Fig. 2.1. Interaction of cavitation bubbles with a shock wave near a solid sphere. Interframe time between (**a**) and (**b**) is 1μs (Tomita et al. [17])

examined in combination with damage experiments. In Sect. 2.4, high temperature phenomena and light generation in sonoluminescence due to the rapid collapse of a bubble or a cavity will be discussed.

2.2 Violent Bubble Collapse and Liquid Jet Formation

A spherical shock wave in a liquid is characterized by its sudden increase of pressure, with an extremely short rise time to reach a maximum, followed by a rapid decrease in the form of an exponential decay. Due to the momentum transfer of the reflected shock front, which accelerates the bubble surface to a velocity twice the particle velocity behind the shock wave, a pressure gradient occurs across the bubble. Therefore the shock wave-induced bubble collapse looses spherical symmetry. In the final stage of the collapse, a liquid jet is formed along the shock wave passage.

Figure 2.2 shows two selected frames of a high-speed shadowgraph of the interaction of a shock wave, incident from the left, with a rising air bubble with the equivalent radius of 1.3 mm. The shock strength in this case was 65 MPa which was generated by exploding a 10 mm silver azide pellet with a pulsed YAG laser. Frame a, corresponds to a snapshot taken just before the interaction and frame b, just after the interaction. The interframe time is 1 μs. Due to the large ratio of the acoustic impedances of water and air, which is approximately 3,500, a part of the incident shock wave reflects at the bubble surface as an expansion wave, propagating back into the original direction. Comparing a with b shows that the density gradient of the expansion wave is

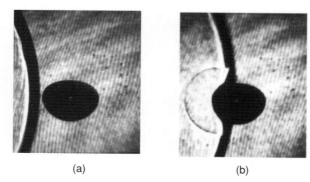

(a)   (b)

Fig. 2.2. Two frames of shadowgraphs of the interaction of a shock wave with a single air bubble, showing (**a**) immediately before the impact and (**b**) after the impact. Interframe time between (**a**) and (**b**) is 1μs. The shock wave with a strength of 65 MPa was generated by exploding a 10 mm silver azide pellet with a pulsed YAG laser and the equivalent radius of the bubble was 1.3 mm (Tomita et al. [22])

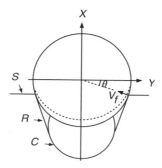

Fig. 2.3. Schematic illustration of shock wave – circular cavity interaction (Dear and Field [3])

completely contrary to that of the incident shock wave. A clear deformation of the impacted bubble surface can be observed. Figure 2.3 shows a conceptual illustration of the interaction of a planar shock wave with a circular cavity [3]. Reflection of the incident shock wave, S, occurs at the free liquid surface to produce a corner wave, C, and a reflected tensile wave, R. Linear theory suggests that the initial deformation of the bubble surface would have a velocity of twice the particle velocity behind the incident shock wave. The velocity imparted to the free surface at the angle θ is given by

$$V_f = 2u_p \sin\theta, \tag{2.1}$$

where u_p is the particle velocity behind the shock wave written as

$$u_p = \frac{p_s - p_0}{\rho_0 U_s}, \tag{2.2}$$

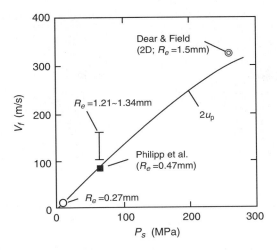

Fig. 2.4. Initial surface velocity V_f of a bubble (or a cavity) against shock wave strength p_s (Dear and Field [3]; Shima et al. [11]; Philipp et al. [12]; Tomita et al. [22])

where p_0 and ρ_0 are the pressure and density in the undisturbed liquid, and U_s and p_s the shock wave velocity and shock wave pressure, respectively. Figure 2.4 collects some experimental data on the initial surface velocity V_f of a bubble (or a cavity) on the symmetry axis (i.e., $\theta = 90°$ in Fig. 2.3), for different shock wave pressures p_s. The solid line represents a theoretical prediction of the particle velocity defined as $V_f = 2u_p$. An open circle denotes a data point from a hydrogen bubble experiment [11], a solid square is associated with the data of the lithotripter-generated shock wave–air bubble interaction [12]. Furthermore a stretched bar indicates the data for the interaction of shock waves with 65 MPa peak pressure with bubbles of the aspect ratio χ , ranging from 1.29 to 1.33 [22]. The aspect ratio is defined here as the ratio of axial and lateral bubble dimension. The result of a two-dimensional experiment, carried out by Dear and Field [3], is also shown in the same figure. It is found that the initial velocity, V_f, of a spherical bubble (or a cavity) increases as increasing shock wave pressure. For nonspherical bubbles with $\chi > 1$, the values of V_f are apt to be larger than for the spherical bubble case. The initial deformation velocity V_f as a function of aspect ratio χ is shown in Fig. 2.5, emphasizing the effect of curvature of the bubble surface. A shock wave with 65 MPa peak pressure did interact with two kinds of gas bubbles, i.e., air and helium bubbles, corresponding to open and filled circles, respectively. One data point by Philipp et al. [12], denoted by a filled square, has also been plotted in the same figure. Again, we notice that the initial deformation velocity increases with increasing surface curvature. Thus the shock-induced bubble deformation is significantly affected by the initial bubble shape, or, in other words, by the curvature of the bubble surface impacted by a shock wave [23]. An overview of

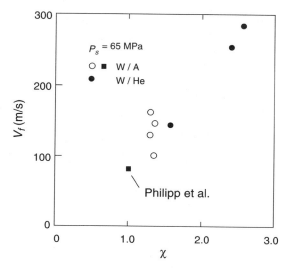

Fig. 2.5. Effect of the curvature of bubble surface, characterized by the aspect ratio χ, on the initial bubble surface velocity V_f when p_s is 65 MPa, where W/A: air bubbles in water (*open circle, filled square*); W/He: helium bubbles in a water (*closed circle*), and a symbol (*filled square*) is a data point by Philipp et al. [12]

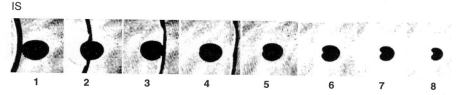

Fig. 2.6. Successive shape change of an air bubble after interacting with a shock wave, where IS means the incident shock wave. Interframe time is 1µs

the collapse process of a bubble by a shock wave is demonstrated in Fig. 2.6, in which a rising air bubble with the equivalent radius of 1.3mm was loaded by a shock wave of 65MPa peak pressure [22]. By taking d_x to be the bubble diameter in the shock direction and d_y the bubble diameter perpendicular to it, we define the bubble aspect ratio as $\chi = d_x/d_y$. Because of the bubble flatness $f(= (d_x - d_y)/d_x)$ being relatively high in this case, i.e., $f = 0.23$, the bubble surface considerably deforms. At 1 µs after the impact (i.e., at the 2nd frame in Fig. 2.6) the velocity of the bubble surface is determined to be 103 ms^{-1} which is as fast as 2.3 times the particle velocity behind the incident shock wave. Figure 2.7 shows the time variation of the upstream and downstream surfaces of the bubble shown in Fig. 2.6 along the shock wave passage. The shock wave completely passes over the whole bubble within 2 µs. Then the whole bubble surface shrinks together, i.e., collapses three-dimensionally. It is easily

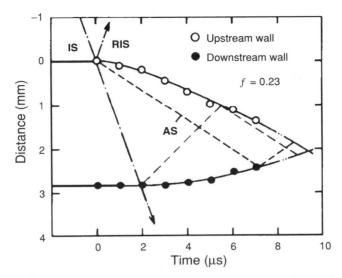

Fig. 2.7. Time histories of upstream and downstream surfaces of an air bubble with the flattening factor of $f = 0.23$, where IS: incident shock, RIS: reflection of incident shock and AS: air shock (Tomita et al. [22])

seen that the upstream surface deforms faster than the downstream surface. The velocity of the former attains $250\,\text{ms}^{-1}$ and that of the latter $130\,\text{ms}^{-1}$. A notation AS implies an air shock propagating inside the bubble. Another air shock originates at the rear side of the bubble wall, traveling back to the upstream bubble surface. Thus air shock waves may be crossing somewhere inside the bubble. If the bubble would collapse spherically during its whole motion without any disturbance, a spherical air shock focusing could occur, producing extremely high values of temperature and pressure at the bubble center.

Grove and Menikoff [24] numerically calculated the planar shock wave-cylindrical air cavity interaction for two specified initial water shock strengths. Their main interest is the anomalous reflection and diffraction of shock waves at interfaces and therefore their results are limited to the times during which the shock wave is interacting with the bubble. For the weak shock case ($p_s/p_0 = 100$, where the Mach number ahead of the shock is $M = U_s/c_w = 1.009$ with c_w being the sound velocity in water), the surface impacted by a shock wave moves with the velocity V_f approximately twice that of the fluid behind the incident shock wave. When the shock wave first hits the bubble, a pair of diffraction nodes is formed. Due to the curvature of the curved cavity surface, the angle between the incident shock wave and contact surface increases as the interaction proceeds forwards and eventually each diffraction node bifurcates into an anomalous reflection. Finally, when the shock wave passes through the back of the cavity, the two nodes

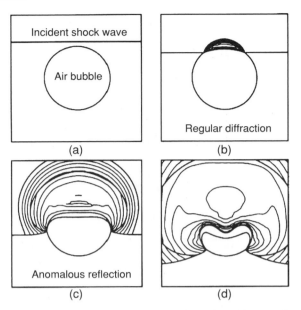

Fig. 2.8. Time sequence of pressure contours after the collision of a shock wave in water with an air bubble. The fluid ahead of the shock wave is at normal conditions of 101.3 kPa and the pressure behind the incident shock wave is 1 GPa (Grove and Menikoff [24]). Permission is granted from Cambridge University Press

formed by the diffraction of the Mach line through the cavity interface coalesce, and then the water shock wave detaches from the cavity. For the strong shock wave case of $p_s/p_0 = 10,000$, the nonlinearities of the flow are revealed and the rarefaction waves spread out over a larger distance. A time sequence of pressure contours is shown in Fig. 2.8. The anomalous reflection and the curvature of the incident shock wave are more clearly visible. The contact surface of the cavity is greatly distorted. As shown in Fig. 2.8c and d, the pressures at the front increasingly develop during the collapse and eventually drive a finite volume of water to form a liquid jet which will impact the back of the cavity. Masubuchi et al. [25] obtained the time evolution of the collapse of a bubble impacted by a shock wave and the calculated diameters d_x and d_y, defined as that taken in the shock direction and one perpendicular to it, are plotted in Fig. 2.9 including a solid line as a theoretical prediction, called Rayleigh's spherical bubble collapse [26]. For a relatively small shock pressure $(p_s/p_0 = 10)$, a bubble collapses almost spherically, whereas a notable deformation of the impacted bubble surface is found for a relatively strong shock pressure $(p_s/p_0 = 1,000)$. Finally a jet is formed and passes across the cavity, striking and penetrating the far side of the cavity surface. Once a liquid jet is formed, the fluid ahead of it is displaced, creating a local shear layer between the jet and the outer fluid. Ding and Gracewski [27] performed numerical

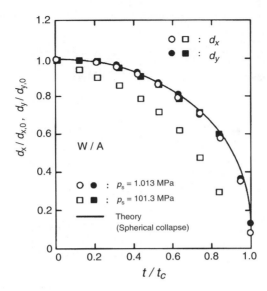

Fig. 2.9. Time variations of the axial and horizontal diameters of an air bubble in water for two different shock strengths, $p_s = 1.013\,\text{MPa}$ (*open circle, filled circle*) and $p_s = 101.3\,\text{MPa}$ (*open square, filled square*): d_x and d_y are the axial and horizontal bubble diameters, respectively (Tomita et al. [22]; Masubuchi et al. [25]) and a solid line depicts the theoretical result of Rayleigh for spherical bubble collapse (Rayleigh [26])

calculations by using a finite volume method for the two-dimensional interaction of a gas cavity with two types of shock strengths, i.e., weak shocks ($p_s < 30\,\text{MPa}$) and strong shocks (p_s ranging from 500 to 2,000 MPa). They concluded that for strong shocks, a jet is created after the shock wave impacts the interface of a cavity and penetrates the opposite side with a velocity as high as $2{,}000\,\text{ms}^{-1}$. For a weak shock, it was proven that the Gilmore model can provide an accurate prediction of cavity collapse. Experimentally, at the beginning of the 1960s, Bowden and Brunton [28] produced high-speed liquid jets as a result of shock wave–cavity interaction. Later numerous investigators have also tackled these phenomena. The density gradient due to liquid jet penetration through the opposite bubble surface can be captured with a Schlieren method. Figure 2.10 shows a high-speed photograph where the interframe time is 1 μs [10]. A hydrogen bubble of 0.38 mm in radius was generated at the top of a needle by means of electrolysis. As clearly seen in the 1st frame, a primary shock wave coming from the right interacted with the bubble. The bubble collapses almost spherically during the initial phase of motion, but gradually looses its spherical shape. At the later stage of the collapse, the impacted surface becomes flat, leading to jet formation. Viewing the 7th frame we notice a dark portion of the liquid on the left side of

Primary Shock Wave

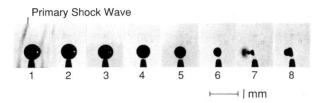

Fig. 2.10. High speed Schlieren photograph indicating the collapse process of a hydrogen bubble loaded by a shock wave. Interframe time is 1μs (Tomita et al. [10])

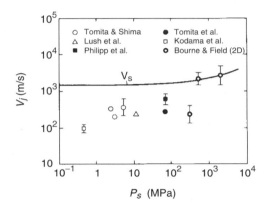

Fig. 2.11. Liquid jet velocities recorded by several investigators (Tomita and Shima [9]; Philipp et al. [12]; Tomita et al. [22]; Bourne and Field [23]; Lush et al. [32]; Kodama et al. [33])

the bubble, which is believed to be the density change due to liquid jet flow. Tulin [29] suggested that when a planar shock wave impinges on a wedge or a conical cavity, the impacted surface deforms considerably, forming an ultra high-speed liquid jet with the velocity of as high as twice the shock speed. This is a mechanism very similar to that of a Munroe jet formation in shaped charges [30]. The velocity of this type of tiny jet particularly depends on the curvature of the liquid surface to be impacted. In the experiment of Bowden and Brunton [28], the liquid surface was approximately spherical and it might perhaps be regarded as a "half bubble." A variety of two-dimensional experiments on shock wave–cavity interaction has been carried out by Field and his collaborators, involving shock-induced collapses of single cavities [23], arrays of cavities [3] and further including various types of configurations of cavities [31]. Cavities were made by punching gelatin sheets. When a jet penetrates the opposite surface of the cavity, a pair of linear vortices is created which moves towards the boundary. Available data of jet velocity, V_j , as a function of shock wave strength, p_s, are compiled in Fig. 2.11 [9, 12, 22, 32, 33], together with two-dimensional data by Bourne and Field [23]. A solid line indicates the relationship between the shock wave velocity V_s and shock wave strength p_s, calculated by making use of the Hugoniot curve of water. Clearly a stronger shock wave produces a liquid jet with much higher velocity, while the jet impact velocity attainable by an isolated bubble collapse is of order of 100–1,000 ms^{-1} at most. Two-dimensional data demonstrate that jet velocity could exceed the shock wave velocity when the shock pressure p_s is above (1,000–2,000) MPa [23]. Even in the case of small scale phenomena, a liquid jet could be generated and bubble fragmentation occurs resulting from the surface instability [34]. Indeed Postema et al. [35] observed a liquid jet caused by the interaction between an ultrasound contrast agent (UCA) bubble-induced shock wave and a microbubble.

Although the flow field induced by shock wave–bubble interaction is essentially nonspherical, it is worthwhile to look at the effect of pressure impulse on the maximum pressure generated by bubble collapse. In fact there are several situations where bubble motion is almost spherical during the collapse process under some conditions. For example, cavitation sometimes occurs in the form of clouds which contain numerous bubbles with different sizes interacting with each other as already demonstrated in Fig. 2.1. A cavitation bubble cloud also appears in ultrasonic cavitation (e.g., Sanada et al. [36]) in which bubble–bubble and shock wave–bubble interactions are normally generated. For relatively small pressure amplitudes, it will be presumably true that bubbles oscillating in such pressure field collapse almost spherically. Indeed, the bubble radius is often much smaller than the acoustic wavelength. Applying a homogeneous model to a cavitation cloud, Kuttruff [37] calculated the shock wave–bubble interaction where bubbles are pulsating or collapsing independently of each other, but their motions are mutually coupled by some sort of interaction. Since, in each cycle of the exciting sound field the smallest bubbles are the first to implode emitting shock waves outwards. This kind of interaction results in an energy transfer from smaller bubbles to larger ones, which undergo particularly violent collapses. An optical image of acoustic cavitation excited from a quadratic oscillator with a frequency of 28 kHz is demonstrated in Fig. 2.12a which was taken with a CCD camera using a Xenon-flash with 2 μs duration as a light source. A filamentary structure of streamers appears at near the pressure antinode arising in the standing wave of the sound field in water. Individual branches of these streamers consist of small bubbles with a variety of sizes, each bubble behaving almost spherically as shown in Fig. 2.12b. A scale of a platinum wire of 50 μm diameter can be seen at the lower position. For weak shock pressures, therefore, a spherical bubble dynamics will determine which factors essentially affect shock-induced bubble collapse. Church [38] discussed the interaction of an ESWL induced shock wave with spherical bubbles initially at equilibrium in an infinite liquid. The intense acoustic wave generated at the focus of the ESWL was modeled as the impulse response of a parallel RLC circuit. The bubble radius vs. time curves are calculated by numerical integration of the Gilmore–Akulichev formulation for bubble dynamics. Similarly the shock wave–inertial microbubble

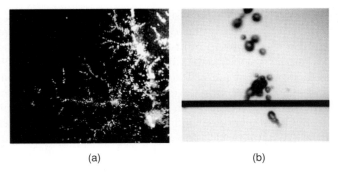

(a)   (b)

Fig. 2.12. Acoustic cavitation: (**a**) a filamentary structure of streamers generated at near the pressure antinode; (**b**) an enlargement of a part of one branch of the streamers. The diameter of a bar in (**b**) is 50 μm

interaction has been theoretically studied by simulating different lithotripters, namely, a modified XL-1, a Dornier HM-3 and an electrohydraulic lithotripter with pressure-release reflector. For calculations of bubble dynamics associated with a modified XL-1 lithotripter the Gilmore formulation was applied, coupled with zeroth-order gas diffusion [39]. Tomita et al. [10] investigated the effect of the pressure profile on the maximum bubble wall pressure, p_{max}, for three different pressure impulses and obtained the result shown in Fig. 2.13. It was found that the maximum pressure value is significantly affected by such factors as the rise time of pressure waves, which is especially important for small bubbles but insensitive to large bubbles, and also the pressure impulse denoted by F_t. Of course a larger impulse is required for a larger bubble to collapse intensively. Finally, it is concluded that an optimum bubble size at which the bubble collapses most violently exists for a definite amount of impulse. Nevertheless we should pay attention to the fact that if a bubble is relatively small, the induced pressure impulse due to bubble collapse will also be small because of the short time scale of the high pressure duration, even though the maximum pressure exceeds the yield strength of common materials. Then, the induced impulse is insufficient to distort the material surface [9–11]. Figure 2.14 shows the variations of p_{max}, t_c being the bubble collapse time, and R_{min}, termed the minimum bubble radius, as a function of the initial bubble radius R_e [11]. A comparison was made between theory and experiment for the data of t_c and R_{min} for two different applied pressure waves. However, because of the difficulty to measure p_{max} accurately, no experimental data were presented for p_{max}. A good agreement is found between the theory and experiment for the t_c–R_e-curve. Therefore it is understood that the theory is a useful tool to estimate a specified equivalent bubble radius at which the value of R_{min} takes a minimum, thus leading to the evaluation of the maximum impulsive pressure. For instance, R_e is about 0.2 mm when $p_s = 5$ MPa.

Recently, the shock wave–cell interaction has received much attention because nano/microbubbles are capable of enhancing noninvasive cytoplasmic molecular delivery in the presence of ultrasound, which enables transient

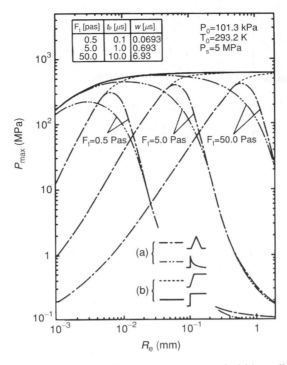

F_t [pas]	t_b [μs]	w [μs]
0.5	0.1	0.0693
5.0	1.0	0.693
50.0	10.0	6.93

Fig. 2.13. Effect of pressure profile on the maximum bubble wall pressure (Tomita et al. [10])

permeabilization of the cell membrane, a phenomenon called sonoporation. To gain maximum impulsive pressures Kodama et al. [20] calculated the dynamics of a bubble subject to an oscillating pressure field with a frequency of 1 MHz by adopting the Keller–Miksis model [40] in order to estimate the limiting domain below which cell permeabilization can occur. At every bubble rebound the liquid surrounding the bubble is rapidly accelerated, eventually producing a shock wave. The peak pressure of a shock wave decays as almost $1/r$ in agreement with the acoustic approximation. The shock wave energy is generally defined as the time integrated energy flux across an area where a shock wave front arrives [41]. It is expressed in the following form based on the acoustic approximation [42]:

$$E_s = \frac{2\pi r_s^2 p_s^2 t_s}{\rho_0 c_w \ln 2},$$
(2.3)

where r_s is the radial distance of the shock front from the origin, p_s the pressure behind the shock front, c_w the sound velocity of undisturbed water and t_s is the FWHM pulse duration of a shock wave (i.e., the time satisfying the condition $p = (p_s + p_0)/2$). Koshiyama et al. [43] investigated the mechanism of molecular delivery into the cytoplasm using a molecular dynamics (MD) simulation.

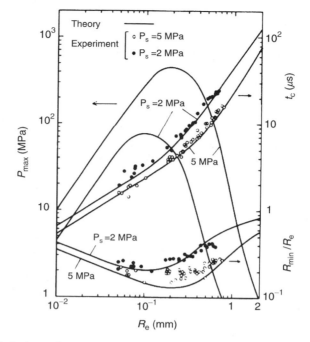

Fig. 2.14. Variations of p_{max}, t_c, and R_{min}/R_e as functions of R_e (Shima et al. [11])

The MD simulation determined the relationship between the uptake of water molecules into the lipid bilayers and the shock wave impulse. The interaction of a shock wave with a lipid bilayer induces a structural change of the bilayer and subsequently increases the fluidity of each molecule. These changes in the bilayer may be an important cause in the use of shock waves to produce transient membrane permeability.

2.3 Shock Wave–Bubble Interaction near Boundaries

Shock impact induced bubble collapse is known as one of the effective physical processes relevant to the generation of high pressures and high temperatures. If this phenomenon occurs in the vicinity of a solid boundary, the resultant pressures could attack against the boundary surface frequently, resulting in cavitation damage. Figure 2.15 is a schematic illustration of the mechanism of cavitation damage proposed by Tulin [29] who has set up a hypothesis that a shock wave due to implosion of one bubble interacts with the other bubble attached to the solid surface, causing an ultra-high-speed liquid jet which eventually impacts on the surface. It is well recognized that during ESWL treatment shock waves are generated and focused to disintegrate calculi in the kidney or gallbladder. Subsequently cavitation bubbles are created

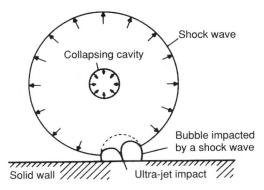

Fig. 2.15. Mechanism of cavitation damage suggested by Tulin [29]

by tensile stresses due to negative pressures following the incident shock pressures. These bubbles with diameters ranging from 10–100 μm will collapse due to the next shock waves. Figure 2.16a shows a high-speed shadowgraph indicating cavitation bubbles generated around a model calculus made of an iron sphere with 10 mm diameter, which was positioned at the second focal point of the reflector [17]. The interframe time is 1μs. As indicated in the 5th frame a shock wave is emitted from bubble A, specified in the 3rd frame, when it starts to rebound and interacts with bubble B located at 1.04 mm apart from bubble A. The time variations of the radii of the two bubbles A and B are shown in Fig. 2.16b. Immediately after the interaction with the shock wave, bubble B stops its motion to expand and starts to shrink. As already observed in Fig. 2.1b, a strong interaction between reflected shock waves and bubbles can generate many microscopic shock waves emitting at each bubble rebound. The curvature of the boundary surface must be one of important factors affecting both bubble motion and shock wave phenomena. Stress waves reflected from the curved rear surface of a calculus focus at a certain position from the rear surface, enhancing its amplitude much more efficiently than that reflecting from a plane interface. The bubble motion and liquid jet behavior are significantly influenced by the surface curvature. These results were confirmed by Tomita et al. [44] who studied the growth and collapse of cavitation bubbles near a curved rigid boundary and found that when a bubble collapses near a convex-shaped boundary, a liquid jet could be developed to achieve its velocity much higher than that of the flat boundary case. A detailed experiment on surface geometry effects in liquid/solid impact has been conducted by Dear and Field [45]. To obtain more detailed information of high pressure generation due to shock wave–bubble interaction, a number of experiments has been carried out in the presence of a solid wall. Using shock waves generated at the instant of underwater spark discharges, shock induced bubble collapses have been investigated with the help of pressure measurements for various kinds of bubble arrangements such as a single air bubble in the vicinity of a solid

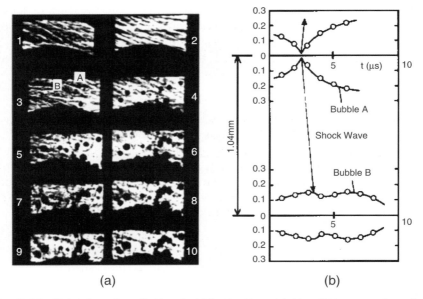

Fig. 2.16. Behavior of cavitation bubbles in the vicinity of the curved surface:
(a) high-speed photographs; (b) time variation of radii of two bubbles, A and B,
specified in (a) (Tomita et al. [17])

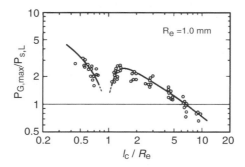

Fig. 2.17. A region where a gas bubble behaves as a source generating more intensive impulsive pressure than the impact pressure induced by shock wave loading
directly on the solid surface in the absence of a gas bubble (Shima et al. [46])

wall [46] and one or two bubbles attached to a solid wall [47,48]. For the case of
air bubbles being positioned near a solid wall and loaded by a spherical shock
wave with 5 MPa peak pressure, the relationship between impulsive pressure
$p_{G,max}/p_{S,L}$, measured with a pressure transducer mounted flush with an upper solid wall, and bubble position l_c/R_e from the wall surface is obtained
as indicated in Fig. 2.17. The shock pressure $p_{S,L}$ at the distance L was measured with a different pressure transducer mounted flush with a lower solid

wall. The distance L from the shock source to the lower solid wall where no bubbles were present was the same as that from the shock source to the upper solid wall. Clearly there exists a region of pressure amplification that is fulfilling the condition of $p_{G,max}/p_{S,L} > 1$, for $l_c/R_e \lesssim 7$. This evidence associated with the shock wave–bubble interaction suggests that damage capability may be elevated as a result of the avalanche effect in bubble collapse. Consequently a gas bubble located in the region of $l_c/R_e \lesssim 7$ behaves as a source capable of enhancing the collapse pressure to a value larger than the shock wave pressure directly impacting on the solid surface in the absence of gas bubbles.

When a liquid jet impacts against the solid surface, a shock wave occurs inside the jet medium due to liquid compressibility and travels back toward the origin of the jet. Figure 2.18 explains this process called "water hammer phenomenon." The water hammer pressure due to the cylindrical liquid jet impact on a flat surface can be expressed by

$$p_w = \frac{\rho_1 c_1 \rho_2 c_2 V_j}{\rho_1 c_1 + \rho_2 c_2},$$
(2.4)

where ρ_1, ρ_2 and c_1, c_2 are the densities and the shock wave velocities of the water and solid, respectively [49]. The water-hammer pressure lasts for an extremely short period restricted by the release wave, which is generated at the contact edge of the jet, to reach the central point. It is given by

$$\tau = \frac{r}{c_1},$$
(2.5)

where r is the radius of the impacting jet and c_1 the shock wave velocity in the liquid. For a rigid boundary, the water-hammer pressure becomes $p_w = \rho_1 c_1 V_j$ because of the condition $\rho_1 \ll \rho_2$, and if a liquid jet impacts on a free surface it can be reduced to 0.5 $\rho_1 c_1 V_j$ since $\rho_1 = \rho_2$. When a liquid jet with a curved tip impacts on a solid surface, the induced pressure inside the jet is not

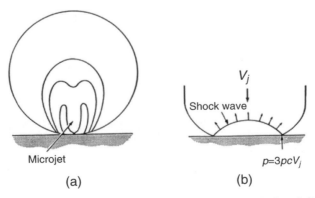

Fig. 2.18. Conceptual illustration of (a) bubble collapse induced liquid jet and (b) subsequent water hammer phenomena

uniform [50] but the "edge" pressures reach their maximum value of $\sim 3\rho c_1 V_j$ just before shock detachment [51]. There are several excellent achievements associated with jet impact phenomena. In a theoretical approach, Lush [52] studied the liquid mass impact against a perfectly plastic solid and Zhong and Chuong [53] numerically calculated the propagation of shock waves in elastic solids, whereas Obara et al. [54] and Bourne et al. [55] carried out valuable experiments on the liquid jet impact on the liquid and solid surfaces associated with cavitation damage and rain erosion problems. Sanada et al. [56] observed the shock wave–bubble interaction by employing double exposure holographic interferometry. Initially an air bubble with the diameter of 1.7 mm was positioned on the epoxy plate and impacted by a shock wave with 39 MPa peak pressure. Figure 2.19 is an optical image of density contours visualized both in water and in solid, which was taken at 21 μs after the ignition. An initial state of the bubble is visible as a nearly hemispherical shadow in the photograph. In addition a flattened dense black image is the bubble collapsing in contact with the epoxy surface. When a liquid jet impacts on the epoxy resin surface, shock waves occur and propagate into both regions of water and solid. The shock wave in the epoxy resin, marked by A in Fig. 2.19, travels with the velocity of $2,500\,\mathrm{ms^{-1}}$, followed by the stress fringes of higher order, marked by B, which originate from the central position where the liquid jet has collided. A pronounced wave, marked by C, can be seen at the top of the bubble collapsing in the water as a part of a circular-shaped wave. This must be the shock wave generated in the liquid jet, emerging from the top of the bubble surface. Supposing the phenomena to be axis-symmetric, we can determine the distribution of the refractive index in the domain of interest with the help of the fringe distributions $S(y)$ at any location which follow from the double exposure holographic interferograms [57]. Using the Lorentz–Lorenz equation we obtain a relationship between the refractive index of water n_i and density ρ_i given by

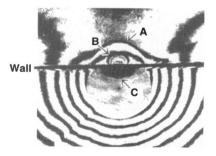

Fig. 2.19. Double exposure holographic interferogram indicating the interaction of a primary shock wave with an air bubble on an epoxy resin, which was taken at 21 μs after the ignition (Sanada et al. [56]). For the explanation, see text. From Sanada et al. [56], reproduced with permission of N. Sanada

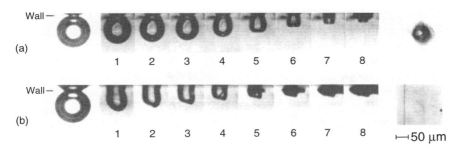

Fig. 2.20. Microjet formation inside an air bubble loaded by a spark-induced shock wave with the conditions of (**a**) $R_e = 0.29$ mm, $p_s = 2.2$ MPa and (**b**) $R_e = 0.3$ mm, $p_s = 2.4$ MPa. The jet in the case (**a**) impacts against the indium surface with a velocity of more than $200\,\mathrm{ms^{-1}}$, subsequently leaving a damage pit on the surface as shown at the right-hand corner. Asymmetric damage (**b**) is marked on the surface when it was loaded by an oblique shock wave. Interframe time is $0.5\,\mu$s (Tomita et al. [58])

$$\rho_i = \frac{n^2 - 1}{R\,(n_i^2 + 2)}, \tag{2.6}$$

where R is a constant value associated with the molar refraction. Eventually we can determine the pressure distribution in the liquid surrounding a bubble by knowing the relationship between the density and pressure. Usually a corrected Tait equation [41] is used as equation of state for water.

In Sect. 2.2, we mentioned the optimum bubble size corresponding with a maximum pressure for a specified pressure impulse (or a shock wave impulse). Verification of this evidence can also be made from a damage experiment with the use of a very soft material, for instance, 99.9 pure indium whose Vickers hardness is 1.6 and the melting point 429K. Figure 2.20 shows two typical high speed photographs of the shock wave–bubble interaction taken with a rate of 2,000,000 frames $\mathrm{s^{-1}}$ [58]. Case (a) ($p_s = 2.9$ MPa, $R_e = 0.29$ mm) is the result for an air bubble initially attached onto the indium surface due to buoyancy and impacted by a spark discharge-induced shock wave. The shock wave was coming from below, i.e., the direction perpendicular to the solid surface. The other case (b) ($p_s = 2.4$ MPa, $R_e = 0.30$ mm) is the situation where a shock wave came from an oblique direction, different from the axis of symmetry, to meet the bubble surface. In both cases, the separate frame at the left-hand side corresponds to the initial state of individual bubbles, and at the right-hand corner a microscopic image of the subsequent damage pit is marked. It should be noted that although the indium surface did undergo incident shock impact many times, no appreciable damage pits produced by shock impact alone when no bubbles were left on the indium surface. After the shock wave loading, a bubble collapses nonspherically due to the pressure gradient in the direction of shock wave passage, resulting in the formation of a high-speed liquid jet which finally pierces the bubble and impacts against

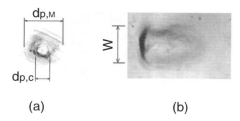

(a)   (b)

Fig. 2.21. Definition of (a) symmetric and (b) asymmetric damage pits

the solid surface. The impact velocity of the liquid jet can be estimated to be more than $200\,\mathrm{ms^{-1}}$ for Fig. 2.20a. The resultant damage pit is about $50\,\mu\mathrm{m}$ in diameter which correlates well with the jet diameter. When a shock wave interacts with a bubble from a direction different from the symmetric axis of the bubble, the impacted bubble surface initially deforms along the shock direction with the velocity derived by (2.1). Afterwards, the jet changes its direction drastically due to the limited flow field bounded by a solid wall in virtue of momentum conservation. Subsequently, an oblique liquid jet impacts on the indium surface, producing an asymmetric damage pit. In Fig. 2.21a definition of damage pit parameters is given for two typical examples: case (a) including a symmetric damage pit defined as $d_{\mathrm{p,c}}$, i.e., the central pit diameter and $d_{\mathrm{p,M}}$, the outer pit diameter, and case (b) indicating an asymmetric damage pit when $p_s = 5\,\mathrm{MPa}$ and $R_\mathrm{e} = 0.16\,\mathrm{mm}$ [10]. As clearly seen in Fig. 2.21b, the asymmetric damage might be caused by both compression and shear forces due to the impact of an oblique liquid jet.

Using numerous numbers of indium specimens and employing shock waves with different strengths, damage experiments were carried out for a wide range of initial bubble sizes. Figure 2.22 is a map of the damage pits consisting of weak pits and intense indentations [58]. A cross symbol indicates the case where no damage occurs, an open circle refers to shallow pits and a solid circle corresponds to the case of intense damage pits. A threshold condition for the damage is denoted by a solid curve. In the region below this no damage occurred. A minimum pressure required for the damage pit formation is about $14\,\mathrm{MPa}$ for initial bubble radii between 0.2 and 0.3 mm.

A dimensionless damaged pit diameter is shown in Fig. 2.23 for three peak pressures of 2, 3.5, and 5MPa [58]. In this figure, the symbols of open and solid circles correspond to $d_{\mathrm{p,M}}$ and $d_{\mathrm{p,c}}$, and a cross means the liquid jet diameter d_j. All of these variables are normalized by the initial bubble radii, R_e. A good correlation between the diameters of the circular dent and liquid jet, which is about 0.08 times the initial bubble diameter, is obviously obtained. By increasing the shock wave pressure for a constant bubble size, various forms of damage appeared on the indium surface. For a shock pressure slightly larger than the limiting condition, a shallow depression was marked with the same diameter of a liquid jet. For a shock wave with higher peak pressure, a more violent bubble collapse occurs. To find out a threshold jet

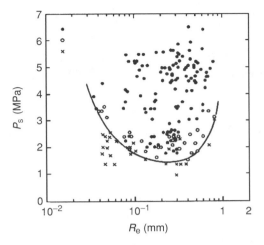

Fig. 2.22. Diagram of shock wave pressure vs. initial bubble radius, including the criteria for damage pit formation (Tomita et al. [58])

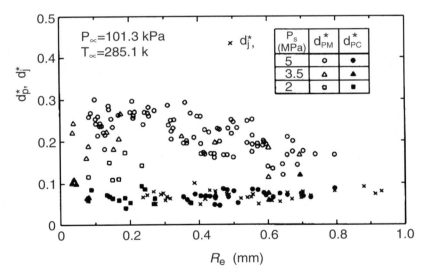

Fig. 2.23. Correlation of pit diameters $d_{p,M}{}^{*}$ and $d_{p,c}{}^{*}$ with jet diameters $d_j{}^{*}$ for various initial bubble radii R_e (Tomita et al. [58])

velocity for causing damage, the relationship between the liquid jet velocity V_j and pit diameter $d_{p,c}{}^{*}$ was examined for bubbles in radii of 0.4–0.79 mm with the shock wave pressure being kept 5 MPa [9]. Figure 2.24 is the result, demonstrating clearly that a jet velocity of over 200 ms^{-1} is needed to cause a damage pit on the indium surface. It is meaningful to mention that no damage occurs when a liquid jet with a velocity of about 100 ms^{-1} impacts

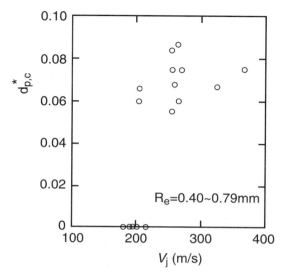

Fig. 2.24. Threshold impact velocity of a liquid jet for causing a damage pit (Tomita and Shima [9])

on an indium surface. This is a normal jet velocity for an isolated bubble collapsing very close to a boundary without shock loading [59]. The water hammer pressure corresponding to the velocity of $200\,\mathrm{ms^{-1}}$ is evaluated to be $300\,\mathrm{MPa}$, followed by the stagnation pressure of $20\,\mathrm{MPa}$ which lasts for a relatively long time after the release of the water-hammer pressure. It is known that for indium the water-hammer pressure is more than 15 times its yield strength and the stagnation pressure is roughly of the same order as the yield strength. From (2.5) the duration of the water hammer pressure is between $10\,\mathrm{ns}$ and $100\,\mathrm{ns}$. Generally, plastic deformation depends on the amount of dislocation which moves with the velocity of transverse wave through a solid layer. For impulsive pressure loading with very short duration, however, there is less time for dislocations to grow, so that a significantly larger pressure is required to deform a material. The optimum bubble size varies with the degree of pressure loading, and the existence of an optimum condition has been confirmed by Philipp et al. [12] who performed an experiment on the interaction between lithotripter-generated shock waves with $65\,\mathrm{MPa}$ strength and air bubbles. Figure 2.25 is their result indicating that the maximum jet velocity of $770\,\mathrm{ms^{-1}}$ occurred at the initial bubble size of $0.55\,\mathrm{mm}$. They explained that the maximum jet velocity is related not only to the temporal profile of the shock wave but also to the effective cross section of the bubble for shock wave energy transfer. Figure 2.26 shows the dimensionless asymmetric pit width, $W/(2Re)$, against the initial bubble radii, R_e, for shock strength of $p_s = 5\,\mathrm{MPa}$ [10]. The outer regions divided by two shaded lines imply no damage zones. It should be noted that the occurrence

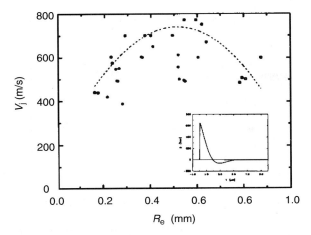

Fig. 2.25. Collapse jet velocity vs. initial bubble radius. The *dashed line* is a quadratic fit of the uppermost values (Philipp et al. [12]). (Courtesy of W. Lauterborn)

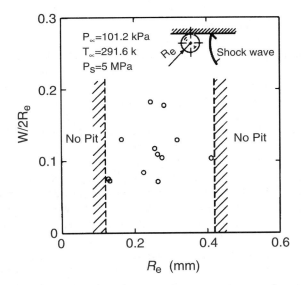

Fig. 2.26. Limited region capable of causing asymmetric damage pits (Tomita et al. [10])

of the damage pits is limited to a narrow region in the neighborhood of the optimum bubble radius, that is for radii ranging from 0.12 mm to 0.42 mm.

Tissue damage associated with the ESWL was well recognized to induce an undesirable complication. Major tissue damage often results from miss-shots of shock focusing. Due to the difference in acoustic impedances, strong shear forces or tensile forces can be generated between tissue and calculi or tissue and water near the second focal point of a reflector. This seems to be one possible cause of tissue damage. The destructive action of cavitation bubbles collapsing near boundaries was also pointed out by Crum [60]. Since arteries are usually located very close to the surface of a kidney or a gallbladder, i.e., within several fractions of a millimeter, the arteries would be damaged due to the impact of the liquid microjets formed as a result of the shock wave–bubble interaction [32, 61, 62]. So, the shock wave–bubble interaction near deformable boundaries is an important subject to explore, because human tissue can be regarded as a viscoelastic material, which is dynamically non-linear, anisotropic, and inhomogeneous. The experiments using either a free surface [55] or an elastic boundary [63] suggested that the bubble collapse-induced pressures depend on the dynamic properties of the surrounding tissue.

A number of research works have been conducted on cavitation bubble dynamics near compliant surfaces [13, 64–67]. Figure 2.27 shows a high-speed photograph of a simulation of tissue damage due to shock wave–bubble interaction by employing gelatin as a boundary. A very fine liquid jet can be seen, which penetrates into the gelatin layer. The final penetration depth of this jet depends not only on the shock wave profile but also on the bubble size and on the physical properties of the gelatin mixture. To elucidate the mechanism of tissue damage due to cavitation bubble collapse, Kodama and Takayama [68] performed an experiment on the interaction of shock waves with air bubbles positioned at either gelatin surfaces or at extirpated livers

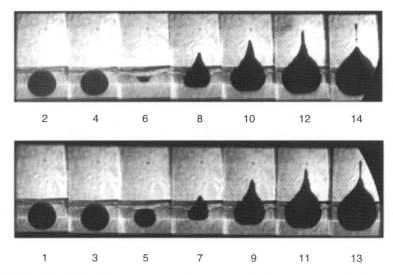

Fig. 2.27. Needlelike jet penetration into a gelatin layer due to shock wave-air bubble interaction. Interframe time is $10\,\mu$s (Lush et al. [32])

of rats. A shock wave generated by laser focusing was applied to interact with air bubbles attached to gelatin surfaces for a study of in vivo bubble dynamics [63]. Furthermore Kodama et al. [69] studied the interaction of an underwater shock wave with two air bubbles attached to a gelatin surface. For exploring an effective technology for revascularization of cerebral embolism, Kodama et al. [70] investigated the interaction of a shock wave with a gas bubble attached to an artificial thrombus, made of either gelatin or clotted blood, inserted into a tube model of a cerebral artery and obtained the penetration depth of the liquid jet as a function of time elapsed from initial jet penetration into the gelatin. Wolfrum et al. [71] conducted an experiment on the interaction of a pressure wave generated by an underwater spark discharge with a Levovist contrast agent bubble, and observed induced rapid phenomena by means of high-speed photography coupled with a microscope. Furthermore they investigated shock wave–microbubble interaction near a cell membrane in order to make clear the mechanism of sonoporation and cell disruption [72]. They found that even low pressure amplitudes are sufficient to cause strong bubble expansion with visible damage to the cells upon subsequent collapse. More recently Ohl and Ikink [19] suggested that shock–bubble interaction induced jetting might act as a micro-syringe, capable of realizing temporary permeabilization and transfection of biological cells.

2.4 Bubble Collapse Induces High Temperature and Sonoluminescence

Light emission, later termed sonoluminescence, was first discovered for multi-bubble systems in 1933. It was observed that when a photographic plate was immersed into a water container, it became black by ultrasonic irradiation [73]. The effect was considered due to light emission accompanied by acoustic cavitation [74]. In fact acoustic cavitation can lead to energy concentration into an extremely small region of fluid, achieving high values of temperature and pressure within the collapsing cavities whose values may be in excess of thousands of degrees Centigrade and more than 1 GPa, respectively. This makes the bubble an intense microreactor with a variety of chemical reactions within and surrounding the bubble. Under certain conditions, the concentrated energy is sufficient to generate light emission, termed sonoluminescence. The sonoluminescence from many bubbles is termed multibubble sonoluminescence (MBSL), while Gaitan [75] first reported experiments on single-bubble sonoluminescence (SBSL). Soon later Barber and Putterman [76] found that the light pulse duration is 50 ps at most and the spectral analysis has revealed that MBSL can be fitted by the blackbody formula with effective temperatures ranging from 10,000 to 50,000 K. However, a more recent measurement by Gompf et al. [77] suggested that the width of the light pulse is actually of the order of a few hundred picoseconds as illustrated in Fig. 2.28 as an example. One of reliable mechanisms of sonoluminescence was pointed out by

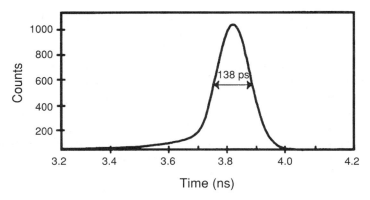

Fig. 2.28. Pulse shape from SBSL [77]. (Courtesy of B. Gompf)

Wu and Roberts [78], who explained that shock waves detach from the bubble surface during the compression of the gas contents and focus at the center. The resultant shock focusing at the bubble center compresses the gas so strongly that light can be emitted, either by ionization and subsequent bremsstrahlung or by blackbody radiation. Later Roberts and Wu [79] discussed the state of the shock wave theory of sonoluminescence in more detail. From the theoretical site Yasui [80] proposed to discuss in detail the internal gas dynamics taking into account chemical reactions and nonequilibrium phase change in a collapsing air bubble under SBSL conditions. Until now there are a number of excellent reviews [81–84] and books [7,18] associated with sonoluminescence. While spherical bubble oscillations are found in acoustic cavitation dynamics for low driving pressures as mentioned before, there exists another situation of nonspherical bubble collapse characterized by liquid jet formation when the driving pressures are high. Under these circumstances strong interactions of shock waves emitted from individual bubbles with various sized bubbles may be frequently encountered. Experimentally numerous results regarding nonspherical bubble collapse and related phenomena have been obtained at the Cavendish Laboratory (e.g., Bourne and Field [85]). Two possible mechanisms may be responsible for the generation of high temperatures resulting from a rapid collapse of a bubble loaded by a shock wave. One is the adiabatic compression of gas inside the bubble. A second possible mechanism is shock heating of a pocket of gas trapped between the jet and the far cavity wall at the moment before impact. Assuming that a gas compresses adiabatically, the temperature inside the bubble is given by

$$T_2 = T_1(V_1/V_2)^{\kappa-1}, \tag{2.7}$$

where T_1, T_2 and V_1, V_2 are the initial and final temperatures and volumes, respectively, and κ is the ratio of the specific heats. Chaudhri and Field [86] concluded that adiabatic heating of a gas bubble impacted by a shock wave was the prime cause for initiation of condensed explosives. Bubbles of argon

($\kappa = 1.67$), helium($\kappa = 1.67$), and air($\kappa = 1.4$) with diameters of about $50\,\mu m$ caused initiation. For butane ($\kappa = 1.08$) with the same diameter no initiation occurred, but bubbles greater than about $300\,\mu m$ were able to cause initiation. Considering the initial temperature of $T_1 = 300\,K$ and taking a reasonable volume ratio $V_1/V_2 = 64$, we obtain $T_2 > 1,500\,K$ for air and $T_2 > 4,800\,K$ for argon and helium, but $T_2 = 418\,K$ for butane which is too low to cause initiation. Generally translational and rotational modes of energy of each molecule are excited within times of $0.1\,ns$ to $10\,ns$, but the transfer of energy to vibrational modes can take place as long as $1\,\mu s$ [87]. To complete heat conduction effectively, a sufficient amount of heat energy is required. This leads to the condition that the high temperature should be maintained during at least several μs, which is possible when a bubble has relatively larger size. Ohl et al. [88] investigated light emission from single collapsing laser induced cavitation bubbles, i.e., single cavitation bubble luminescence (SCBL) for both cases of spherical and nonspherical bubble collapses. They found that luminescence can occur at the bubble center for spherical bubble collapse and at the bubble site of jet impact for nonspherical bubble collapse. Baghdassarian et al. [89] reported several important results obtained by using a spectrometer, expressing that for small bubbles in water the spectrum exhibits a similar trend to that of a black-body at about 7,800K, whereas for bubbles larger than about 1mm maximum radius the OH* molecular band at $310\,nm$ becomes apparent. In the larger bubbles the surface instabilities at the collapse point are induced, often causing the bubble to split into two smaller bubbles just before the collapse. In this case double flashes from two splitting bubbles evidentially occur.

Bourne and Field [90] investigated the shock-induced cavity collapse and luminescence with the help of high-speed photography. Figure 2.29 shows a schematic illustration of the luminescence events. The collapse of a two dimensional cavity was triggered by a 1.88 GPa shock wave, yielding the formation of a liquid jet which impacted at the downstream cavity wall before the incident shock had arrived at the same position. The source of the initial bright luminescence, J, is believed to be the violent shock heating of a pocket of gas trapped between the jet and the far cavity wall at the moment before impact. The principal ignition mechanism appears to be the impact of the jet. Hydrodynamic compression of the material at the downstream wall causes the temperature rise necessary to trigger the onset of reaction. A similar phenomenon has recently been obtained by Ohl et al. [88]. In Fig. 2.29, two flashes denoted by L correspond to the position of the lobes of trapped gas isolated by jet penetration of the downstream wall. Spectral analysis suggests that the cause is due to radiation from the thermally excited ionized gas contents, or chemiluminescence arising from the reactions involving photons [85]. These reaction steps might include recombination of radicals. For water a continuum spectrum of wavelengths shifted to the blue has been observed. The spectral-intensity distribution can be fitted to that of a black-body radiator with a color temperature of 8,800K [85].

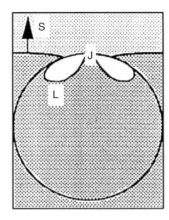

Fig. 2.29. Schematic indicates the relative positions of the flashes J and L. The luminescence J results from gas trapped between the jet tip and downstream cavity wall just before the jet impacts on the downstream wall and the luminescence L from two lobes of entrapped gas (Bourne and Field [90])

Conclusions

Shock wave–bubble interaction and related phenomena have been reviewed. An important characteristic of these phenomena is that energy is concentrated in a small region of liquid, leading to high values of pressure and temperature though only during an extremely short interval of time. At the final stage of the collapse of a bubble a high-speed jet is formed and at the rebound shock waves are emitted. When these phenomena occur either in the neighborhood of the boundary surface or attached to it, high impulsive pressures are inevitably generated. Shortness in duration of these impulsive pressures may be significantly important for microscopic cleaning and fine destruction or disruption of materials which are relevant to a variety of elementary processes in engineering and in medical applications. It is found that there is an optimum condition of bubble collapse impacted by a shock wave at which the maximum pressure or highest jet velocity can be achieved, depending on the characteristics of shock wave profile and bubble geometry as well. Recently the interaction of minute bubbles with shock waves emitted from nano/microsized bubbles under ultrasonic wave has been attracted by many researchers in connection with drug delivery. On the other hand, a violent collapse of a bubble induces a high temperature of the gas inside the bubble, leading to sonoluminescence. An interesting suggestion has recently been made by Ohl et al. [88] who assumed that the shape of the light emission could be an image of the bubble scattering the light from its interior. Heat transfer during this process will cause various kinds of chemical and physical reactions. To elucidate the mechanism of light emission a more essential and detailed basic study should be made in the future by employing femtosecond laser pulses which can provide new phenomena [91,92]. However a puzzle concerning the instability of bubble surface at the very late stage of the collapse still remains to be solved especially for bubbles with much smaller size. Under some conditions the light generated inside the bubble might be unable to pass from the interior to the surroundings of the bubble due to diffused reflection at the irregular bubble surface, creating an opaque bubble.

References

1. Shima, A., Takayama, K., Tomita, Y., Miura, N.: Acustica **48**, 293 (1981)
2. Ellis, A.T.: On jets and shock waves from cavitation. In: Proceedings of the 6th Symposium on Naval Hydrodynamics, Washington, DC, U.S.A., Oct. pp. 137–161 1966
3. Dear, J.P., Field, J.E.: A study of the collapse of arrays of cavities. J. Fluid Mech. **190**, 409–425 (1988)
4. Testud-Giovanneshi, P., Alloncle, A.P., Dufresne, D.: J. Appl. Phys. **67**, 3560 (1990)
5. Hansson, I., Mørch, K.A.: J. Appl. Phys. **51**, 4651 (1980)
6. Lauterborn, W.: Cavitation and Inhomogeneities in Underwater Acoustics. Springer, Berlin Heidelberg New York (1980)
7. Leighton, T.G.: The Acoustic Bubble. Academic, London (1994)
8. Brennen, C.E.: Cavitation and Bubble Dynamics. Oxford University, New York (1995)
9. Tomita, Y., and Shima, A.: J. Fluid Mech. **169**, 535 (1986)
10. Tomita, Y., Shima, A., Sugiu, T.: Mechanisms of impulsive pressure generation and damage pit formation by bubble-shock wave interaction. In: Murai H. (ed.) Proceedings of the Interational Symposium on Cavitation, pp. 77–82 Institute of High Speed Mechanics, Tohoku University, Sendai, (1986)
11. Shima, A., Tomita, Y., Sugiu, T.: AIAA J. **26**, 434 (1988)
12. Philipp, A., Delius, M., Scheffczyk, C., Vogel, A., Lauterborn, W.: J. Acoustic Soc. Am. **93**, 2496 (1993)
13. Blake, J.R., Gibson, D.C.: Ann. Rev. Fluid Mech. **19**, 99 (1987)
14. Philipp, A., Lauterborn, W.: J. Fluid Mech. **361**, 75 (1998)
15. Steinberg, D.J.: J. Stone Dis. **5**, 49 (1993)
16. Vogel, A., Hentschel, W., Holzfuss, J., Lauterborn, W.: Ophthalmology **93**, 1259 (1986)
17. Tomita, Y., Obara, T., Takayama, K. and Kuwahara, M.: Shock Waves: Int. J. **3**, 149 (1994)
18. Srivastava, R.C., Leutloff, D., Takayama, K., Grönig, H.: Shock Focussing Effect in Medical Science and Sonoluminescence. Springer Berlin Heidelberg New York (2002)
19. Ohl, C.D., Ikink, R.: Phys. Rev. Lett. **90(21)**, 214502-1 (2003)
20. Kodama, T., Suzuki, M., Tomita, Y., Koshiyama, K., Yano, T., Fujikawa, S.: Interaction of impulsive pressures of cavitation bubbles with cell membranes during sonoporation. In: Clement G.T., Hynynen, K. (eds.) 5th International Symposium on Theraptutic Ultrasound, American Institute of Physics, New York, 2005. Paper No T067

21. Tezel, A., Mitragotri, S.: Biophys. J. **85**, 3502 (2003)
22. Tomita, Y., Yamada, K., Takayama, K.: Surface deformation of a gas bubble impacted by a shock wave. In: Takayama K. (ed.) Proceedings of Symposium on Shock Waves, JAPAN'93, pp. 257–260 Institute of Fluid Science, Tohoku University, Sendai, (1993) (in Japanese)
23. Bourne, N.K., Field, J.E.: J. Fluid Mech. **244**, 225 (1992)
24. Grove, J.W., Menikoff, R.: J. Fluid Mech. **219**, 313 (1990)
25. Masubuchi, H., Takahira, H., Akamatsu, T.: Numerical analysis of the bubble collapse by ALE method. In: Kato H. (ed.) Proceedings of 7th National Symposium on Cavitation, pp. 83–87, Japanese Science Council, Tokyo, (1992) (in Japanese)
26. Lord Rayleigh: Phil. Mag. **34** , 94 (1917)
27. Ding, Z., Gracewski, S.M.: J. Fluid Mech. **309**, 183 (1996)
28. Bowden, F.P., Brunton, J.H.,: Proc. R. Soc. Lond. A. **263**, 433 (1961)
29. Tulin, M.P.: On the Creation of Ultra-Jets. In: L.I.Sedov 60th Anniversary Volume: Problems of Hydrodynamics and Continuum Mechanics (Soc. Ind. Appl. Math., Philadelphia,1969) pp. 725-747
30. Birkhoff, G., MacDougall, D.P., Pugh, E.M., Taylor, G.I.: J. Appl. Phys. **19**, 563 (1948)
31. Bourne, N.K., Field, J.E.: J. Appl. Phys. **63**, 1015 (1988)
32. Lush, P.A., Tomita, Y., Onodera, O., Takayama, K., Sanada, N., Kuwahara, M., Ioritani, N., Kitayama, O.: Air bubble-shock wave interaction adjacent to gelatine surface. In: Kim Y.W. (ed.) AIP Conference Proceedings 208, Current Topics in Shock Waves: Proceedings 17th International Symposium on Shock Waves and Shock Tubes, pp. 831–836 American Institute of Physics, New York, 1990
33. Kodama, T., Tomita, Y., Shima, A.: Interaction of a bubble attached to a gelatine wall with a shock wave (A study of tissue damage caused by bubble collapse), Trans. Jpn. Soc. Mech. Eng., Ser. B. **59**, 1431–1435 (1993) (in Japanese)
34. Brennen, C.E.: J. Fluid Mech. **472**, 153 (2002)
35. Postema, M., van Wamel, A., Lancee, C.T., de Jong, N.: Ultrasound Med. Biol. **30**, 827 (2004)
36. Sanada, N., Takayama, K., Onodera, O., Ikeuchi, J.: Observation of shock waves induced by vibratory cavitation test, Trans. Jpn. Soc. Mech. Eng., Ser. B. **50**, 2275–2284 (1984) (in Japanese)
37. Kuttruff, K.H.: J. Acoust. Soc. Am. **106**, 190 (1999)
38. Church, C.C.: J. Acoust. Soc. Am. **86**, 215 (1989)
39. Zhu, S., Zhong, P.: J. Acoust. Soc. Am. **106**, 3024 (1999)
40. Keller, J.B., Miksis, M.: J. Acoust. Soc. Am. **68(2)**, 628 (1980)
41. Cole, R.H.: Underwater Explosion. Princeton University Press, (1948)
42. Tomita, Y., Tsubota, M., An-naka, N.: J. Appl. Phys. **93**, 3039 (2003)
43. Koshiyama, K., Kodama, T., Yano, T., Fujikawa, S.: Molecular dynamic simulation of cell permeabilization induced by shock wave impulse. In: Ikohagi T.(ed.) 4th International Symposium on Advanced Fluid Information and 1st International Symposium on Transdisciplinary Fluid Integration, pp. 36–37 Institute of Fluid Science, Tohoku University, Sendai, 2004
44. Tomita, Y., Robinson, P.B., Tong, R.P., Blake, J.R.: J. Fluid Mech. **466**, 259 (2002)
45. Dear, J.P., Field, J.E.,: J. Appl. Phys. **63**, 1015 (1988)

46. Shima, A., Tomita, Y., Takahashi, K.: Proc. Instn. Mech. Engrs. **198C**, 81 (1984)
47. Tomita, Y., Shima, A., Takahashi, K.: Trans ASME I: J. Fluids Eng. **105**, 341 (1983)
48. Tomita, Y., Shima, A., Ohno, T.: J. Appl. Phys. **56**, 125 (1984)
49. Dear, J.P., Field, J.E., Walton, A.J.: Nature **332**, 505 (1988)
50. Hwang, J.-B., Hammitt, F.G.: J. Fluids Engng., Trans. ASME, Ser. I. **99**, 396 (1977)
51. Field, J.E., Lesser, M.B., Dear, J.P.: Proc. R. Soc. Lond. A. **401**, 225 (1985)
52. Lush, P.A.: J. Fluid Mech. **135**, 373 (1983)
53. Zhong, P., Chuong, C.J.: J. Acoust. Soc. Am. **94**, 19 (1993)
54. Obara, T., Bourne, N.K., Field, J.E.: Wear **186–187**, 388 (1995)
55. Bourne, N.K., Obara, T., Field, J.E.: Proc. R. Soc. Lond. A. **452**, 1497 (1996)
56. Sanada, N., Ikeuchi, J., Takayama, K., Onodera, O.: Interaction of an air bubble with a shock wave generated by a micro-explosion in water. In: Murai H. (ed.) Proceedings of the International Symposium on Cavitation, pp. 67–72 Institute of High Speed Mechanics, Tohoku University, Sendai, 1986
57. Abe, A., Takayama, K.: Spherical shock waves from the open end of a shock tube. In: Takayama K.(ed.) Proceedings of the 1988 National Symposium on Shock Wave Phenomena, pp. 41–56 Institute of Fluid Science, Tohoku University, Sendai, 1989
58. Tomita, Y., Shima, A., Takayama, K.: Formation and limitation of damage pits caused by bubble-shock wave interaction. In: Takayama K. (ed.) Proceedings of the 1988 National Symposium on Shock Wave Phenomena, pp. 149–160 Institute of Fluid Science, Tohoku University, Sendai, 1989
59. Tomita, Y., Shima, A.: Acustica **71**, 161 (1990)
60. Crum, L.A.: J. Urol. **140**, 1587 (1988)
61. Lush, P.A., Takayama, K., Tomita, Y., Obara, T.: Cavitation and induced shock wave-bubble interaction as a cause of human tissue damage in ESWL. In: Grist T. (ed.) Proceedings of the 3rd International Conference on Cavitation, pp. 55–64 Institution of Mechanical Engineers, London 1992
62. Lush, P.A., Tomita, Y., Obara, T.: Simulation of human tissue damage using air bubble collapse adjacent to gelatine layers. In: Takayama K. (ed.) Proceedings of the 18th International Symposium on Shock Waves, pp. 1181-1186 Springer-Verlag 1992
63. Kodama, T., Tomita, Y.: Appl. Phys. B. **70**, 139 (2000)
64. Shima, A., Tomita, Y., Gibson, D.C., Blake, J.R.: J. Fluid Mech. **203**, 199 (1989)
65. Brujan, E.-A., Nahen, K., Schmidt, P., Vogel, A.: J. Fluid Mech. **433**, 251 (2001)
66. Tomita, Y., Kodama, T.: Some aspects of the motion of two laser-produced cavitation bubbles near a free surface. In: King A.C., Shikhmurzaev Y.D. (eds.) IUTAM Symposium on Free Surface Flows, pp. 303–310 Kluwer, Dordrecht, 2001
67. Tomita, Y., Kodama, T.: J. Appl. Phys. **94**, 2809 (2003)
68. Kodama, T., Takayama, K.: Ultrasound in Med. Biol. **24**, 723 (1998)
69. Kodama, T., Takayama, K., Nagayasu, N.: J. Appl. Phys. **80**, 5587 (1996)
70. Kodama, T., Takayama, K., Uenohara, H.: Phys. Med. Biol. **42**, 2355 (1997)
71. Wolfrum, B., Mettin, R., Kurz, T., Lauterborn, W.: Appl. Phys. Lett. **81**, 5060 (2002)

72. Wolfrum, B., Kurz, T., Mettin, R., Lauterborn, W.: Phys. Fluids **15**, 2916 (2003)
73. Marinesco, M., Trillat, J.J.: Compt. Rend. **196**, 858 (1933)
74. Frenzel, H., Schultes, H.: Z. Phys. Chem. **27b**, 421 (1934)
75. Gaitan, D.F.: An experimental investigation of acoustic cavitation in gaseous liquids, Ph.D. dissertation, University of Mississippi (1990)
76. Barber, B.P., Putterman, S.J.: Nature **352**, 318 (1991)
77. Gompf, B., Gunther, R., Nick, G., Pecha, R., Eisenmenger, W.: Phys. Rev. Lett. **79**, 1405 (1997)
78. Wu, C.C, Roberts, P.H.: Phys. Rev. Lett. **70**, 3424 (1993)
79. Roberts, P.H., Wu, C.C.: The Shock-Wave Theory of Sonoluminescence. In: Srivastava, R.C., Leutloff, D., Takayama, K., Grönig, H. (eds.) Shock Focussing Effect in Medical Science and Sonoluminescence, pp. 1–27 Springer, Berlin Heidelberg New York 2002
80. Yasui, K.: Phys. Rev. E. **56**, 6750 (1997)
81. Matula, T.J.: Phil. Trans. R. Soc. Lond. A. **357**, 225 (1999)
82. Ohl, C.D., Kurz, T., Geisler, R., Lindau, O., Lauterborn, W.: Phil. Trans. R. Soc. Lond. A. **357**, 269 (1999)
83. Brenner, M.P., Hilgenfeldt, S., Lohse, D.: Rev. Mod. Phys. **74**, 425 (2002)
84. Lohse, D.: Nature **434**, 33 (2005)
85. Bourne, N.K., Field, J.E.: Phil. Trans. R. Soc. Lond. A. **357**, 295 (1999)
86. Chaudhri, M.M., Field, J.E.: Proc. R. Soc. Lond. A. **340**, 113 (1974)
87. Zeldovitch, Ya B., Raizar, Yu P.: Physics of Shock Waves and High Temperature Hydrodynamic Phenomena, Vol.1. Academic, New York (1966)
88. Ohl, C.D., Lindau, O., Lauterborn, W.: Phys. Rev. Lett. **80**, 393 (1998)
89. Baghdassarian, O., Chu, H.–C., Tabbert, B., Williams, G.A: Spectrum of luminescence from laser-induced bubbles in water and cryogenic liquids. In: Brennen, C.E. (ed.) Proceedings of the 4th International Symposium on Cavitation, California Institute of Technology, Pasadena, 2001 Paper No.A2.001
90. Bourne, N.K., Field, J.E.: Proc. R. Soc. Lond. A. **435**, 423 (1991)
91. Lauterborn, W., Kurz, T., Schenke, C., Lindau, O., Wolfrum, B.: Laser-induced bubbles in cavitation research. In: King, A.C., Shikhmurzaev, Y.D. (eds.) IUTAM Symposium on Free Surface Flows, Birmingham, pp. 169–176 Kluwer, Dordrecht (2001)
92. Brzoska, J.-B.: Private communication on "Bubble fusion: feasibility with the help of femtosecond light pulses"(2005)

Shock Induced Cavitation

Valery K. Kedrinskii

3.1 Introduction

For decades, fundamental studies of bubble cavitation were primarily focused on the state of a real liquid in terms of its inhomogeneity, on the mechanism of bubble cluster formation and liquid strength as well as on medical aspects of shock wave applications. The macroscopic structure of the liquid is such that even when the liquid is carefully purified by distillation or deionization there are always microinhomogeneities, which act as cavitation nuclei. Experiments show that microbubbles of free gas, solid microparticles or their conglomerates can play a role as cavitation nuclei. It means that the state of a real liquid can be determined as a multiphase one and corresponding mathematical models describing wave propagation and cavitation processes in liquids should be constructed. These models should take into account the character of the medium to have a possibility to analyze the structure and the parameters of the wave field as well as the limiting values of the tensile stress allowed by a cavitating liquid. The main results will be presented on the following problems.

Real liquid state. The nature of microinhomogeneities, their parameters, density, and size spectra were studied using an original technique as the combination of the shock tube and the scattering indicatrix method.

Formation mechanism of bubble clusters. Two approaches to the explanation of this effect (bubble form instability and/or successive saturation of cavitation zone by bubbles attaining visible size) will be discussed.

Mathematical model of cavitating liquid. Rarefaction waves (phases) in a real liquid are described using the two-phase mathematical IKW- or so called pk-models [1]. "IKW" is the model formulated by Iordansky S. (1960), Kogarko B.(1961) and Leen van Wijngaarden (1964). "pk-model" is the system of equations of second order for two main variables "p" (average pressure) and "k" (relative value of average volumetric gas concentration) determining the state dynamics of bubbly liquid (suggested by the author in 1968).

These models allowed one to solve:

the problem of the dynamic strength of a liquid which will be considered within the framework of the axisymmetrical problem of an underwater explosion near a free surface;

the problems of tensile stress relaxation and ultimate tensile stresses allowed by a cavitating liquid.

Effect of "frozen" profile of mass velocities in a cavitation zone. The effect is due to the fast relaxation of tensile stresses behind the unloading wave.

Cavitation threshold. The methods of light scattering on microinhomogeneities and of a capacitance technique for recording the dynamics of a free surface of the liquid under reflection of a shock wave are applied.

Shock waves, bubbles, and biomedical problems. Microbubbles and cavitation clusters, their interaction with shock waves (SW), micro bubble pulsations play both positive (a kidney stone disintegration) and undesirable (destruction of tissues and cells) roles in the processes of damage, destroying and treatment.

3.2 Real Liquid State (Nucleation Problems)

In physical acoustics, fundamental statements on bubble cavitation were primarily focused on the state of a real liquid in terms of its homogeneity and on the mechanism of bubble cluster formation and on the notions of liquid strength. Determination of the nature of the microinhomogeneities, their parameters, density, and size spectra are the basic problems of the analysis of the state of a real liquid. The most reliable results are provided by a combination of light scattering and an electromagnetic shock tube.

A typical experimental setup is shown schematically in Fig. 3.1: a pulse magnetic field is generated by an electromagnetic source in a narrow gap between a membrane and a plane helical coil on which a high-voltage capacitor is discharged [2]. It should be noted that the parameters of the high-voltage circuit are chosen to ensure that the discharge is aperiodic and to eliminate pressure fluctuations in the shock wave generated in the liquid by a moving membrane driven by the magnetic field.

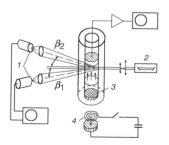

Fig. 3.1. Diagram of electromagnetic shock tube : (1) photomultipliers, (2) an He–Ne laser, (3) a membrane, (4) a plane helical coil in the high-voltage discharging circuit

This method can produce shock waves with amplitudes of up to 10–20 MPa and durations of the positive phase of about 3–5 µs. Distilled water is placed in the transparent working section of the shock tube. The light source is a He–Ne laser, with a beam diameter of 1.5 mm, that is transmitted at a depth of 3 mm from the free surface of the liquid. The scattered light is collected by photomultipliers whose position relative to the direction of the laser beam is adjusted to a specific problem [2]. The signals from the photomultipliers are fed into AD converters and a PC. A variable-capacitance transducer registers the displacement of the free surface during reflection of the shock wave.

The angular distribution of the intensity of scattered light (the so-called *scattering indicatrix*) has characteristic maxima, which correspond to concrete sizes of microinhomogeneities. This fact is clearly illustrated by Fig. 3.2 (curve 1 for the angle > 32° the scale was changed by an order of magnitude). The figure also presents the calculation results obtained using the known relationships from [3].

The light scattering technique was used in [4] to study microcracks within 200–2,000 Å in the experiments on pulse loading of Plexiglas specimens. The scattering indicatrix was close to the Rayleigh value and consisted of one petal. The dynamics of submicron cracks was reconstructed from the change of the relation of intensities measured in two fixed directions within the zero maximum. In the case of microbubbles, the indicatrix has several petals, whose intensity, number, and position depend on the size of scattering particles. Thus, their dynamics can be reconstructed from the number of petals recorded by the photomultiplier at a fixed observation angle and the requirements to the measurement accuracy of the intensity can be relaxed. Application of this

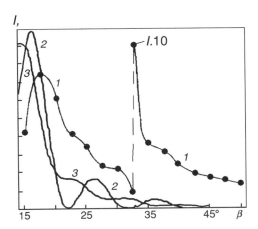

Fig. 3.2. Experimental scattering indicatrix for distilled water (1), calculated indicatrices for particles of radius 1.5 µm (2), and for a mixture of particles of radii 1.3, 1.5, and 1.7 µm (3)

method yielded a number of important results on the state of microinhomo-geneities in distilled water.

Static measurements reported in [2] revealed the stable existence of microinhomogeneities and proved that, in addition to the zero maximum, the scattering indicatrix has a clear maximum at $\beta = 17.5°$ and two smooth maxima at $\beta = 28°$ and $37°$ (Fig. 3.2, curve 1). Their position unambigu-ously determines the radius of scattering particles as $a \approx 1.5\,\mu$m. One may assume that the indicatrix petals are smoothened due to the size dispersion of the microparticles. To check this assumption, the integral indicatrix was calculated for the mixture of particles with radii 1.3, 1.5, and 1.7 μm (Fig. 3.2, curve 3). The calculations confirmed smoothening of the second and the third maxima to the same extent as in the experiments. Static experiments (unper-turbed liquid) with distilled water and a wavelength $\lambda = 0.63\,\mu$m show that the radii of the nuclei are approximately $1.5 \pm 0.2\,\mu$m and are nearly monodis-perse. We note that this result may be associated with a certain selectivity of the recording system, which limits its capability of determining the true distribution.

The experimental results of [5] for settled water were generalized in [6] to the simple relation $\sqrt{N_i}V_i = C$, where i refers to the bubble group of certain size, N_i is the number density of bubbles for the corresponding group in the given size interval, V_i is the bubble volume, and $C \simeq 10^{-9}\,\text{cm}^{3/2}$. Obviously this relation does not describe the entire distribution, which from physical considerations should have a maximum and asymptotically approach zero as the volume of bubbles approaches zero and at infinity. A distribution with these properties (see Fig. 3.3b curve) can be presented as [7]

$$N_i = N_0 \frac{(V_i/V_*)^2}{1 + (V_i/V_*)^4}. \tag{3.1}$$

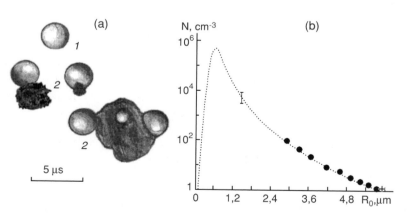

Fig. 3.3. The structure of microinhomogeneities (**a**) and the spectrum of cavitation nuclei (**b**): (1) free gas bubbles and (2) combination structures

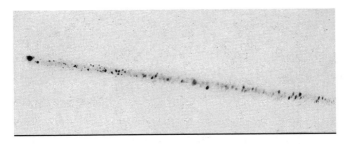

Fig. 3.4. Tracks of diffraction spots crossing the laser beam due to natural convection of water in the cuvette

This distribution involves two unknown parameters: the total number of particles per unit volume N_0 and the normalization parameter V_*, which can be taken as the volume of a bubble of radius R_* corresponding to the distribution maximum. If the above experimental value $R \simeq 0.85\,\mu\text{m}$ is taken as R_* and the "tail" of the distribution (1) fits the data of [5] (filled circle), [8] (+) for $R_i \geq 3\,\mu\text{m}$, the total density of microinhomogeneities is $N \simeq 1.5 \cdot 10^5\,\text{cm}^{-3}$, which is consistent with the experimental data 10^5–$10^6\,\text{cm}^{-3}$ obtained by measuring the tracks of diffraction spots from light scattered at inhomogeneities of any nature (Fig. 3.4).

Of principal importance is the question on the nature of inhomogeneities in a liquid. An original technique combining the shock tube and the scattering indicatrix method was proposed in [2]. The technique made it possible to solve the problem unambiguously. The idea is based on the essential dependence of the intensity of scattered light on the size of inhomogeneities. We selected two observation angles $\beta = 10°$ and $15°$. The intensity distribution of the light scattered at these angles is presented in Fig. 3.5a. An arrow marks the initial size of the inhomogeneities $(1.5\,\mu\text{m})$.

If the inhomogeneities are the microbubbles of a free gas, when the shock wave passes along the liquid, the cavitation nuclei will collapse and the intensity of scattered light should vary in different ways for the selected registration angles (Fig. 3.5a), i.e., increase with respect to the background for $\beta = 10°$ (1) and decrease for $\beta = 15°$ (2) [2]. In the experiment, a laser beam, 4 mm in diameter, was focused along the tube diameter (Fig. 3.1) at a depth of 6 mm beneath the free surface of water. The oscillograms shown in Fig. 3.5b ($\beta = 10°$ for beam (1) and $\beta = 15°$ for beam (2)) have three characteristic stages: oscillations due to the particle dynamics in the shock wave, reconstruction of unperturbed state, and the section of increasing intensity of oscillations and subsequent decline. The incident shock wave consists of three attenuating pressure pulses with an amplitude of 6–7 MPa and a passage period of about $3\,\mu\text{s}$. The intensity fluctuations of scattered light coincide with the period of passage of the pulses and characterize the dynamics of the average radius of scattering particles.

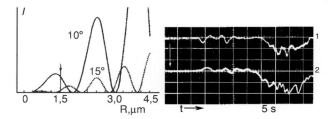

Fig. 3.5. Two curves of the intensity of scattered light *I(R)*, where *(R)* is the size of microinhomogeneities, and two oscillograms on *I(t)* dynamics of the scattering light behind the shock- and rarefaction-wave fronts for the viewing angles $\beta = 10°$ (1) and 15° (2)

As follows from Fig. 3.5b, the experimental arrangement actually demonstrates an antiphase change of signals from the photomultiplier (the first section of oscillograms) related to the decrease in size of cavitation nuclei, which is a direct prove of the existence of free gas microbubbles among the cavitation nuclei. The third stage of the oscillograms accounts for the synchronous growth of microbubbles in the rarefaction waves (after reflection of the shock wave from the free surface) from the initial sizes up to $4, \ldots, 5\,\mu$m. In the case of growth to these sizes, the upper beam registers two explicit bursts, while the lower registers three intensity bursts against the background of growth. Then bubbles collapse, which is proved by the intensity decay.

Two problems associated with the state of a real liquid are the stabilization of nuclei and their number density N, which are directly associated with the mechanism of bubble cluster formation in rarefaction waves. Several models have been proposed to solve the first problem:

— Fluctuating cavities: $R_* = \sqrt{kT/\sigma}$ (Frenkel, 1945)
— Hydrophobic particles with nuclei in the interstices: $R_* = 2\sigma/p_0$ (Harvey, 1944)
— Surface organic films (Herzfeld and Fox, 1954)
— Ionic mechanism (Blake, 1949, and Akulichev, 1966)
— Solid-particle nuclei (Plesset, 1969)
— Microbubbles when the influence of heat fluxes, Stokes forces, and buoyancy forces is balanced: $R_* = (\nu^2 kT/\rho_0 g^2)^{1/7}$ (Kedrinskii, 1985)
— Combination structures ("microparticles-microbubbles," Fig. 3.3a; Besov, Kedrinskii and Pal'chikov, 1991)

3.3 Bubble Clusters

3.3.1 Formation Mechanisms

All of the above types of microinhomogeneities with their stabilization properties can exist in a real liquid and hence determine a wide spectrum of sizes, from nanometers to tens of microns (Fig. 3.3). The second problem is to determine the number density of nuclei N_0 in the cavitation zone and their volume concentration k_0. Results on acoustic diagnostics of free gas microbubbles obtained in particular by Strasberg [8], indicate a very low value of N_0 and contradict to experimental data observed in developed cavitation clusters. The experiments on a development of cavitation in the focal zone of an ultrasonic concentrator ($f = 550\,\text{kHz}$), registered using high-speed photography, have shown that immediately after application of the field only a single bubble appears in the frame [9].

Strasberg's results seem to be confirmed, but after ten periods a dense cloud of cavitation bubbles is formed in the vicinity of the focus. This fact became a base to suggest an avalanche-like population of the zone of developing bubble cavitation by nuclei [9] caused by the shape instability and by the fragmentation of the bubbles under intense collapse. The fragments act as new cavitation nuclei. Then the process repeats over and over again: the number of nuclei in the ultrasonic field is increasing continuously. Thus, one can conclude that the *process of cavitation development consists of a peculiar ultrasonic pumping of the liquid by nuclei*. This mechanism is quite probable and can be physically justified for cavitation in ultrasonic fields. However, two facts fail to fit the scheme: (1) the velocity of fragments must be high enough so that they could quickly and uniformly distribute in space (usually a dense cluster of fragments attached to one another appears instead of one bubble) and (2) a dense cluster of bubbles is also observed in the field of the *single rarefaction pulse* (for example, in the case of a shock wave of underwater explosion reflected from a free surface) when the effect of the pumping by nuclei is not possible.

It should be noted that the density of nuclei in a cavitation zone must be associated with microinhomogeneities of any nature, including solid nuclei and their combinations with gas nuclei (Fig. 3.3a, 2) at which gas–vapor bubbles can develop under the effect of tensile stresses.

A fundamentally new mechanism for the generation of cavitation zones was proposed by the author in [10]. The essence of the mechanism is as follows: (a) it is assumed that a real liquid contains a spectrum of nuclei of radii 10^{-7}–10^{-3} cm and the constant density of the bubble number of 10^5–10^6cm^{-3} (see Fig. 3.3b); (b) the concept of a visible (detectable) size of the cavitation nuclei is introduced; (c) the apparent multiplication of cavitation nuclei in relatively weak ultrasonic fields is explained by the successive saturation of the zone by bubbles that need different times for attaining a visible size depending on their initial size (Fig. 3.3b); (d) for strong rarefaction phase the entire spectrum of nuclei can reach visible size simultaneously; and (e) behind the front of a strong rarefaction wave the density of saturation of the cavitation zone by nuclei reaches a maximum, the spectrum transforms into a monodisperse structure.

3.3.2 Mathematical Model of Cavitating Liquid

The experimental data mentioned support the unambiguous interpretation of a real liquid as a two-phase medium, in spite of the negligibly small initial gas content (volume concentration of $k_0 \simeq 10^{-8}$–10^{-12}). In this case, it is logical to assume that the development of bubbly cavitation excited by rarefaction wave (RW) changes the state of a two-phase medium and, hence, the parameters of wave field too. It is said that the profile and the parameters of RW are transformed. The mechanism of RW transformation is similar to the one of shock wave transformation in bubbly media. The latter is the base to use two-phase mathematical models for describing the state dynamics of a cavitating liquid and wave processes in it [11, 12].

The *IKW-model* is a combination of conservation laws for the average density ρ, pressure p, and mass velocity v, and a subsystem describing the dynamics of the bubbly medium state. But the *pk-model* as system of equations for two basic characteristics, pressure p and reduced volume concentration $k = (R/R_0)^3$ of the gas–vapor phases, is the most convenient form to describe processes in cavitating media:

$$\Delta p - c_0^{-2}\frac{\partial^2 p}{\partial t^2} = -\rho_0 k_0 \frac{\partial^2 k}{\partial t^2}, \tag{3.2}$$

$$\frac{\partial^2 k}{\partial t^2} = \frac{3k^{1/3}}{\rho_0 R_0^2}(p_g - p) + \frac{1}{6k}\left(\frac{\partial k}{\partial t}\right)^2. \tag{3.3}$$

Here $p_g = p_0 \cdot k^{-\gamma}$. This system of equations was suggested for the first time by the author in 1968 [13] to describe the more complicated process of shock wave attenuation in a liquid with bubbles of different sizes. The advantage of the system (3.2)–(3.3) is that it enables analytical solutions, if the appropriate physical conditions are formulated in the problem. For example, the experimental data on the propagation of a plane one-dimensional shock wave in bubbly media demonstrate the effect of splitting of the incident shock wave into two waves [13]: a precursor (a shock wave pulse transformed by the plane bubbly layer) and a reradiation of the bubble layer. I would like to remind that the Rayleigh equation (in the IKW-model) is only one of the elements of the subsystem describing the state of bubbly liquid. Because equation (3.2) for p takes account some functions of this subsystem (its right part is determined on the base of the state equation of the bubbly liquid) equation (3.3) as the "generalized" Rayleigh equation plays the key role in the processes both pressure field transformation and of formation of cavitation clusters.

On the base of experimental data, one can assume that as a precursor forms, bubbles are only radially accelerated. It means that $k \simeq 1, \partial k/\partial t \simeq 0$ and, consequently, the equation for the concentration k becomes

$$\frac{\partial^2 k}{\partial t^2} \simeq \frac{3}{\rho_0 R_0^2}(p_0 - p).$$

which makes it possible to obtain from (2) the Klein–Gordon equation that describes the formation of a precursor

$$\triangle p - c_0^{-2}\frac{\partial^2 p}{\partial t^2} = \alpha^2 (p - p_0).$$

One can see that the Klein–Gordon equation involves a *similarity parameter*

$$\alpha^2 = 3k_0/R_0^2,$$

which depends on the characteristics of the bubbly medium.

3.3.3 Comparison with Experiments

Comparison with experimental data [13] shows that the process of shock wave attenuation in the bubbly medium is characterized by *the similarity criterion*

$$\beta = \alpha \cdot l \quad or \quad \beta = \sqrt{3k_o} \cdot \frac{l}{R_o},$$

where l is the thickness of the bubbly layer. It means that the shock wave amplitude attenuates on one and the same law $p(\beta)$ for different combinations of all three parameters k_0, R_0, l.

The essentially changing state of the medium with microinhomogeneities behind the rarefaction-wave front (or in an ultrasonic field [14]) makes it possible to neglect the gas content in the expanding bubbles of a cavitation cloud and the inertia term $k_t^2/6$ in (3.3). Then the dynamics of volume concentration in a bubble cluster can be described by the equation

$$\frac{\partial^2 k}{\partial t^2} \simeq -\frac{3k^{1/3}}{\rho_0 R_0^2} p. \tag{3.4}$$

Changing the left-hand part of (3.2), neglecting the compressibility of the liquid component of the medium (carrier phase), and introducing the spatial variable $\eta = \alpha r k^{1/6}$ under the assumption of the validity of $|p_{\eta\eta}\eta_r^2| \gg |p_\eta\eta_{rr}|$ and $k \gg |rk_r/6|$ (where $\alpha = \sqrt{3k_0}/R_0$), we obtain an equation of the Helmholtz type for the pressure in a cavitating liquid

$$\Delta p \simeq p. \tag{3.5}$$

Such a "heuristic" approach justifies its validity in the mathematical modeling of the transformation of shock waves in bubbly systems [13]. The simultaneous solution of (3.4) and (3.5) determines the parameters of the tensile stress field and the dynamics of the cavitation process [11,15,16]. It should be noted that comparison of these approximations with the numerical solutions of the complete system of equations and with experimental data proved that this model allows one to estimate the main characteristics of wave processes in cavitating liquids. This conclusion can be confirmed, in particular, by the calculation results for the dynamics of the visible cavitation zone presented in Fig. 3.6a for the times $t = 16, 32, 48$, and $64\,\mu s$. The region in which bubbles at a given moment exceed the minimum visible size is shown darkened. Figure 3.6b presents the frames of high-speed photography for the same times: the

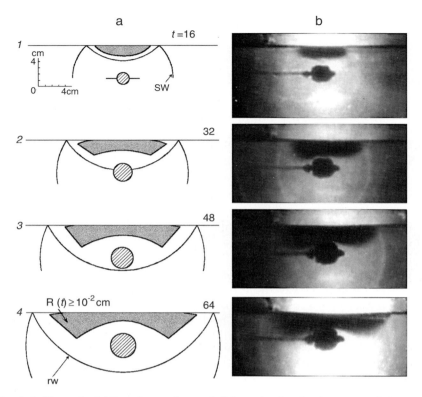

Fig. 3.6. Numerical (**a**) and experimental (**b**) results for the dynamics of the visible cavitation zone (charge was exploded on 5.3 cm depth, time is presented in µs)

cavitation zone is developing in the rarefaction wave caused by SW reflection from free surface at underwater explosion of a 1.2-g charge at a depth of 5.3 cm.

A comparison of the numerical and experimental data shows that the two-phase *pk-model* provides a quite satisfactory description of the dynamics of the cavitation zone.

3.3.4 Dynamic Strength of Liquid

It is known that the tensile stresses at the front of a rarefaction wave are continuous and reach maximum values after a final time Δt_*, which can be defined as the slope of the RW-front. The "slope of the front" has a dimension of time, it is the finite time interval during which the module of negative pressure in the rarefaction wave front increases from the hydrostatic level to the maximum value. The solution of the axisymmetrical problem for the cavitation development caused by an underwater explosion of HE (high explosive) charge near a free surface shows that this fact is essential in determining the

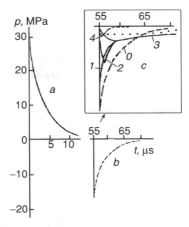

Fig. 3.7. Profiles of shock (**a**) and rarefaction (**b**) waves at underwater explosion calculated for certain point ($r = 4.5$ cm) on the symmetry axis between a charge and free surface; on Fig. (**c**) the rarefaction wave profiles calculated for different physical conditions and experimental profile (*dots*) are compared

limiting stress the liquid can withstand [16]. During the time Δt_* the reduced volume concentration k of the gas–vapor phase increases by several orders of magnitude with respect to k_0 and significantly changes the state of the medium and the applied stress field. Note, that the upper limit of the k value is restricted by the condition that the distance between bubbles must exceed the bubble diameter. As a result, the maximum negative pressure amplitudes in a cavitating liquid can decrease by several orders of magnitude [16] with respect to those in an ideal one-phase model.

   Figure 3.7 shows the profiles of a shock wave (a) and its rarefaction phases (b, c) calculated for the underwater explosion of a 1.2 g charge at a depth of 18.5 cm at a point with the coordinate $r = 4.5$ cm at the symmetry axis. The rarefaction phase profiles (b, c) shown by dashed lines correspond to the one-phase model. Figure 3.7c enables to compare the experimental data on rarefaction phase profile (dots) and profiles calculated at different conditions within the framework of the two-phase model (3.4)–(3.5). The curve 0 is for the one-phase model, the curves 1–3 were obtained for $R_0 = 5\,\mu$m, $k_0 = 10^{-11}$, and $\Delta t_* = 0$, 1, and 5 μs, respectively, the curve 4, for $k_0 = 10^{-10}$ and $\Delta t_* = 1$ μs. The results 2 and 4 have different volume concentration. One can see that the experimental points (dotted line) are located near the curve 3, which justifies the feasibility of estimating the wave field parameters.

3.3.5 Tensile Stress Relaxation (Cavitation in a Vertically Accelerated Tube)

Interesting experimental and numerical estimates of the relaxation time of tensile stresses in a cavitating liquid are given by the example of cavitation

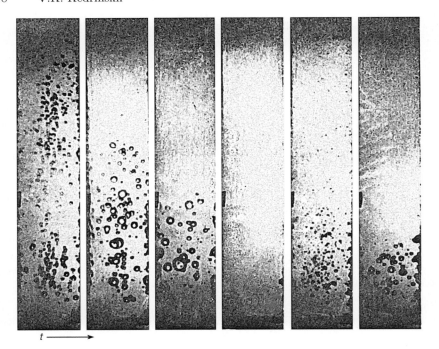

$t \longrightarrow$

Fig. 3.8. Dynamics of the cavitation zone in a liquid that experiences downward acceleration pulse in the tube

produced near the bottom of a vertical tube filled with liquid when the tube is suddenly accelerated downward by an impact (Fig. 3.8) [14].

According to the experimental data obtained by the author together with I. Hansson and K. Morch at the Danish Technical University, a zone of intense bubble cavitation develops at the bottom of the tube (Fig. 3.8). For a sufficiently high acceleration amplitude, this cavitation zone can transform into a continuous gas–vapor layer, which will result in detachment of the liquid column from the bottom. The experiment was modeled numerically using the one-dimensional system (3.4)–(3.5) for the following boundary condition at the tube bottom ($z = 0$):

$$\frac{\partial p}{\partial z} = -\rho_0 a(t). \tag{3.6}$$

Here $a(t)$ is the acceleration of the tube. It was assumed for simplicity that the liquid occupies the entire half-space $z \geq 0$. Solving (3.5), we get an analytical dependence for $p(k)$

$$p = -\rho_0 \frac{|a(t)| \exp\left(-\alpha k^{1/6} z\right)}{\alpha k^{1/6}}, \tag{3.7}$$

which, being substituted in (3.4), reduces the equation for the volume concentration near the tube bottom to the form

$$\frac{d}{dt}\left(k^{-1/6}\frac{dk}{dt}\right) = \frac{3\,a(t)}{\alpha R_0^2}.$$

One can remind that the gas pressure inside the bubbles is neglected, which is justified for the purpose of estimating the relaxation time, the above equation can be solved analytically, given the explicit dependence $a(t)$. We estimate the transformation of the pulse profile, as a model of reflection of an underwater shock wave from a free surface, assuming that the acceleration and hence the unloading pulse have an exponential form. In this case, the last equation turned out to have the analytical solution

$$k^{5/6} = 1 + \frac{5t_* a_{max}}{2\alpha R_0^2}[t - t_*(1 - \exp(-t/t_*))].$$

Using (3.7), we obtain main estimates for the relaxation processes in a cavitating liquid. If the maximum tensile stress is applied instantaneously and the acceleration has the form $a(t) = a_{max}e^{-t/t_*}$, we can estimate:

— The relaxation of tensile stresses by the time $t = t_*$

$$\frac{p}{p_{max}} \simeq \frac{1}{e}\left[\frac{2\alpha R_0^2 e}{5t_*^2 a_{max}}\right]^{1/5}.$$

— The time of relaxation of stresses when it reaches the p_{max}/e value

$$t_{cav} \simeq 14.3(k_0 R_0^2 a_{max}^{-2})^{1/4}.$$

Simple estimates based on these dependencies for $k_0 = 10^{-10}$, $R_0 = 1\,\mu m$, $a_{max} = 5 \times 10^7\,cms^{-2}$ (corresponding to $-30\,MPa$), and $t_* = 10\,\mu s$ prove that

— The amplitude of the rarefaction wave in the cavitating liquid decreases 20-fold as compared to $p/p_{max} = 1/e$ in the one-phase liquid
— The relaxation time of tensile stresses in the cavitating liquid is less than $0.1\,\mu s$. If the slope of the front of tensile stresses is described as a linear growth of the amplitude of the applied acceleration $a(t) = a_{max}(t/t_*)$, one can easily find for the ultimate tensile stress:

$$p \simeq -\rho_0\left(\frac{a_{max}^2 R_0^3}{3k_0}\right)^{2/5} t_*^{-2/5}.$$

Following this estimate, the cavitating liquid at $t = t_*$ accepts only $-3\,MPa$ instead of $-30\,MPa$ prescribed by the one-phase model. We used the following parameters: $k_0 = 10^{-11}$, $R_0 = 0.5\,\mu m$, and $t_* = 1\mu s$ for the front slope. The calculations in [15] yielded a quantity of the same order, which shows the reliability of the expressions for estimating both the relaxation time and the limiting tensile stresses.

3.4 Methods of Hydrodynamic Pulse Tubes and Experimental Technique

3.4.1 Hydrodynamic Tube of Rarefaction

The complete system of equations for the cavitating liquid, supplemented by kinetic equations of the heat transfer for the calculation of the gas pressure inside a bubble was applied to the problem of a pulse rarefaction tube. This tube is an analog of the classical shock tube. It consists of a high-pressure chamber filled with a liquid studied and a low-pressure chamber which contains a gas. They are separated by a diaphragm. When the latter is suddenly broken the rarefaction wave begins to propagate into the compressed liquid sample. So we have the classical gas dynamic problem of the decay of an arbitrary discontinuity with an essentially unsteady and nonlinear stage of rarefaction wave formation in a liquid sample.

In (3.3), the pressure in gas p_g is determined from the known equation

$$\frac{dp_g}{dt} = 3(\gamma - 1)\frac{q}{4\pi R^3} - \frac{3\gamma p_g}{R}\frac{dR}{dt},$$

where

$$q = 4\pi R^2 \lambda_g \, Nu \, \frac{T_0 - T}{2R}, \quad T = \frac{p_g}{(\gamma - 1)c_v \rho_g} = T_0\left(\frac{R}{R_0}\right)^3 \frac{p_g}{p_{go}},$$

$$Nu = \sqrt{Pe} \quad \text{at} \quad Pe > 100, \quad Nu = 10 \quad \text{at} \quad Pe < 100,$$

$$Pe = 12(\gamma - 1)\frac{T_0 R |S|}{\nu |T_0 - T_g|}.$$

The structures dynamics of the wave field in a cavitating liquid shown in Fig. 3.9 [17] was calculated for the following conditions: $\nu = 0.001\,\mathrm{cm^2 sec^{-1}}$, $c_v = 0.718 \cdot 10^7\,\mathrm{cm^2\,(s^2 \cdot deg)^{-1}}$, $\lambda_g = 2470\,\mathrm{g\,cm\,(s^3 \cdot deg)^{-1}}$, $k_0 = 10^{-4}$, and $R_0 = 50\,\mu\mathrm{m}$. The rarefaction wave formation at initial moments ($t = 3.3, 10, 20, 30$, and $40\,\mu\mathrm{s}$) after the discontinuity decay is shown in Fig. 3.9a. Two wave profiles with regard to heat exchange (curve 1) and for the adiabatic version (curve 2) are compared for the times 90μs in Fig. 3.9b and 440μs in Fig. 3.9c. One can see that the wave field is split into two characteristic parts: a precursor formed by a centered rarefaction wave and propagating with the speed of sound over the unperturbed liquid and the main disturbance in the form of a wave with an oscillating front and propagating with the equilibrium speed characteristic for a two-phase bubbly medium. We note the analogy with the separation of shock waves into a precursor and a wave packet observed by the author [13] for shock wave in bubbly media.

3.4.2 Hydrodynamic Shock Tube, Pulse X-Ray Method and Resolution of Cavitation Zone Dynamics

The optically opaque stage of the cavitation process in intense rarefaction waves was studied using three pulse x-ray devices and a two-diaphragm

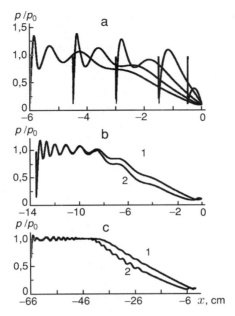

Fig. 3.9. Calculation of the wave pattern in the high-pressure chamber of the hydrodynamic rarefaction tube: (**a**) initial moments after the discontinuity decay, (**b**) 90 μs later, (**c**) 440 μs later; 1 with regard to heat exchange and 2 for adiabatic process

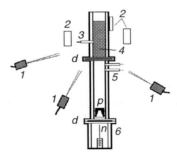

Fig. 3.10. Two-diaphragm shock tube with the x-ray pulse technique: (1) x-ray apparatuses, (2) photo film, (3) pressure gauge, (4) working section with the liquid studied, (5) optical sensor for measuring the piston velocity, and (6) high-pressure chamber

hydrodynamic shock tube (Fig. 3.10). The tube consists of three sections: a high-pressure chamber 6, an evacuated acceleration channel (with the driving piston p and two separating diaphragms d at the ends), and a working section 4 made of duralumin and containing the liquid under study. The lower diaphragm was ruptured by an electromagnetic system with a needle n. The velocity of the piston before the rupture of the upper diaphragm separating the channel from the liquid was measured by fiber-optic sensors 5. To arrange rapid-series frame-by-frame radiography, the system was synchronized with the three x-ray devices started with different given time lags. The dynamics of the inner structure of the cavitation zone (cluster) was studied using pulsed x-ray devices PIR-100/240 designed in the Lavrentyev Institute of Hydrodynamics (Novosibirsk). The spectrum of the γ-quantum energy of the devices was adjusted to the resolution of structural inhomogeneities in the few-centimeters-thick layers of a cavitating liquid: the average radiated energy was 70 keV, the maximum energy was up to 200 keV, and the duration of a single burst in the pulse half-height was 80 ns.

A shock wave in the liquid sample was produced by an impact of the piston. Its positive phase lasted for several tens of microseconds and the amplitude varied within 5–30 MPa. Reflection of the shock wave from the free surface of the liquid resulted in intense cavitation development, as shown by the x-ray photographs in Fig. 3.11 (for different times from zero to 1 ms). The first frame corresponds to the arrival of the shock wave front to the free surface (the diameter of the liquid column was 30 mm). We see that already after 600 µs (frame 4) the cavitation zone reached dense packing with coarse gas–vapor cells. Computer processing of the density distribution of the x-ray negatives enabled analysis of the process dynamics without interfering the medium by sensors.

Scanned and computer-processed x-ray negatives of the cavitation zone developing behind the front of the shock wave reflected from the free surface were used to study the dynamics of the average density $\bar{\rho}$ of the cavitation zone and the time t^* required for the cavitating zone to reach the bulk density of bubbles as a function of the deformation rate of the medium $\dot{\varepsilon}$ [18] (Fig. 3.12):

$$\bar{\rho} = \rho_0(1 + \dot{\varepsilon}t)^{-1}.$$

The value of t^* estimated as $t^* \simeq (\dot{\varepsilon})^{-1}$ is rather convenient for analysis when the profile and parameters of the shock wave incident on the free

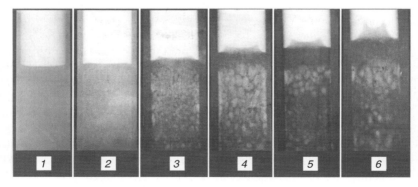

Fig. 3.11. X-ray images of the structure dynamics of the cavitation zone (the interval between frames is 200 µs)

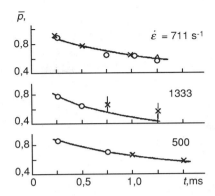

Fig. 3.12. Dynamics of the mean density of the cavitation zone based on the computer-aided analysis of x-ray photographs

surface are known. Following the experimental data, the relaxation time of the medium to the foam-like state for $\dot{\varepsilon} \simeq 1,330\,\mathrm{s}^{-1}$ is approximately $700\,\mu\mathrm{s}$, while for $\dot{\varepsilon} \simeq 500\,\mathrm{s}^{-1}$ it is about $2,000\,\mu\mathrm{s}$, which is consistent with the above dependence.

3.4.3 "Frozen" Profile of Mass Velocities in a Cavitation Zone

The dynamics of the distribution of mass velocities throughout the cavitation zone at initial stages of the process was studied experimentally using the successive flash x-ray recording of a sample with special lead foil traces under loading. Some traces were made of $25 \times 0.5 \times 0.04\,\mathrm{mm}$ bands attached by small pieces of flimsy to lavsan bands, other $1 \times 1\,\mathrm{mm}$ traces were attached by thin threads suspended independently in parallel to the sample axis [18], [19]. In all experiments, the process was recorded under identical conditions at fixed times with preliminary registration of initial position of the traces. Six photos were obtained for each experiment: three for the initial state from different observation angles and three at certain moments.

The immovable marks at the external side of the shock tube fixed the initial position of the traces with respect to the tube and their shift over the given times. X-ray tubes were located at the arc of a circle with a pitch of $60°$ in the plane perpendicular to the tube axis. This enabled control over the spatial arrangement of the traces and their departure from the tube walls and provided the surveillance of the zone from different sides. The mass velocity of the medium was determined from the displacement of traces for a given time interval between the x-ray frames, including the initial state. The plane of linear traces was arranged horizontally to observe the dynamics of the velocities in the cross-section of the cavitation zone. The reliability of the trace method was checked by means of an electromagnetic sensor used to measure the mass velocity profile in short shock waves [20]. The sensor was

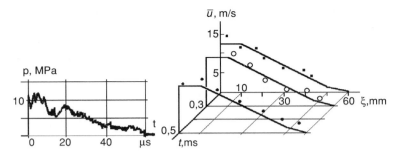

Fig. 3.13. The effect of "freezing" of the mass velocity profile $\bar{u}(\xi)$ in the cavitating liquid (*right*); shock wave profile $p(t)$ (*left*)

made of an aluminum-backed lavsan film whose size was close to that of the linear traces. The film was placed in the field of a permanent magnet.

In the run of experiments, the average amplitude of shock waves was 11.4 ± 0.6 MPa, the duration was 60–70 µs, and the wave profile near the front had a "step" (Fig. 3.13, left). The position of the linear traces in the photos was determined by the center of lead bands. Although the radial distribution of the velocity was not measured, at later times experiments revealed a bend in linear traces, which qualitatively characterized the radial structure of the flow and yielded the velocity profile typical of a viscous liquid flow. To show the depth distribution of mass velocities, the measurement results of two identical experiments are shown by dots in the coordinates "velocity–Lagrangian coordinate–time" (Fig. 3.13, right). The point of origin is related to the free surface.

These results were compared with calculations following the model of instantaneous relaxation of tensile stresses behind the rarefaction wave front [21] in the following statement. Let a shock wave, whose profile is given as a certain monotone function $p_{\text{sh}} = p_{\text{fr}} f(x)$, where p_{fr} is the pressure in the shock-wave front, reaches the free surface (coordinate $x = x_0$) at the time $t = 0$. At $t > 0$ a rarefaction wave propagates with the speed of sound c_0 in the liquid with a monodisperse distribution of microinhomogeneities as free gas microbubbles of $1\,\mu\text{m}$ radius and nuclei concentration of $N = 10^4\,\text{cm}^{-3}$. An instantaneous relaxation of tensile stresses occurs behind the wave front. Then following the analysis of Chernobaev [20], we write the system of equations describing a flow of a cavitating liquid in Lagrangian mass coordinates

$$\frac{\partial x}{\partial t} = u, \qquad \frac{\partial 1/\bar{\rho}}{\partial t} = \frac{1}{\rho_0}\frac{\partial u}{\partial \xi}, \qquad \frac{\partial u}{\partial t} = 0,$$

$$p = p_0, \qquad \bar{\rho}^{-1} = \rho_0^{-1} + v_{\text{bubl}},$$

where ξ is the Lagrangian coordinate, u is the mass velocity, $\bar{\rho}$ and ρ_0 are the density of the mixture and the liquid, respectively, v_{bubl} is the volume of cavitation bubbles per mixture mass unit, and p_0 is the atmospheric pressure. If we take $\xi = x_1$, within the accepted approximation that at the moment of arrival of the rarefaction wave the pressure in the point x_1 instantaneously decreases from $p_1 = p_{fr} f(2\xi - x_0)$ to p_0, the mass velocity u_1 becomes equal to $2p_1/\rho_0 c_0$, and its distribution (by virtue of the law of conservation of momentum $\partial u/\partial t = 0$) in the cavitation zone becomes equal to $u = u(\xi)$ and is determined by the profile of the initial shock wave:

$$u(\xi) = \frac{2p_{fr}}{\rho_0 c_0} f(2\xi - x_0).$$

Now one can easily find the deformation rate of the cavitating liquid $\dot{\varepsilon} = \partial u/\partial \xi$

$$\dot{\varepsilon} = \frac{2p_{fr}}{\rho_0 c_0} f_\xi(2\xi - x_0)$$

and, after integrating the equation of continuity, obtain the equation for determining the function $\bar{\rho}(t)$

$$\frac{1}{\bar{\rho}} = \frac{1}{\rho_0}\left[1 + \dot{\varepsilon}(\xi)\left(t - \frac{\xi}{c_0}\right)\right].$$

The subscript ξ at the function f denotes a derivative with respect to the Lagrangian coordinate.

Based on the experimental pressure profile (Fig. 3.13, left), we present the model velocity distributions for the appropriate times in Fig. 3.13, right. One can see that the calculated curves are adequate to the experimental data, support the tendency to the preservation of the velocity profile for different times, and suggest "freezing" of the mass velocity profile in the cavitating liquid. At later times and greater depths from the free surface one can notice the deviation of calculated curves from the experimental data due to the idealization of the pressure profile or the approximate character of the model neglecting the inhomogeneity of the macrostructure of the cavitation zone.

3.5 Comparison of the Initial Stages of Disintegration of Solids and Liquids

3.5.1 Liquids

As we emphasized already, cavitation is a discontinuity in the tensile stress field of a liquid (the initial stage of transformation of its state), accompanied by the growth of gas–vapor bubbles on cavitation nuclei. From the viewpoint of physical acoustics, one of the parameters that characterizes the cavitation strength of a liquid is the cavitation threshold. It means that the tensile stresses exceed a critical value and, as a result, a growth of cavitation nuclei starts, changing rather abruptly the dynamics of the free surface [22].

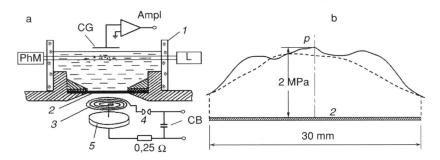

Fig. 3.14. Schematic diagram of experimental setup (**a**) for shock wave generation in liquids by magnetic field; wave front profiles (**b**) in the vicinity of membrane (*solid line*) and of free surface (*dotted line*)

The pressure profile of the incident shock wave can be reconstructed from the dynamics of the free surface of water for the precavitation reflection period. The statistic scattering of the experimental cavitation threshold values can be considerably decreased using the capacitance technique for recording the dynamics of the free surface of the liquid under reflection of a shock wave from it [22, 23].

Consider the setup and wave profiles in the liquid column (Fig. 3.14). If the liquid column in the working section is 30 mm high, due to diffraction effects the linear dimension of a cross-section of the wave beam (between the lateral maxima) generated by the membrane reduces from 20–25 mm near the membrane (Fig. 3.14, wave front profile, solid line) to 10–15 mm near the free surface of the liquid (Fig. 3.14, dotted line). The latter can be estimated on the width of the cavitation zone (Fig. 3.15, (1)) near the free surface – the black horizontal line in the center of the frame (1). One can see that a free-wave zone arises in the specimen. Thus, within a reasonable registration time span one can neglect the effect of the working chamber walls on the process, for example, when studying the cavitation zone, near the free surface under the reflection of a shock wave from it (Fig. 3.15). Fig. 3.15 (1) presents one frame of the cavitation zone development between free surface and membrane 2 (1 μsec exposure time).

The thin structure of the cavitation zone and wave field was resolved using practically instantaneous photographs with 3 nsec exposure time (Fig. 3.15, (2)). Here the free surface is the upper boundary of the frame. We can see cavitative bubbles of different sizes, cumulative jets and a system of shock waves radiated by bubbles. The bubbles in the cavitation zone have different sizes and hence they are in different stages of pulsations. The latter gives the base to conclude that the effect of bubble collapse with cumulative jet formation is a result of the interaction of shock waves generated by bubble groups which are characterized by a relatively high frequency of pulsation and more "slow" bubbles. Comparison of frames in Fig. 3.15 shows that obviously

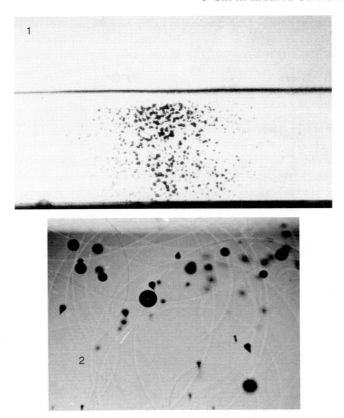

Fig. 3.15. Cavitation zones: (1) $1\,\mu\text{sec}$ exposure time, (2) $3\,\text{nsec}$ exposure time

an exposure time of $1\,\mu\text{sec}$ (frame (1)) is enough in order to obtain a smooth image and a practically uniform cavitation zone.

The dynamics of the cluster, as found in [23], determines the inadequate (in terms of a uniform medium) behavior of the free surface of the liquid, which depends essentially on the amplitude of the incident shock wave and which is characterized by an unjustifiably (at first sight) prolonged reaction of the surface to the loading. As follows from experimental data [23], the shift of the surface is registered at a longer interval than the shock wave duration. According to the experimental data for distilled water [23], three effects can be distinguished (for $3\text{--}4\,\mu\text{s}$ incident SW duration): prethreshold ($p_{\text{sh}}{=}2.9$ MPa) when no cavitation is observed, threshold ($p_{\text{sh}}{=}3.4$ MPa) when the first reaction to the shock wave loading is manifested, and cavitation ($p_{\text{sh}}{=}5.9$ MPa), which results in significant changes of the dynamics of the free surface.

The velocity of the free surface V corresponding to the completed reflection of the shock wave from it can be associated with the specific volume ϕ (gas content) of the formed bubbles by the simple relation $\phi = V/c$, where c is the speed of sound in water [24]. Practical observations show that for threshold loading the value of ϕ is within 10^{-4}–10^{-3} and their results allow one to make conclusion about an increase in the liquid strength with decreasing time of exposure to the pulse loading.

3.5.2 Solids

A similar effect is observed under disintegration of solids. The closest phenomenon to cavitation in liquids is spall fracture in solids, when the material strength increases as the loading time decreases. For the case of threshold pulses, a weak dependence of the failure time on the amplitude of the initial loading pulse is established. The effect is called a dynamic branch. In [25], [26], a structure-time failure criterion was proposed that allows one to calculate the strength increase observed in the experiments and to explain the phenomena in brittle materials observed in experiments on spallation [25]. A similar criterion is used for materials displaying plastic properties, which enables an effective prediction of the behavior of the dynamic yield stress

$$\frac{1}{\tau} \int_{t-\tau}^{t} \left(\frac{\sigma(t')}{\sigma_c} \right)^{\alpha} \mathrm{d}t' < 1, \tag{3.8}$$

where σ_c is the static yield stress, τ is the incubation time related to the dynamics of the dislocation process.

The parameter α characterizes the sensitivity to the stresses that lead to irreversible deformation. It is greater than or equal to one in solids and up to several tens in some materials, for example, $\alpha = 10$–30 for steels and alloys. It should be noted that the condition of irreversible growth of cavitation bubbles under critical loading that causes disintegration of the sample is applied to liquids as well [27]. To analyze the initial stage of cavitation damage the criterion (3.8) was used taking into account that in solid mechanics tensile stresses are positive, while in fluid mechanics tensile stresses are negative.

The contribution of compressibility tends to that the relation for liquids becomes

$$\frac{1}{\tau} \int_{t-\tau}^{t} \mathrm{sign}(\sigma(t')) \cdot \left(\mathrm{abs} \left(\frac{\sigma(t')}{\sigma_c} \right) \right)^{\alpha} \mathrm{d}t' < 1. \tag{3.9}$$

If we assume σ_c equal to 0.1 MPa for distilled water, since the intense growth of the cavitation nuclei under the effect of tensile stresses starts with neutralizing the forces of surface tension $\sigma_c = 2\sigma_w/r$, where σ_w is the surface tension in water, $r = 1.5\,\mu\mathrm{m}$ is the radius of the cavitation nuclei [28]. We select τ and α using experimental points. The load applied in the experiments we approximate using the formula

$$\sigma(t) = -P_A \cdot \sin\left(\frac{\pi\tau}{T} \right) \cdot \mathrm{e}^{-t/T_1} \tag{3.10}$$

The applied load $\sigma(t)$ induces a pressure wave moving towards the surface. If the surface coordinate is zero and the time is reckoned from the moment when the wave arrives at the surface, the wave can be written as $\sigma(t + x/c)H(t + x/c)$. Upon reflection from the surface the wave acquires the form $-\sigma(t - x/c)H(t - x/c)$ and the pressure in the liquid will be determined by the sum of two waves. Here $H(t)$ is the Heaviside step function, c is the velocity of propagation of the wave in the liquid ($c = 1,500\,\mathrm{m\,s^{-1}}$). The calculations show that the experimental loading curves are well described by the formula (10) at $T_1 = 2.85 \times 10^{-6}\,\mathrm{s}$. To calculate the critical strength, we substitute normalized values of pressure for all times for which pressure is nonzero in the integral of (3.9). Then we find the time and the coordinate at which the maximum value of the integral is achieved.

The sought value of P_m is the value at which, for a given time and position, the LHS of (3.9) becomes unity. The calculations show that the best agreement with the experiment is observed for α within the range from 0.4 to 0.5. In particular, at $\alpha = 0.5$ and $\tau = 19\,\mu\mathrm{s}$ the experimental data turned out to practically coincide with the theoretical values of strength as a function of the pulse duration $P_m(T)$, calculated from (3.9).

Thus, it was established experimentally that the cavitation strength of water grows as the duration of the loading pulse decreases; the appropriate dependence is nonlinear. Application of the structure-time criterion makes it possible to calculate the experimentally observed increase in the cavitation threshold with decreasing pulse duration. The data give evidence to the fundamental character of the structure-time approach that provides an adequate reflection of the dynamics of failure of solids and the initial disintegration stage of liquids [29].

3.6 Shock Waves, Bubbles, and Biomedical Problems

3.6.1 General Questions and Statements

Numerous and experimental investigations confirm that the physics of cavitation is directly related to the wide spectrum of phenomena studied within the frame work of medical applications of shock waves for special purposes. Microbubbles and cavitation clusters, their interaction with shock waves (SW) and SW generation, microbubble pulsations and cumulative microjets play both positive (a kidney stone disintegration) and undesirable (destruction of tissues and cells) roles in the processes of damage, destroying, and treatment.

An overview of the basic physical principles of shock waves, and the history and basic research behind shock-wave use in medicine for disintegration of kidney and ureteral stones, in orthopaedics and traumatology was presented in [30] and [31]. One noted that the theory of shock-wave therapy for orthopaedic diseases involves the stimulation of healing processes in tendons, surrounding tissue and bones. This is a completely different approach from that of urology, where shock waves are used for disintegration. It is well known that lithotripsy is a common effective treatment for kidney stones. However, focal volumes are often larger than stones, and surrounding tissue is often injured. In [32] a new dual-pulse lithotripter consisting of two opposing, confocal and simultaneously triggered electrohydraulic sources has been suggested to accelerate stone fragmentation and to reduce cell lysis in vitro.

Model gypsum stones and human erythrocytes were exposed to dual pulses or single pulses. The results of tests in vitro have shown that at the focus, model stones treated with 100 dual pulses at a charging voltage of 15 kV broke into eight times the number of fragments as stones treated with 200 single pulses at 18 kV. Using an experimental system that mimics stone fragmentation in the renal pelvis, the role of stress waves and cavitation in stone comminution in shock-wave lithotripsy (SWL) was investigated in [33]. Spherical plaster-of-Paris stone phantoms ($D = 10$ mm) were exposed to 25, 50, 100, 200, 300, and 500 shocks at the beam focus of a Dornier HM-3 lithotripter operated at 20 kV and a pulse repetition rate of I Hz.

This apparent size limitation of the stone fragments produced primarily by stress waves (in castor oil) is likely caused by the destructive superposition of the stress waves reverberating inside the fragments, when their sizes are less than half of the compressive wavelength in the stone material. On the other hand, if a stone is only exposed to cavitation bubbles induced in SWL, the resultant fragmentation is much less effective than that produced by the combination of stress waves and cavitation. It is concluded that, although stress wave-induced fracture is important for the initial disintegration of kidney stones, cavitation is necessary to produce fine passable fragments, which are most critical for the success of clinical SWL.

Stress waves and cavitation work synergistically, rather than independently, to produce effective and successful disintegration of renal calculi in SWL. It should be mentioned that an analogous effect was predicted as a result of numerical studies in [34]. Several mechanisms of kidney stone fragmentation in extracorporeal shock wave lithotripsy (ESWL) were under discussion also in [35]. As a new mechanism, the circumferential quasistatic compression or "squeezing" by evanescent waves in the stone has been introduced. Cleveland et al. [36] performed experiments on lithotripsy shock waves inside pigs with small hydrophones. A hydrophone was positioned around the pig kidney following a flank incision. It appeared that a combination of nonlinear effects and inhomogeneities in the tissue broadened the focus of the lithotripter. The shock rise time was of the order of 100 ns, substantially more than the rise time measured in water, and was attributed to a higher absorption in tissue.

A novel, less invasive, shock wave source has been developed, that can be introduced into an arbitrary position in a human body percutaneously [37]. The shock wave source consists of an explosive, an optical fiber, a balloon catheter, and a Nd:YAG laser, which generates a spherical explosive shock wave. The destructive potential of the present source for injuring tissue was confirmed and a subsequent cell elongation and splitting in the direction of the shock wave propagation has been observed. The studies in [38] were undertaken to examine the liquid jet impact effect on fibrinolysis in a tube model of an internal carotid artery. The shock wave was generated by detonating a silver azide pellet weighing about a few µg located in a balloon catheter. Thrombi were formed using fresh human blood and gelatin. The fibrinolysis induced by the liquid jet impact with urokinase was explored as the percentage of the weight loss of the thrombus.

The interaction of air bubbles attached to gelatin surfaces, extirpated livers or abdominal aortas of rats with underwater shock waves was investigated in [39] to help clarify the tissue-damage mechanism associated with cavitation bubbles induced during extracorporeal SWL. The bubble attached to gelatin or to a rat's liver surface migrates away from the surface with an oscillatory growth/collapse behavior after the shock-wave interaction. The penetration depth of the liquid jet into the gelatin and the radius of the subsequent damage pit on the surface depend on the initial bubble radius. The effect of extracorporeal shock waves on hemoglobin release from red blood cells was recently found to be minimized under minute static excess pressure [40].

The experiments carried out have shown that a dominant mechanism of shock wave action is a shock wave–gas bubble interaction. A method for real-time in vitro observation of cavitation on a prosthetic heart valve has been developed. Cavitation of four blood analog fluids (distilled water, aqueous glycerin, aqueous polyacrylamide, and aqueous xanthan gum) has been documented for a Medtronic/Hall(TM) prosthetic heart valve [41]. The observations were made on a valve that was located in the mitral position, with the cavitation occurring on the inlet side after valve closure on every cycle. Stroboscopic videography, was used to document the cavity life cycle. Bubble cavitation was observed on the valve occluder face.

For each fluid, cavity growth and collapse occurred in less than one millisecond, which provides strong evidence that the cavitation is vaporous rather than gaseous. The cavity life time was found to decrease with increasing atrial pressure at constant aortic pressure and beat rate. The area of cavitation was found to decrease with increasing delay time at a constant aortic pressure, atrial pressure, and bear rate. Cavitation was found to occur in each of the fluids, with the most cavitation seen in the Newtonian fluids (distilled water and aqueous glycerin). As a rule, in studies of cells or stones in vitro, the material to be exposed to shock waves is commonly contained in plastic vials. It is difficult to remove air bubbles from such vials.

The attempt to determine whether the inclusion of small, visible bubbles in the specimen vial has an effect on SW-induced cell lysis was made [42]. It was found that even small bubbles led to increased lysis of red blood cells and that the degree of lysis increased with bubble size. Thus, bubble effects in vials could involve the proliferation of cavitation nuclei from existing bubbles. Whereas injury to red blood cells was greatly increased by the presence of bubbles in vials, lytic injury to cultured epithelial cells was not increased by the presence of small air bubbles. This suggests different susceptibility to SW damage for different types of cells. Thus, the presence of even a small air bubble can increase SW-induced cell damage, perhaps by increasing the number of cavitation nuclei throughout the vial, but this effect is variable with cell type.

Impulsive stresses in repeated shock waves administered during ESWL cause injury to kidney tissue. In a study of the mechanical input of ESWL, the effects of focused shock waves on thin planar polymeric membranes immersed in a variety of tissue-mimicking fluids have been examined [43]. A direct mechanism of failure by shock compression and an indirect mechanism by bubble collapse have been observed. Thin membranes are easily damaged by bubble collapse. After propagating through cavitation-free acoustically heterogeneous media (liquids mixed with hollow glass spheres, and tissue) shock waves cause membranes to fail in fatigue by a shearing mechanism.

Shocks with large amplitude and short rise time (i,e., in uniform media) cause no damage. Thus the inhomogeneity of tissue is likely to contribute to injury in ESWL. A definition of dose is proposed which yields a criterion for damage based on measurable shock wave properties. The role of shock waves in the treatment of soft tissue pain is at present unknown. There is a potential for further therapeutic applications of shock waves since shock waves exert a strong biological effect on tissue which is mediated by cavitation. Experiments using shock waves for tumor therapy have shown some promising results [44], yet devices which generate other waveforms than lithotripters are probably better suited.

Shock waves cause a transient increase in the permeability of the cell membrane, and this might lead to further applications of shock waves. Characteristics of the underwater shock waves and of ultrasound focusing were studied by in [45] by means of holographic interferometric flow visualization and polyvinylidene difluoride (PVDF) pressure transducers. These focused pressures, when applied to clinical treatments, could effectively and noninvasively disintegrate urinary tract stones or gallbladder stones. However, as it was noted, despite clinical success, tissue damage occurs during ESWL treatments, and the possible mechanism of tissue damage is briefly described.

Miller and Song studied lithotripter shock waves with cavitation nucleation agents and showed that they can produce tumor growth reduction and gene transfer in vivo [46]. Cavitation nucleation agents (CNA) can greatly enhance DNA transfer and cell killing for therapeutically useful applications of nonthermal bio-effects of ultrasound (US) and shock waves (SW).

3.6.2 Some Results on Modeling of ESWL Applications

H.Grönig in his survey [47] gave a brief review of ESWL and distinguished the works of Jutkin, Goldberg (1959), Forssman (1975), Chaussy et al., Reichenberger and Naser, Ziegler and Wurster (1986), Muller (1987), Delius and Brendel, Takayama et al., and Coleman and Saunders (1988), in which the structure and parameters of the wave field in the focus zone were studied in detail. But the cavitation effects in them are the result of the interaction of shock waves with artificially produced gas bubbles. Such bubbles do not have any connection to cavitation, as a new medium state, which considerably affects the wave-field parameters [12] and the conditions at which bubble collapse with cumulative microjet formation is possible.

Physical aspects of wave focusing in the context of ESWL are analyzed by Sturtevant [48], who raised the question on the mechanism of disintegration of kidney stones and on the role of the rarefaction wave in it. According to [48], strong rarefaction waves following the shock wave front lead to a reduced duration of its positive phase, the sharply peaked waveform, and to limitation of the amplitude. When such a wave penetrates into the target, it induces shear stresses leading to spalls. Kuwahara [49] cited experimental data for the disintegration process recorded by the high-speed video camera. Following his data, when a shock wave falls on the target, first an intense cavitation zone develops around the stone; thereafter, the stone deforms and disintegrates. In Kuwahara's opinion, the disintegration mechanism is governed by the cavitation bubbles.

Similar effects were observed by Kitayama et al. [50], who noted that after wave focusing, the pressure around the target reduces sharply, which leads to the formation of a cavitation zone. It is assumed that over this time cracks grow inside the stone, then they are widened to a certain maximum size, and finally a collapse occurs. Then the second stage of expansion starts when the cracks grow further and the stone disintegrates. This is only a suggestion, since there is no observational evidence for this process, because the object is shielded by the ambient cavitation cloud. Grunevald et al. [51] proposed an acoustic scheme for calculating the pressure field in the shock wave focus that takes into account the shift of the source.

Delius [52] considered three possible mechanisms of tissue damage by shock waves: thermal effects, direct mechanical effects, and the indirect effect called cavitation. His estimates show that for typical wave forms generated in the systems one can expect a temperature increase by about $2°C$ for a pulse frequency of $100\,Hz$. Delius related the damage to the formation of cumulative microjets produced either by asymmetrical collapse of the bubble near the target solid wall or by the interaction of the cavitation bubble with the shock wave [53]. It should be noted that this is one of the typical mechanisms of the cavitation erosion in ultrasonic fields, but under ultrasonic erosion there arise cumulative microjets whose frequency is of the order of tens of kilohertz, which does not happen in lithotriptor systems due to the low-frequency generation of the sequence of shock waves and the complex structure of the cavitation zone near the target that has the form of a dense cloud of bubbles [50].

Figure 3.16 presents three intervals (1–3) of the formation of a bubble cluster on the stone surface under shock-wave focusing (Delius' experiments, 1990). The time between frames in each interval is $40\,\mu s$ and the time lag between successive intervals is $200\,\mu s$. The shock wave was generated by a Dornier Lithotriptor with an amplitude of $65\,MPa$ in the front and a rarefaction phase of maximum amplitude $-6\,MPa$. A stone was positioned onto a metal foil in the geometrical focus of the lithotriptor. Typical shock wave

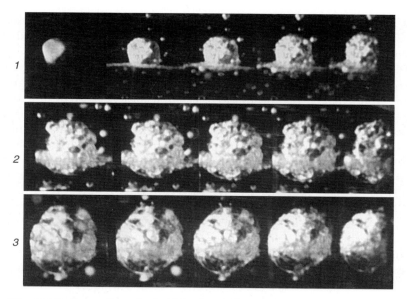

Fig. 3.16. Formation of a bubble cluster formation on the stone surface

profiles with the rarefaction phases and their interaction with micro-bubbles of micron sizes are presented in [54,55].

Prat [56] concluded that a shock-induced cavitation has a potential for selective and noninvasive destruction of deep stomach tumors. For this purpose, an outgoing pulse is a sequence of rarefaction and compression phases. Experimental investigations of the "pressure–density" diagram in human tissues and in tissue-simulating gelatin and numerical simulation aimed at predicting the level of high pressure generated in tissues and its effect on the tissues, are reported in [57]. In particular, the paper describes hysteresis effects found under blood examination when the density of the medium did not come back to the initial values after removal of the load. It should be noted that a similar effect for distilled water was first described in [28]. Its nature appeared to be determined by the two-phase state of the medium containing microinhomogeneities that played the role of cavitation nuclei and changed their structure under the effect of the wave field.

Probably such a mechanism determines the nature of hysteresis found in blood [57]. On the other hand, some researchers, for example, in [58], doubt the key role of cavitation in the disintegration of kidney stones. Sato et al. [59] proposed to use soft-flash radiography to visualize shock-induced cavitation zone in living tissues. Earlier, the method of flash x-rays was used for the same purposes in studying the dynamics of irreversible cavitation processes in distilled water with high-density gas-vapor phase [60,61]. In [62] different types of reflectors were experimentally studied in order to generate spherical shock waves in water and focus them in certain types of oils.

The analysis of the role of SW and bubbly cavitation should be continued, but the explanation for the disintegration would remain as mentioned above: cumulative jets and (or) pulse pressure arising under collapse of cavitation bubbles and/or shear stresses produced by ultrashort strong shock waves inside the target and resulting in the development of microcracks in it.

References

1. Kedrinskii, V.K.: Hydrodynamics of Explosion: Experiments and Models, Springer, Berlin Heidelberg New York (2005)
2. Besov, A.S., Kedrinskii, V.K., Pal'chikov, E.I.: Investigation of initial stage of cavitation by diffraction optical technique. Pis'ma ZhTF, **10(4)**, (1984)
3. Shifrin, K.S.: Light Diffusion in Turbid Media. Gostekhizdat, Moscow-Leningrad (1951)
4. Volovets, L.D., Zlatin, N.A., Pugachev, G.S.: Pis'ma Zh. Exp. Teor. Fiz., **4**, (1978)
5. Hammitt, F.G., Koller, A., Ahmed, O., Pjun, J., Yilmaz, E.: Cavitation threshold and superheat in various fluids. In: *Proceedings of Conference on Cavitation*, pp 341–354 Mechanical Engineering Publication Limited, London 1976
6. Kedrinskii, V.K.: Peculiarities of bubble spectrum behavior in the cavitation zone and its effect on wave field parameters. In: *Proceedings of Conference Ultrasonics International* vol. 85 Gilford, London (1985)
7. Kedrinskii, V.K.: On relaxation of tensile stresses in cavitating liquid. In: Proceedings of 13th *International Congress on Acoustics*, vol. 1, pp 327–330 Dragan Srnic Press Sabac (1989)
8. Strasberg: Undissolved air cavities as cavitation nuclei. In: Cavitation in Hydrodynamics National Physics Laboratory, London (1956)
9. Sirotyuk, G.: In: Rozenberg, L.D. (ed.) Strong Ultrasonic Fields, part 4, pp 75–81 Nauka, Moscow (1968)
10. Kedrinskii V.K.: On multiplication mechanism of cavitation nuclei. In: Shaw, E. (ed.) *Proceedings of 12th International Congress on Acoustics* Toronto (1986)
11. Kedrinskii, V.K.: Zh. Prikl. Mekh. i Tekh. Fiz., **16**, 5 (1975)
12. Kedrinskii, V.K.: Zh. Prikl. Mekh. i Tekh. Fiz., **34**, 3 (1993)
13. Kedrinskii, V.K.: Zh. Prikl. Mekh. i Tekh. Fiz., **9**, 4 (1968)
14. Hansson, I., Kedrinskii, V., Morch, K.: J. Phys. D Appl. Phys., 15 (1982)
15. Kedrinskii, V.K.: Zh. Prikl. Mekh. i Tekh. Fiz., **19**, 4 (1978)
16. Kedrinskii, V.K.: Acta Astron. **3**, 7–8 (1976)
17. Kedrinskii, V.K., Plaksin, S.: Rarefaction wave structure in a cavitating liquid. In: *Problems of Nonlinear Acoustics: Proc. of IUPAP-IUTAM Symposium on Nonlinear Acoustics*, ed. V.K. Kedrinskii, part 1 (Novosibirsk 1987)
18. Berngardt, A.R., Kedrinskii, V.K., Pal'chikov, E.I.: Zh. Prikl. Mekh. i Tekh. Fiz. **36**, 2 (1995)
19. Berngardt, A.R.: Dynamics of the cavitation zone under impulsive loading of a Liquid. PhD Thesis, Novosibirsk (1995)
20. Besov, A.S., Kedrinskii, V.K., Pal'chikov, E.I.: On threshold cavitation effects in pulse rarefaction waves. In: Pravica, P. (ed.) Proceedings of 13th International Congress on Acoustics, vol 1, pp. 355–358 Dragan Srnic Press, Sabac 1989

21. Chernobaev, N.N.: Modeling of shock-wave loading of liquid volumes. In: Adiabatic Waves in Liquid–Vapor Systems, G.E.A. Meier, P.A. Thompson (eds.), Proceedings IUTAM Symposium on Springer, Berlin Heidelberg New York (1989), pp. 361–370

22. Besov, A.S., Kedrinskii, V.K. Palchikov, E.I.: Pis'ma Zh. Exp. Teor. Fiz., 15, 16 (1989)

23. Besov, A., Kedrinskii, V.: Dynamics of bubbly clusters and free surface at shock wave reflection. In: Blake, J., Boulton-Stone, J., Thomas, N. (eds.) *Proceedings International Symposium on Bubble Dynamics and Interface Phenomena*, pp 93–103 Birmingham, 6-9 Sept, 1993, Kluwer (1994)

24. Cornfeld, M.: Elasticity and Strength of Liquids Inostrannaya Literatura, Moscow 1951, p 46

25. Morozov, N.F., Petrov, Yu.V., Utkin, A.A.: Dokl. Akad. Nauk **313**, 2 (1990)

26. Morozov, N.F., Petrov, Yu.V.: Problems of Dynamics of Failure of Solids St. Petersburg University Publishers, St. Petersburg, p 132 (1997)

27. Pernik, A.D.: Cavitation Problems, Sudostroenie, Leningrad (1966)

28. Besov, V.K., Kedrinskii, A.S., Matsumoto, Y. et al: Microinhomogeneity structures and hysteresis effects in cavitating liquids. In: Proceedings of 14th International Congress on Acoustics, Beijing (1992)

29. Besov, A.S., Kedrinskii, V.K., Morozov, N.F., Petrov, Yu.V., Utkin, A.A.: Dokl. Akad. Nauk **378**, 3 (2001)

30. Thiel, M., Nieswand, M., Dorffel, M.: Minimally invasive therapy and allied technologies **9**, 3–4, 247–253 (AUG 2000)

31. Thiel, M.: Clinical Orthopaedics and related research **387** (JUN 2001) pp. 18–21

32. Sokolov, D.L., Bailey, M.R., Crum, L.A.: Ultrasound in medicine and biology **29**, 7 (JUL 2003) 1045–1052

33. Zhu, S.L., Cocks, F.H., Preminger, G.M., Zhong, P.: Ultrasound in med. bio. **28**(5), 661–671 (MAY 2002)

34. Kedrinskii, V.K.: On a mechanism of target disintegration at shock wave focusing in ESWL.: In: Kuhl, P.K., Crum, L. (eds.) Proceedings of 16th International Congress on Acoustics, Seattle, USA, vol 4, pp 2803–2804 University of Washington, 1998

35. Eisenmenger, W.: Ultrasound in medicine and biology **27**, 5, 683–693 (MAY 2001)

36. Cleveland, R.O., Lifshitz, D.A., Connors, B.A., et al: Ultrasound med. bio. **24**(2), 293–306 (FEB 1998)

37. Kodama, T., Uenohara, H., Takayama, K.: Ultrasound med. bio. **24**(9), 1459–1466 (NOV 1998)

38. Kodama, T., Tatsuno, M., Sugimoto, S., et al: Ultrasound medicine biology **25**, 6 (JUL 1999) 977–983

39. Kodama, T., Takayama, K.: Ultrasound med. bio. **24**(5), 723–738 (JUN 1998)

40. Delius, M., Ueberle, F., Eisenmenger, W.: Ultrasound med. bio. **24**(7), 1055–1059 (SEP 1998)

41. Zapanta, C., Lisziszka, E.G., Lamson, T.C., et al: J. Biomech. Eng.-Trans. ASME **116**(4), 460–468 (Nov 1994)

42. Williams, J.C., Stonehill, M.A., Colmenares, K., et al: Ultrasound in med. bio. **25**(3), 473–479 (MAR 1999)

43. Howard, D., Sturtevant, B.: Ultrasound in med. bio. **23**(7), 1107–1122 (1997)

44. Delius, M.: Zentralblatt fur Chirurgie **120**(4), 259–273 (1995)

45. Takayama, K.: Jpn. J. Appl. Phys., Part 1 - regular papers, Short Notes and Review Papers, **32**(5B), 2192–2198 (MAY1993)
46. Miller, D.L., Song, J.M.: Ultrasound med. and bio. **28**, 10, 1343–1348 (OCT 2002)
47. Grönig, H.: Past, present and future of shock focusing research. In: Proceedings of International Workshop on Shock Wave Focusing, Sendai, pp 1–38 (1989)
48. Sturtevant, B.: The physics of shock focusing in the context of ESWL. In: Proceedings of International Workshop on Shock Wave Focusing, Sendai, pp 39–64 (1989)
49. Kuwahara, W.: Extracorporeal shock wave lithotripsy. In: Proceedings of International Workshop on Shock Wave Focusing, Sendai, pp 65–89 (1989)
50. Kitayama, O., Ise, H., Sato, T., Takayama, K.: Non-invasive gallstone disintegration by underwater shock focusing. In: Groñig, H. (ed.) Proceedings of 16th International Symposium on Shock Tubes and Waves, VCH, Aachen, pp 897–904 (1987)
51. Grunevald, M., Koch, H., Hermeking, H.: Modeling of shock wave propagation and tissue interaction during ESWL. In: Groñig, H. (ed.) Proceedings of 16th International Symposium on Shock Tubes and Waves, VCH, Aachen, pp 889–895 (1987)
52. Delius, M.: Effect of lithotriptor shock waves on tissues and materials. In: Hamilton, M., Blackstock, D. (eds.) Proceedings of 12th ISNA, Frontiers of Nonlinear Acoustics, ESP Ltd, London, pp 31–46 (1990)
53. Kedrinskii, V.K., Soloukhin, R.I.: J. Appl. Mecha. Tech. Phys., **2**(1) (1961)
54. Church, C., Crum, L.: A theoretical study of cavitation generated by four commercially available ESWL. In: Hamilton, M., Blackstock, D. (eds.) Proceedings of 12th ISNA, Frontiers of Nonlinear Acoustics, ESP Ltd, London, pp 433–438 (1990)
55. Church, C.: J. Acoust. Soc. Am. **86**, 1 (1989)
56. Prat, F.: The cytotoxicity of shock waves: cavitation and its potential application to the extra-corporeal therapy of digestive tumors. In: Brun (ed.), Proceedings of 19th International Symposium on Shock Waves, Dumitrescu, Marseille (1993)
57. Nagoya, H., Obara, T., Takayama, K.: Underwater shock wave propagation and focusing in inhomogeneous media. In: Brun (ed.) Proceedings of 19th International Symposium on Shock Waves, Dumitrescu, vol 3, Marseille, pp 439–444 (1993)
58. Stuka, C., Sunka, P., Benes, J.: Nonlinear transmission of the focused shock waves in nondegassed water. In: Brun (ed.) Proceedings of 19th International Symposium on Shock Waves, Dumitrescu, vol 3, Marseille, pp 445–448 (1993)
59. E. Sato, et al: Soft flash x-ray system for shock wave research. In: Brun (ed.) Proceedings of 19th International Symposium on Shock Waves, Dumitrescu, vol 3, Marseille, pp 449–454 (1993)
60. Berngardt, A., Bichenkov, E., Kedrinskii, V., Pal'chikov, E.: Optic and x-ray investigation of water fracture in rarefaction wave at later stages. In: Pichal M. (ed.) Proceedings of IUTAM Symposium on Optical Methods in the Dynamics of Fluids and Solids, Prague (1984), Springer, Berlin Heidelberg New York, (1985)
61. Bayikov I., Berngardt, A. Kedrinskii, V. Pal'chikov, E.: Zh. Prikl. Mekh. i Tekh. Fiz., **25**(5) (1984)
62. Isuzugawa, K., Fujii, M., Matsubara, Y., et al: Shock focusing across a layer between two kinds of liquid . In: Proceedings of 19th Intetnational Symposium on Shock Waves, ed by Brun, Dumitrescu, (Marseille, 1993)

Shocks in Cryogenic Liquids

Masahide Murakami

4.1 Cryogenic Fluids and Superfluid Liquid Helium

A unified definition of cryogenic fluids has not been clearly given so far. Nevertheless, for convenience's sake in this chapter, it may be defined as fluids of which normal boiling points (NBP) are below liquid nitrogen temperature, 77 K. The saturated vapor pressure and the triple points and the critical points are shown for typical cryogenic fluids in Fig. 4.1. In this temperature range, cooling of superconductive magnets and devices (both of metal and high temperature superconductors) and superfluidity of superfluid helium are of particular interest. Among these cryogenic fluids, liquid nitrogen (NBP = 77 K) is the most popular cryogenic fluid that is frequently used as coolant at cryogenic temperature such as in high temperature superconductor cooling and freezing, and as a substitute of other cryogenic fluids because of its easy and cheap procurement. Liquid oxygen (NBP = 90 K) is used as an oxidizer of liquid rockets and is an oxygen source for manned flight. Liquid hydrogen (NBP = 20 K) is a very important potential energy source and has been frequently used for liquid rocket fuel. Liquid helium exists at still lower temperatures, which in fact appears in two liquid phases, super (He II) and normal (He I) liquid heliums. Liquid helium has been used as a coolant for metal superconductor and is an object of low temperature physics experiments.

Shock wave related phenomena in cryogenic fluids may be caused by a quench of a superconductive magnet, when violent boiling occurs, and in a sudden thermal shrink of the surrounding structure during cryogenic cooling. The thermal shock wave is a phenomenon characteristic only for superfluid helium. In addition, cryogenic fluids may be utilized for the research of cavitation, for enhanced thermodynamic effects and for shock wave and high Mach number flows in low sound speed fluids. Some of these topics will be described in more detail later.

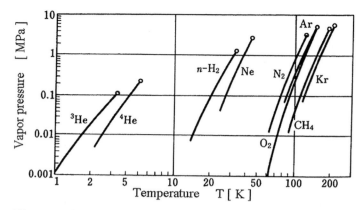

Fig. 4.1. Saturated vapor pressure curves for cryogenic liquids

4.1.1 Cryogenic Fluids

Shock wave related phenomena in cryogenic fluids are not fundamentally different from those in normal fluids with the thermal shock wave in He II as an exception. However, some characteristic features of thermofluid dynamic phenomena of cryogenic fluids may be pointed out: One of them is, of course, the fact that they appear in an environment at very low temperature. They usually occur in a rather restricted closed space inside of a cryostat as illustrated in Fig. 4.2, which imposes some limitations on the size and the degree of freedom of the movement of measuring instruments. For the purpose of flow visualization, specially designed vacuum-tight windows that are composed of multiple glass plates for thermal insulation to minimize parasitic radiation heat leak from the room temperature environment must be attached to the cryostat. Furthermore, such conventional measuring instruments and tools in thermofluid dynamics experiments, as a hot wire and a manometer cannot be utilized. Most commercially available sensors for the temperature and the pressure measurements cannot be applied in a cryogenic environment. Even if some of them can be used, their operation will be outside of the guarantee of a product and they are easy to break against thermal cycles and on-site calibration will be required. It is rather difficult to find suitable tracers for flow visualization.

Cryogenic fluid properties have a high sensitivity to temperature. Even a small variation in temperature may cause large variations in physical properties, which are sometimes large enough to cause an essential change of the physical state. It is another common characteristic feature of cryogenic fluids that their latent heat of evaporation is very small. Thus, in cryogenic fluids evaporation tends to occur easily. These features may be summarized in short that thermodynamic effects are enhanced in cryogenic fluids.

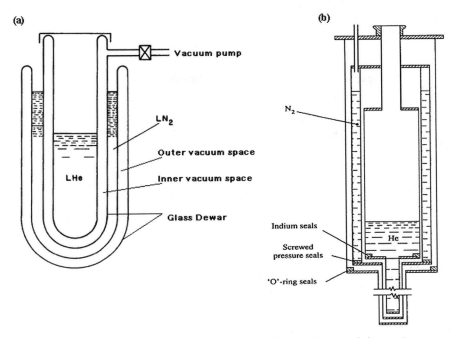

Fig. 4.2. Cryostats for cryogenic experiment. (**a**) glass Dewar, (**b**) metal cryostat

Some research topics related to shock waves in cryogenic fluids except superfluid liquid helium are briefly introduced here. The development of a cryogenic shock tube by the group of Caltech. led by Professor Liepmann should be especially mentioned as a very pioneering work in this field. The original purpose of this development was an attempt to construct a shock tube to realize an ultra-high Mach number shock tube by cryogenic cooling of test gas by liquid helium down to about 2 K [1, 2]. Later, the work progressed to the development of a research facility for superfluid liquid helium thermofluid dynamics [3]. Liepmann presented a comprehensive review about the cryogenic shock tube [4, 5]. A new type cryogenic shock tube, that is a diaphragm-free cryogenic shock tube equipped with the M–O valve, was developed by Maeno [6] for the research of thermofluid dynamics and shock waves of cryogenic fluids. Cavitation in cryogenic fluids is one of such shock wave related phenomena. Generation and collapse of bubbles and subsequent shock wave emission were investigated. Some critical point phenomena, usually observed at very high temperature and high pressure in the cases of conventional fluids can be studied under moderate pressure condition in cryogenic fluids. A quench of a superconducting magnet may be one of the most serious phenomena in the application of cryogenic engineering, which may cause compression waves in the coolant, that is normally liquid helium, and thermal shocks in structural components as a result of suddenly occurring violent boiling. Solidified gases may be of use for the researches of ablation.

4.1.2 Superfluid Liquid Helium (He II)

Superfluid helium is one of the main topics in this chapter, and special focus is placed on its particular feature in fluid dynamics including shock waves. First, the phase diagram of helium at very low temperature, below 7 K, will be introduced to characterize the physical state of the superfluid helium that is the lower temperature liquid of the two kinds of liquid phases of helium. Then, the two-fluid model to elucidate the particular superfluid phenomena is presented. Some of the typical superfluid phenomena are explained on the basis of the two-fluid model. The highlights among them are the first and second sounds. However, even under moderate heat transfer conditions, the superfluid helium is often in the breakdown state, where quantized vortices are generated and some dissipative phenomena appear, though the superfluid phenomena are not completely destroyed. In the state where tangled mass of high-density quantized vortices is generated, serious effects of them arise. This state is called the superfluid turbulent state.

Superfluidity and Two-Fluid Model

It is the superfluid phenomena of ^{4}He that is treated in this section. The superfluidity of ^{3}He that is an isotope of helium is excepted from this section. The phase diagram of ^{4}He at low temperatures is shown in Fig. 4.3. We notice some singular properties of helium at cryogenic temperature in this figure. There are two liquid phases (He I and He II) separated by the lambda-line (λ-line). The liquid phase, He II, does exist even at the temperature of absolute zero, which explains why it is called a permanent liquid, and does not have a triple point at which three phases, vapor, liquid, and solid, coexist. Furthermore, it is another singular property that an anomaly of the specific heat in the form of the character λ appears at the lambda point temperature (2.17 K), because of which the He I–He II transition is called the λ-phase transition. This phase transition is a second-order phase transition in which no latent heat participates. The liquid phase on the lower temperature side of the λ-line is called He II showing superfluidity, while the higher temperature phase, He I, is just a normal viscous fluid. The fact that the superfluid phenomena appear only at temperatures near absolute zero suggests that superfluidity originates from a quantum effect. The low mass of the helium atoms and the extremely weak forces between them makes the quantum effect dominate. Because of the smallness of the atoms, the zero-point energy is large and thus the total energy of the liquid reaches a minimum at a considerably greater atomic volume than the potential-energy minimum. Therefore the interatomic forces are strong enough to produce the liquid phase at a low enough temperature,

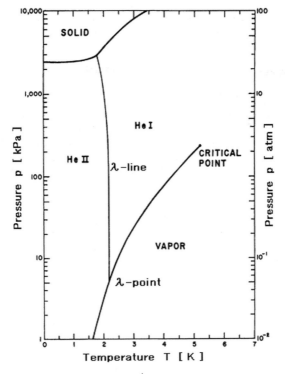

Fig. 4.3. Phase diagram of ^{4}He at cryogenic temperatures

but the high zero-point energy keeps the density of the liquid rather small. The permanent liquid state originates in these physical facts.

The thermofluid dynamic behavior of a superfluid can be well explained on the basis of the two-fluid model. The model was phenomenologically derived from experimental observations, and is related to the theory of Bose–Einstein condensation. A superfluid is considered as a mixture of two fluids, i.e., the superfluid component with zero viscosity and the normal fluid component as an ordinary viscous fluid. Furthermore, the superfluid component is assumed to be a fluid that does not participate in the entropy. In other words, the superfluid component has zero entropy because the motion of the component corresponds to the collective motion of a phase being macroscopically in the ground state in the sense of Bose–Einstein condensation. In the ideal super-fluid state the superfluid component does not exchange momentum with the normal fluid component, resulting in frictionless flow. Attention is called here about the following fact that this description for the two fluid components is just an expedient and that it is meaningless to ask whether individual helium atoms belong either to the superfluid component or to the normal component. The relative concentration of the two fluid components is mainly a function of temperature and thus it is subject to heating or cooling. Heating He II creates excitations, which is the normal component, and as the result the concentration of the normal component increases there. Under this condition a large difference in chemical potential exists, which in turn induces a flow of the superfluid component towards the heating site. A kind of internal counterflow between the two components is created so as to fulfill the total mass conservation, because no motive force to drive the total fluid is generated. This flow is called the thermal counterflow, where a one-directional heat flow is created because only the normal component carries the entropy, though no total mass flow is induced. In such a situation a macroscopic amount of heat can be transported without any appreciable temperature gradient and total mass flow, and therefore it is called the superthermal conduction. The effective thermal conductivity becomes about two orders of magnitude larger than for high-purity metals at low temperatures, and, of course, far larger than the thermal conductivity of He I. In fact, the superthermal conduction is not a thermal conduction phenomenon but a convective one in a form of internal convection with a huge effective thermal conductivity. This phenomenon is sometimes more important than a frictionless flow in engineering applications.

The frictionless flow of He II through a narrow channel, as the most well-known superfluid phenomenon, can be understood on the basis of the two-fluid model. In narrow flow passages with a width of the order of $1\,\mu\text{m}$ or less the normal fluid component cannot flow because it sticks to the wall of the flow passage due to viscosity. On the other hand, the superfluid component can freely flow through even such narrow channels because of its inviscid nature. As a result, it looks that He II can flow through a very narrow channel in a frictionless manner. However, this frictionless flow results from zero viscosity not of the total fluid He II, but of the superfluid component. It was reported that a persistent flow in a torus-shaped channel filled with very fine fibrous foam was experimentally realized [7, 8]. A flow channel through which the superfluid component can flow but in which the normal fluid component is in the clamped state, is called a superleak. A superleak is commonly produced in a tube packed tightly with fine powder, the spaces between which form complex passages of very narrow width, typically $\sim$100 nm.

Another example of particular thermophysical behavior of He II is the second sound. In fact, there are four kinds of sounds in He II. For the first sound, the super and normal fluid components oscillate together in phase with a common velocity equal to the center of mass velocity as in an ordinary sound wave. Thus, the first sound is a longitudinal pressure wave involving its compression and rarefaction motions. The second sound arises from a relative motion of the two fluid components with no center of mass motion. In this sound mode, the two fluid components oscillate through each other. This relative motion between the two fluid components produces compression and rarefaction of the entropy, which can be detected as a propagating temperature wave. The third sound, a kind of capillary or surface tension wave, is generated

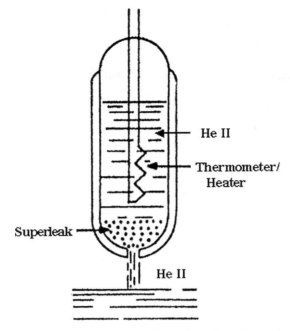

He II

Thermometer/
Heater

Superleak

He II

Fig. 4.4. Apparatus for thermomechanical and mechanocaloric effects

on the free surface of a He II thin film, where the normal fluid component motion will be effectively damped because of its viscous action in an extremely thin film. The fourth sound is generated through narrow porous channels, where the normal fluid is effectively clamped by its viscosity. Shock wave research for superfluid He II is mostly related to the nonlinear forms of the first and second sounds.

The physically most important and most interesting superfluid phenomena appear when some temperature variation is imposed, which is caused by zero-entropy property. Here, the thermomechanical effect that was first observed by Daunt and Mendelssohn [9] is discussed, referring to Fig. 4.4. A vacuum insulated glass container, plugged at the bottom with fine emery powder working as a superleak, is arranged so that it can be raised or lowered into a bath of He II. A resistance thermometer based on the temperature dependency of the electric resistance is mounted in the container. It was found that when the container was lowered into the He II, liquid flowed into it through the powder and the temperature dropped about 10^{-2} K, and when liquid flowed out the temperature rose. In both cases only the superfluid component flowed through the superleak. It was also demonstrated with the same apparatus that this phenomenon was reversible. On the other hand, when the He II inside the container was electrically heated by the resistance thermometer, liquid flowed into the container. As the results of these findings, the assumption that the entropy of He II was associated only with the normal fluid component was experimentally confirmed in both qualitative and quantitative senses. In a more strict sense, the former is called the mechanocaloric effect, and the latter the thermomechanical effect.

The flow phenomena of He II seem to obey Euler's equation in situations such as the Torricelli flow, where He II flows out of an open container through a hole at the bottom due to gravity, and such as the propagation of a sound wave. It is rationalized that in these situations the normal and superfluid components move together in phase and thus no superfluid phenomena appear. However, in other experiments Euler's equation apparently give an inadequate description of the fluid dynamics of He II. One of the most striking phenomena to demonstrate this was the thermomechanical effect. The landmark experiment of Kapitza in 1941 [10] provided a clear and quantitative unification of the thermomechanical and mechanocaloric effects. Based on these experimental results, the idea of the two-fluid model was first presented by Tisza in 1938 [11]. A consistent set of the dynamical equations for He II was presented by Landau in 1941 [12], inspired by various experimental results. It is now called the Landau Two Fluid equations [13].

Two-Fluid Equations

The most fundamental assumption of the two-fluid equations is that two independent velocity fields are required to describe the superfluid He II hydrodynamic phenomena. Thus the He II fluid dynamic behavior can be characterized by eight variables, ρ, s, $\boldsymbol{v}_{\mathrm{n}}$ and $\boldsymbol{v}_{\mathrm{s}}$. The two fluid components are distinguished by the suffix s for the super fluid component and n for the normal fluid component. The equations can be written in various approximation levels. In the following, the form derived by Khalatnikov [14] is shown, where the terms including the relative velocity between the super and normal fluid components up to the second order of smallness sufficient to describe shock waves are retained. The normal viscosity is neglected, because it is not essential in the process of shock waves. These equations are applicable to the fluid dynamic phenomena in the ideal superfluid level. If the relative velocity exceeds a critical value, then quantized vortices are generated in He II and some excess dissipation appears. Under this condition the equations must be modified so as to include the effects of quantized vortices. The details of the effects of a tangled mass of high-density quantized vortices are discussed Sect. "Superfluid Breakdown and Quantized Vortices."

In what follows, the equations of continuity (4.1), total momentum for the ith component (4.2), entropy conservation (4.3), the motion of the superfluid component (4.4), and the equation for the chemical potential (4.5), are presented, respectively.

$$\frac{\partial}{\partial t}\left\{\rho+\frac{1}{2}w^2\rho^2\frac{\partial}{\partial p}\left(\frac{\rho_{\mathrm{n}}}{\rho}\right)\right\}+\nabla\cdot\boldsymbol{j}=0, \tag{4.1}$$

where, $j = \rho_s v_s + \rho_n v_n$.

$$\frac{\partial j_i}{\partial t} + \frac{\partial \Pi_{ik}}{\partial r_k} = 0, \tag{4.2}$$

where, $\Pi_{ik} = \rho_s v_{si} v_{sk} + \rho_n v_{ni} v_{nk} + p\delta_{ik}$.

$$\frac{\partial}{\partial t}\left[\rho s + \frac{\rho w^2}{2}\left\{\frac{\partial}{\partial T}\left(\frac{\rho_n}{\rho}\right) + \rho s \frac{\partial}{\partial p}\left(\frac{\rho_n}{\rho}\right)\right\}\right] + \nabla \cdot (\rho s v_n) = 0. \tag{4.3}$$

$$\frac{\partial v_s}{\partial t} + \nabla \left(\mu + \frac{v_s^2}{2} - \frac{\rho_n w^2}{2\rho}\right) = 0. \tag{4.4}$$

$$d\mu = -sdT + \left(\frac{1}{\rho}\right)dp - \left(\frac{\rho_n}{2\rho}\right)dw^2. \tag{4.5}$$

It should be noted that the pressure p and the temperature T are chosen as the independent thermodynamic variables. However, μ, ρ, ρ_n, ρ_s and s generally depend on the relative velocity $w(w = |\mathbf{w}| = |\mathbf{v}_n - \mathbf{v}_s|)$. For large values of w this dependence can only be determined by a microscopic theory. Here we restrict ourselves to quadratic terms in w, which is a valid approximation for moderately large w such as in shock waves. Hence we may neglect the dependence of ρ_n and ρ_s on w. The choice of the equations for the momentum has some degrees of freedom. Here, those of the total momentum, (4.2), and of the superfluid momentum, (4.4), are selected. Any two among the three equations for the momentum that is the total momentum, the superfluid momentum and the normal fluid momentum may be selected. The terms including w^2 in (4.1), (4.3)–(4.5) express the variations of the thermophysical quantities induced by the relative velocity between the super and the normal fluid components. The second term of (4.3) reflects the fact that the entropy is carried only by the normal fluid component. The first term of the right hand side of (4.5) presents the thermomechanical effect, which is very large in He II. It is understood from (4.5) that peculiar superfluid phenomena appear in the cases where w is not zero and heat flow causing some temperature variation is imposed to He II.

Superfluid Breakdown and Quantized Vortices

The above equations describe superfluid dynamic phenomena in the ideal superfluid level, where no dissipative effects participate in movement. However, if the relative velocity w between the super and the normal components exceeds a critical value w_{cr}, that varies from several $cm\,s^{-1}$ to a few $m\,s^{-1}$ according to the thermophysical situation, quantized vortices of which circulations are quantized are generated in He II and some excess dissipation appears. This state is called the superfluid breakdown state. Note that the superfluid breakdown does not mean the superfluidity completely disappears but the superfluidity is accompanied by excess dissipation as the result of interaction with vortices. This state of He II is analogous to the mixed state of type II superconductor where quantized magnetic fluxes partially penetrate. In the mixed state, the electrical resistance increases with the magnetic field, though the resistance is still smaller than that of normal conductor state. The size of each quantized vortex line is comparable with the interatomic spacing and the size of a vortex core is of the order of 10^{-1} nm (=1Å). In the state where the relative velocity w becomes considerably larger than the critical value, the vortex density becomes very high so that they become entangled closely, and the vortex line density, that is the total length of vortices per unit volume, becomes as high as 10^{10} m m^{-3}. This state is called the superfluid turbulent state where the superfluid component interacts with the normal fluid component through the action of vortices. This effect is very large in particular on the temperature field and brings about some dissipative effects on heat transfer.

The development of vortices is rather rapid as compared with the time scale of conventional superfluid flow phenomena. In these situations, the vortex development may be assumed to be a momentary phenomenon, and thus the vortex density is always in an equilibrium state. However in shock wave phenomena, as the characteristic time is usually shorter than that of vortex development, the time variation of the vortex density must be considered.

In order to take into account the effect of tangled mass of quantized vortices on He II thermofluid dynamic phenomena, the two fluid equations must be modified. The introduction of the mutual friction term proposed by Gorter and Mellink [15] to the equation of motion may be the most convenient way:

$$\boldsymbol{F}_{\mathrm{ns}} = A\rho_{\mathrm{n}} \left(w - w_{\mathrm{cr}}\right)^2 \left(\boldsymbol{v}_{\mathrm{s}} - \boldsymbol{v}_{\mathrm{n}}\right). \tag{4.6}$$

Here the coefficient A is an empirical constant given by Gorter and Mellink. This mutual friction term was derived on the basis of empirical laws that the vortex density is in proportion to $(w - w_{\mathrm{cr}})^2$, and the mutual friction is proportional to the product of the vortex density and the relative velocity $(\boldsymbol{v}_{\mathrm{s}} - \boldsymbol{v}_{\mathrm{n}})$. The term $\boldsymbol{F}_{\mathrm{ns}}$ is added to the left hand side of (4.4). The effect of the mutual friction appears more prominent in the temperature field rather than the flow resistance. This treatment mostly yields satisfactory results in the case of steady heat transfer, but rather poor results in high-speed phenomena such as shock waves because then the time development of the vortex density should be taken into consideration. In this case, the equation for the vortex development must be simultaneously solved with the two fluid equations. The equation for the vortex development proposed by Vinen [16, 17] has been frequently used:

$$\frac{\partial L}{\partial t} + \nabla \cdot (V_{\mathrm{T}} L) = \mathrm{X}_1 \frac{B\rho_{\mathrm{n}}}{2\rho} w L^{3/2} - \mathrm{X}_2 \frac{h}{\pi m} L^2. \tag{4.7}$$

Here, L is the vortex line density, V_{T} is the drift velocity of the vortex tangle, X_1 and X_2 are empirical constants given by Vinen, and B is a constant for the definition of the mutual friction term. The mutual friction term is also written in terms of L as follows;

$$\boldsymbol{F}_{\mathrm{ns}} = \left(\frac{\kappa}{3}\right)\left(\frac{\rho_{\mathrm{n}}}{\rho}\right) BL\left(\boldsymbol{v}_{\mathrm{s}} - \boldsymbol{v}_{\mathrm{n}}\right), \tag{4.8}$$

where κ is the unit of quantized circulation ($= h/m$), and h and m are Planck's constant and the mass of a helium atom. It is clear from (4.6) and (4.8) that the mutual friction force does not work in the cases of $\boldsymbol{v}_{\mathrm{n}} = \boldsymbol{v}_{\mathrm{s}}$ where two fluid components move together in phase as the mutual friction force is a kind of internal force acting between the super and the normal fluid components. The numerical solutions for shock wave problems based on the two fluid equations (4.1)–(4.5) and Vinen's vortex density development equation (4.7) seem fairly good, but still suggest that this treatment has some shortcoming with respect to a little bit too fast decay of vortices.

First Sound and Second Sound [13]

In general, the validity of a new theory must be checked by the comparison of the prediction based on the theory with experimental results. It was confirmed by the capability of predicting the existence of the second sound and by the excellent agreement of the prediction of the propagation speeds of the first and second sounds with experimental results. In what follows, the propagation speeds of the first and second sounds are derived from the one-dimensional form of the linearized two fluid equations. It is assumed that the velocities of both fluid components are small and the deviation of each physical quantity from quiescent equilibrium quantity is also small, and thus the equation system can be linearized. Each physical quantity is expressed in the form of a sum of small deviation denoted by prime and the equilibrium value denoted by suffix 0 as follows:

$$v_{\mathrm{s}} = v'_{\mathrm{s}}, \quad s = s' + s_0, \quad - - -.$$

Further, ρ' and s' are expressed in terms of p' and T' with the aid of such thermodynamic relations as:

$$\rho' = \left(\frac{\partial \rho_0}{\partial p_0}\right)_{\mathrm{T}} p' + \left(\frac{\partial \rho_0}{\partial T_0}\right)_{\mathrm{p}} T',$$

$$s' = \left(\frac{\partial s_0}{\partial p_0}\right)_{\mathrm{T}} p' + \left(\frac{\partial s_0}{\partial T_0}\right)_{\mathrm{p}} T'.$$

Then, the following equations can be derived:

$$\left(\frac{\partial \rho_0}{\partial p_0}\right)_{\mathrm{T}} \frac{\partial^2 p'}{\partial t^2} - \Delta p' + \left(\frac{\partial \rho_0}{\partial T_0}\right)_{\mathrm{p}} \frac{\partial^2 T'}{\partial t^2} = 0, \tag{4.9}$$

$$\left(\frac{\partial s_0}{\partial p_0}\right)_{\mathrm{T}} \frac{\partial^2 p'}{\partial t^2} + \left(\frac{\partial s_0}{\partial T_0}\right)_{\mathrm{p}} \frac{\partial^2 T'}{\partial t^2} - \frac{\rho_{s0} s_0^2}{\rho_{\mathrm{n}0}} \Delta T' = 0, \tag{4.10}$$

where Δ expresses the Laplacian operator. By substituting a planar wave solution in (4.9) and (4.10) the dispersion relation is found. Two speeds of sound appear:

$$u_1 = \sqrt{\left(\frac{\partial p}{\partial \rho}\right)_s}, \tag{4.11}$$

$$u_2 = \sqrt{\frac{Ts^2\rho_s}{C\rho_n}}. \tag{4.12}$$

Here, the suffix 0 is omitted for simplicity, and the following thermophysical relations are taken into account:

$$C_p \approx C_v = C,$$

$$\left(\frac{\partial p}{\partial \rho}\right)_T \approx \left(\frac{\partial p}{\partial \rho}\right)_s = \frac{\partial p}{\partial \rho}.$$

The small disturbances propagating with speeds u_1 and u_2 are designated as first and second sounds, respectively. The variation of the two speeds of sound with the temperature is shown in Fig. 4.5. They are different by one order of magnitude.

The physical nature of the two sounds is examined in the case of a plane sound wave. In such a sound wave, the velocities v'_s, v'_n and the small disturbances p', T' of the pressure and the temperature may be considered proportional to each other.

$$v'_s = av'_n, \quad p' = bv'_s, \quad T' = cv'_s.$$

For the first and the second sound waves, the approximate solutions for a, b, and c are:

$$a_1 = 1 + \frac{\beta\rho}{\rho_s s}\frac{u_1^2 u_2^2}{(u_1^2 - u_2^2)}, \quad b_1 = \rho u_1, \quad c_1 = \frac{\beta T u_1^3}{C(u_1^2 - u_2^2)}, \tag{4.13}$$

$$a_2 = -\frac{\rho_s}{\rho_n} + \frac{\beta\rho}{\rho_n s}\frac{u_1^2 u_2^2}{(u_1^2 - u_2^2)}, \quad b_2 = \frac{\beta\rho u_1^2 u_2^3}{s(u_1^2 - u_2^2)}, \quad c_2 = -\frac{u_2}{s}. \tag{4.14}$$

Here, β is the thermal expansion coefficient, which magnitude is very small for He II. If the terms including β are ignored due to their smallness, the approximate solution is so simplified that the physical nature of the two sound waves are easily understood. For the first sound, one has

$$v'_s = v'_n, \tag{4.15}$$
$$T' = 0. \tag{4.16}$$

This approximate solution means that in a rough approximation for the first sound the super and the normal fluid components move in-phase and no temperature disturbance is induced. So the first sound is understood to be a pure

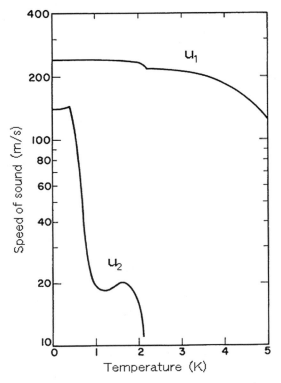

Fig. 4.5. Speeds of sound in liquid helium, the first sound u_1 and the second sound u_2

pressure or density wave. On the other hand, to the same approximation level, the second sound solution results in:

$$j = \rho_s v'_s + \rho_n v'_s = 0, \tag{4.17}$$
$$p' = 0. \tag{4.18}$$

In the second sound, the two fluid components move in antiphase so that the total mass flux density is zero, and no pressure disturbance is excited. The resultant motion leads to a propagating compression and rarefaction in the entropy, because the entropy is carried only by the normal fluid component. The compression and rarefaction in the entropy result in a temperature oscillation. Thus the resulting wave motion is detected in the form of a propagating temperature wave. The essential physical nature suggests that the first sound is efficiently generated by a periodic movement of a boundary wall in the normal direction to the surface, and the second sound by a periodic oscillation of the surface temperature of a wall induced by periodic heating and cooling.

4.2 Shock Waves in He II

It is concluded in the above section that two kinds of sound waves can propagate in He II. If a disturbance has a finite amplitude, nonlinearity shows up during its propagation and the wave develops into a shock wave. The physical properties of the two sound waves are very much different. A compression shock wave just like shock waves in ordinary fluids originates from the first sound mode, while a thermal shock having a discontinuity in the temperature, that is a temperature shock, originates from the second sound mode.

4.2.1 Compression Shock Wave

The research in the field of compression shock waves in liquid helium had been initiated as an extension of first sound experiments. The pioneering research using a cryogenic shock tube was started by H. Liepmann [1]. This shock tube facility was then modified so that experiments on superfluid shock wave phenomena could be performed, where a gasdynamic shock wave impinges onto the free surface of He II and a compression shock wave penetrates into He II and, at the same time, a thermal shock wave is generated at the free surface and also propagated into He II. On the other hand, in a pulsed heating process in He II, a thermal shock wave is dominantly generated, but only an infinitesimally weak pressure wave is induced. And thus this scheme is inadequate for the generation of a strong compression shock wave. A method of direct generation of a compression shock wave by some mechanical means may be possible, but these kinds of method have never been tried so far.

Cryogenic Shock Tube [2, 3]

The primary purpose of the development of the cryogenic shock tube at Caltech. was an attempt to construct a very-high-Mach-number shock tube by reducing the temperature of test gas T_1. The shock Mach number M_s is approximately given by the formula for the condition $T_4/T_1 \gg 1/4$ as,

$$M_s = 4 \left(\frac{T_4}{T_1} \right)^{1/2} , \qquad (4.19)$$

where T_1 and T_4 are the temperatures of the undisturbed test gas and the driver gas, respectively. In the environment of helium vapor coexisting with He II at $T_1 = 2\,\mathrm{K}$ and a driver helium gas $T_4 = 300\,\mathrm{K}$, creating a shock wave with $M_s = 40$ is not difficult in the facility. Even in such high Mach number shock waves no effect of ionization and dissociation of test gas are induced in the cryogenic shock tube. The cryogenic shock-tube constructed by Liepmann and Laguna [4] is shown in Fig. 4.6. For the operation of this shock tube, helium gas is used as the working fluid with the driver section at room temperature. The low-pressure section immersed in liquid helium in the liquid

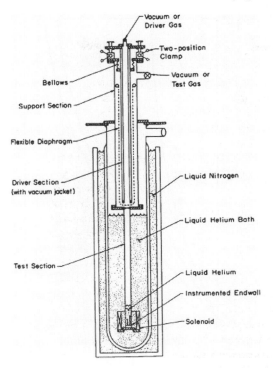

Fig. 4.6. Cryogenic shock-tube developed at Cal. Tech. [4]

helium bath is directly cooled by liquid helium. The solenoid magnet installed at the bottom part of the low-pressure tube was used for the adjustment of the working temperature range of the superconductive temperature sensor detecting the temperature signals of He II shocks during their propagation. A thin Mylar film is used as a diaphragm for the initial separation of driver gas from test gas. The construction is such that a repetition of the shock tube operation is possible avoiding potential problems of frost contamination of the inside wall of the low-pressure tube at cryogenic temperature.

Shown in Fig. 4.7 is the comparison of the shock Mach number between the ideal shock tube theory and the experimental result of the cryogenic shock tube. It is evident in this figure that the ideal shock tube theory becomes more satisfactory for lower temperature operation. The best agreement is found in the case where helium vapor was set as the test gas at 2.3 K and at the highest initial pressure ratio. Helium vapor can be regarded as an ideal gas with the specific heat ratio of 5/3. It is also suggested that the boundary layer in the cryogenic shock tube is so thin that it can be ignored. This results from the extremely small viscosity of helium vapor at cryogenic temperatures. These features of the cryogenic shock tube experiment suggest such interesting research topics as the ultra high Reynolds number flow, shock wave propagation

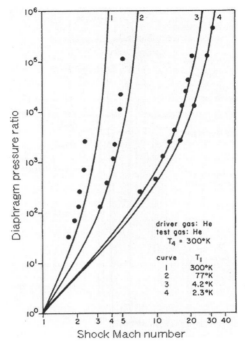

Fig. 4.7. Comparison of the shock Mach number between the ideal shock tube theory and the experimental result of the cryogenic shock tube [4]

through a gas with density stratification, evaporation phenomena, and shock propagation through He II caused by the impingement of a gas dynamic shock wave on the free surface.

Compression Shock Wave Induced by Gas Dynamic Shock Wave Impingement on Free Surface of He II

The matching of the shock wave impedance between liquid helium and vapor is far better than for any other fluids, and thus a shock wave in a vapor phase can well penetrate the free surface of He II. In this situation, only a compression shock wave propagates through ordinary liquid such as He I, while two shock waves, a compression shock wave and a thermal shock wave, propagate through He II. These two shock waves are quite different in physical nature, and can be easily distinguished by their propagation speeds because they are different by one order of magnitude. The shock propagation process for He II is shown in Fig. 4.8 in the form of an x–t diagram for two cases of the initial temperature, 1.522 K (Fig. 4.8a) and 2.095 K (Fig. 4.8b), respectively [4]. In these figures, the abscissa is the distance measured from the initial liquid surface normalized by the initial liquid depth, and the ordinate

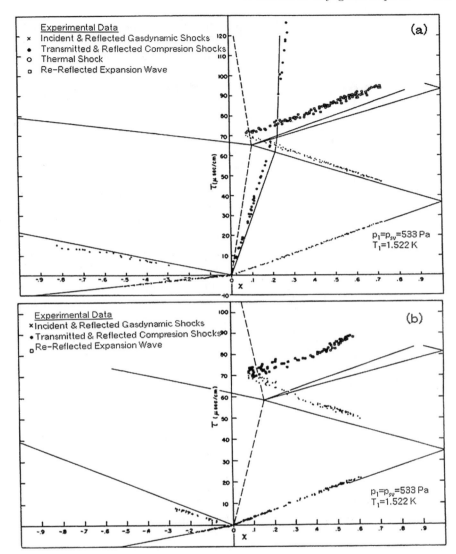

Fig. 4.8. Wave diagram of shock waves obtained by impingement of a gasdynamic shock wave on the He II free surface. The solid lines are the computational result based on the Khalatnikov approximation in He II and classical shock-wave theory in the gas. ×: incident and reflected gasdynamic shocks, (*filled circle*): penetrated and re-reflected compression shocks, (*open circle*): thermal shock wave, (*open square*): rarefaction wave generated upon the reflection of the reflected compression shock wave at the free surface. (a) $T_1 = 1.522\,\mathrm{K}$, $P_1 = 533\,\mathrm{Pa}$; (b) $T_1 = 2.095\,\mathrm{K}$, $P_1 = 4.13\,\mathrm{kPa}$ [4]

is the time divided by the initial liquid depth. These data were acquired by superconductive temperature sensors fixed in the He II in the cryogenic shock tube shown in Fig. 4.6. The initial states of liquid helium were He II for both cases, but the states behind compression shocks were He II in the case of Fig. 4.8a and He I for Fig. 4.8b. The pressure behind the transmitted compression shock wave, the region 6 of the wave diagram shown in Fig. 4.9, is considerably higher for Fig. 4.8a than for Fig. 4.8b. Therefore, in the case of Fig. 4.8b the thermal wave is missing because He II changed to He I due to the lambda-phase transition within the shock transition zone. This result suggests that a shock-induced lambda-phase transition occurs in the cryogenic shock tube. It is evident in the results shown in Fig. 4.8 that the experimental results for the propagation speed of the penetrated compression shock wave are qualitatively in good agreement with the result from the two-fluid equations. The disagreement of the propagation speeds of the reflected shock and the thermal shock is connected with the heat flow in region 6 of the wave diagram shown in Fig. 4.9 and evaporation in region 5 in Fig. 4.9. These discrepancies become bigger as the shock intensity increases. Though solidification of He II by shock compression above 25 bar. would be possible in this kind of shock tube, no attempt has been made so far. This kind of shock-compression would induce a highly non-equilibrium solidification process.

4.2.2 Thermal Shock Wave

The second sound pulse as a wave of the temperature also develops into a shock wave as a result of the nonlinearity, in the cases where the amplitude of the pulse is finite in magnitude. A thermal shock wave, being independent of

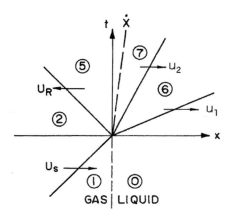

Fig. 4.9. Schematic x–t diagram of the shock experiment conducted in the cryogenic shock tube with He II in the bottom portion of the test section. The gasdynamic shock wave impinges on the free surface of He II, and a reflected shock wave appears and a compression shock wave and a thermal shock wave are simultaneously induced in He II. U_S; incident gasdynamic shock wave, U_R; reflected gasdynamic shock wave, u_1; penetrating compression shock wave, u_2; induced thermal shock wave, the broken line; movement of the free surface

a compression shock wave, can be generated by the incidence of a gasdynamic shock wave on the free surface of He II as well as by pulsed heating of He II. The research of thermal shock waves was initiated by Osborne [18] and Dessler et al. [19]. As a result of the approximation of Khalatnikov for He II wave phenomena, the two-fluid equations were proven correct up to second order.

Thermal Shock Wave Profile

Nonlinear waveform deformation of a finite amplitude second sound disturbance can be understood on the basis of the two-fluid equations. The first order correction to the phase velocity of the second sound is derived from the second-order shock wave theory and is written in terms of the temperature variation as;

$$u_2 = u_{20} \left(1 + \frac{1}{2} B \frac{\Delta T}{T} \right), \tag{4.20}$$

where u_{20} is the speed of second sound relative to the quiescent equilibrium state. The constant B is the coefficient that determines the waveform evolution and is therefore denoted as the steepening coefficient, defined as:

$$B = T \frac{\partial}{\partial T} \left[\ln \left(\frac{u_{20}^3 c_p}{T} \right) \right]. \tag{4.21}$$

The graph of the steepening coefficient vs. T is given in Fig. 4.10. As seen from Fig. 4.10, the steepening coefficient is a strong function of the temperature. It is negative above 1.88 K, which is called the B-nodal point, and

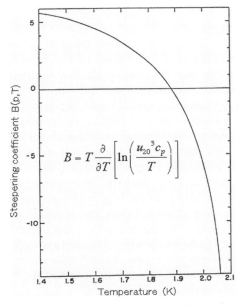

Fig. 4.10. Temperature dependence of the steepening-coefficient at saturated vapor pressure [22]

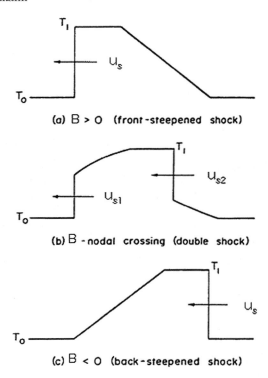

Fig. 4.11. Three shock forms generated by rectangular heat pulses. Arrows indicate the direction of propagation of each shock front. (**a**) front-steepened shock, (**b**) double shock, (**c**) back-steepened shock

positive below 1.88 K. In the temperature range of a positive steepening co-efficient, that is the range of T lower than 1.88 K, the high temperature part of a disturbance propagates faster than the low temperature part and may even overtake the latter, forming a shock in temperature as is schematically shown in Fig. 4.11a. This is called the front-steepened shock. On the other hand, in the temperature range of negative steepening-coefficient, T higher than 1.88 K, the lower temperature part of the disturbance catches up with the higher temperature part, and thus a shock is formed at a backside of the wave (Fig. 4.11c). It looks just like an expansion shock that is not physically allowed to exist in regular fluids and the temperature discontinuously drops after the passage of the shock wave. It is very interesting to see that at the equilibrium temperatures slightly below the B-nodal point a double shock may be generated, where discontinuities in the temperature appear at both the leading and trailing wave fronts of a thermal pulse (Fig. 4.11b). However, for the correct quantitative prediction of the double shock structure, higher or-der quantities up to w^2 must be retained in the two-fluid equations presented in Sect. Two-Fluid Equations.

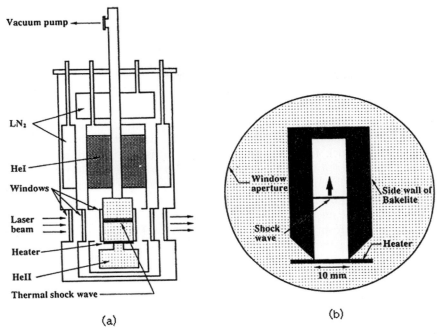

Fig. 4.12. Schematic of the superfluid thermal shock tube facility. (**a**) cryostat with optical windows, (**b**) cross-sectional view of the superfluid thermal shock tube part

Thermal Shock Wave Experiment

In addition to the cryogenic shock tube for the investigation of compression shock waves, the superfluid thermal shock tube was also used as an experimental facility for the thermal shock wave, where a thermal shock was generated by a pulsed heating in He II. The schematic illustration in Fig. 4.12 is an example of the superfluid thermal shock tube used by Iida et al. [20]. The superfluid thermal shock tube is installed in the cryostat with optical windows for holographic visualization. The detailed description of the cryostat is given in next section. The shock tube is composed of a thin film heater at the bottom to generate a thermal pulse and of adiabatic sidewalls. The shock tube width of only 10 mm is sufficiently wide because the sidewall viscous boundary layers are very thin. The length of the present shock tube is restricted to 60 mm due to the size of the cryostat, though it had better be longer because a thermal pulse propagates far longer than 10 cm. A rather large depth of 6 cm is required for the integrated optical path length to be large enough for laser holographic interferometry. A thermal pulse which soon changes into a thermal shock wave, is generated by Joule heating from the thin film heater for a heating time varying between several μs and several ms. If the heating time is longer than several ms, boiling may occur. The condition for the occurrence of boiling is determined not only by the heating time but also by the magnitude of the heat flux. In general, a larger heat flux than $10\,\mathrm{W\,cm}^{-2}$ with a heating time over $10\,\mu\mathrm{s}$ will cause boiling in He II. The strength of a thermal shock wave, which is the temperature amplitude, is given in terms of the heat flux q_p in cases without boiling as,

$$\Delta T = \frac{q_p}{\rho u_{20} c_p}. \tag{4.22}$$

In the cases where the temperature amplitude is higher than $20\,\mathrm{mK}$, the possibility that boiling takes place is high. If a very strong thermal shock is required, a shock focusing technique can be used to avoid boiling. The essential feature of a thermal shock is made clear through the temperature measurement and thus a superconductive temperature sensor with a high resolution and a quick time response [21–23] and some visualization technique [20, 24, 25] may be an adequate choice.

Waveforms of frontal thermal shock waves in He II at $T_b = 1.70\,\mathrm{K}$ measured at a location from the heater, $z = 10\,\mathrm{mm}$ by a superconductive temperature sensor are shown in Fig. 4.13 [23, 26]. This figure shows the variation of a waveform with the peak heat flux q_p and the heating time t_H for an initial heat pulse. For a small heat flux q_p the wave height of a thermal pulse is proportional to q_p and the pulse duration is equal to t_H. It is, however, evident that in the cases of large q_p and t_H the waveforms are highly deformed due to the effect of quantized vortices as described in Sect. Superfluid Breakdown and Quantized Vortices. In this case the wave height diminishes, furthermore, the rear side falls slowly forming a long tail and therefore the length of the thermal pulse becomes longer. It seems that the waveform deformation increases with q_p and t_H for large values of q_p and t_H, and finally the waveform turns into

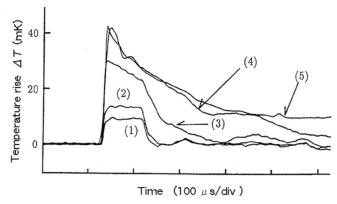

Fig. 4.13. Waveform variation of a frontal thermal shock wave in He II measured at $z = 10\,\mathrm{mm}$ with the heat flux q_p and the heating time t_H for an initial heat pulse. $T_b = 1.70\,\mathrm{K}$. (1); $q_p = 7.1\,\mathrm{W\,cm}^{-2}$, $t_H = 100\,\mu\mathrm{s}$, (2); $q_p = 9.0\,\mathrm{W\,cm}^{-2}$, $t_H = 100\,\mu\mathrm{s}$, (3); $q_p = 20.0\,\mathrm{W\,cm}^{-2}$, $t_H = 100\,\mu\mathrm{s}$, (4); $q_p = 40.0\,\mathrm{W\,cm}^{-2}$, $t_H = 100\,\mu\mathrm{s}$, (5); $q_p = 40.0\,\mathrm{W\,cm}^{-2}$, $t_H = 500\,\mu\mathrm{s}$

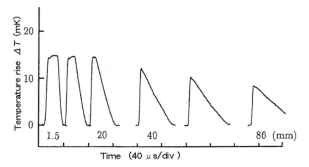

Fig. 4.14. Waveform deformation of a frontal thermal shock during propagation measured by a superconductive temperature sensor. $T_b = 1.69\,\mathrm{K}$, $q_p = 9.0\,\mathrm{W\,cm^{-2}}$, $t_\mathrm{H} = 30\,\mu\mathrm{s}$

a universal form almost independently of q_p and t_H as can be seen from the waveforms (4.4) and (4.5) in Fig. 4.13. Shown in Fig. 4.14 [23] is the waveform deformation of a frontal thermal shock during its propagation measured at a number of locations in the case of $T_b = 1.69\,\mathrm{K}$, $q_p = 9.0\,\mathrm{W\,cm^{-2}}$, $t_\mathrm{H} = 30\,\mu\mathrm{s}$. In this case, q_p is small and the effect of vortices is not clear yet. In the very initial phase immediately after the pulse emission, the wave maintains an initial trapezoidal form just similar to the time variation of the heat input. Then the front turns to a shock and the backside to an expansion waveform. The length of the wave peak becomes short with the propagation and then the waveform results in a triangle. Then, the peak of the triangular waveform decreases with the propagation though the area of the triangular waveform is kept nearly constant because of energy conservation in the nearly ideal superfluid state. All these features result from the nonlinear wave behavior of a second sound thermal pulse. In most cooling applications using He II, the thermal situation is rather in the breakdown state than in an ideal superfluid state because of the large heat flux as discussed in Sect. Superfluid Breakdown and Quantized Vortices.

Effect of High Density Quantized Vortices on Thermal Shock Wave

The thermofluid dynamic state of He II, where the line density of quantized vortices is very high, as high as $10^{10}\,\mathrm{mm^{-3}}$, is called the superfluid turbulent state. In this state the quantized vortices strongly affect the second sound thermal shock wave, but only weakly affect the compression shock wave in the form of weak wave decay. The knowledge acquired from experiments on the effects of high-density quantized vortices to thermal shocks is summarized as follows. The major effects of quantized vortices are waveform deformation and attenuation. They appear more remarkably in the rear portion within a thermal pulse, because vortex development starts at the arrival of the wave front as a triggering action and takes a time, say of the order of one ms. A long tail with a rather high temperature is formed in the rear part of a wave due to the slow decay of vortices. On the other hand, the effect also strongly appears in the vicinity of a heater. It is because the superfluid component that flows towards the heater interferes with the diffusive spread of vortices that have a tendency to move with the superfluid component. Thus the density becomes very high around the heater surface area and the heat flows only diffusively through the layer of high vortex density. This causes local heat accumulation there, which leads to a temperature rise and possibly to boiling. Under this condition, the fraction of heat transported by the second sound wave mode becomes small and the major part of the heat is consumed in the boiling process. Therefore, the wave formed by the small amount of heat approaches to a limiting waveform that is almost independent of the heat flux and the heating time.

4.2.3 Superfluid Shock Tube Facility

In this section, the Superfluid Shock Tube Facility constructed at theUniversity of Tsukuba is described [27,28]. This facility was designed as a versatile experimental facility for He II shock wave phenomena, and is still underdevelopment. The details of the hardware are introduced, and some of the important experimental results are presented. Some key experimental techniques used in these experiments and also some general technical information necessary in cryogenic thermofluid dynamic experiments is added in this section.

Description of the Superfluid Shock Tube Facility

The schematic drawing and the overview photograph of the Superfluid Shock Tube Facility are shown in Fig. 4.15. It consists of the two major parts, a gasdynamic shock tube with a total length of about 2.9 m and a He II cryostat with optical windows for visual observation. A special feature of this facility is the diaphragm-free design with an M–O valve for the high-pressure chamber section of the gasdynamic shock tube. The M–O valve, invented by Professors. Maeno and Oguchi [6], is a quick opening valve replacing a diaphragm to separate the high pressure gas from the low pressure gas adopted in most shock tubes. The diaphragm-free design allows us easy repetition of experiments without disturbances by frost contamination inside the tube in the cryogenic section because the shock tube is completely airtight without diaphragm replacement. The vertical part of the low-pressure tube with a length of 1.6 m is inserted into the cryostat and is maintained at cryogenic temperature of about 2 K in the bottom part immersed in He II in the He II vessel, and at about 80 K around the top part. The test section of the low-pressure tube is filled with He II and the rest of the space in the low-pressure tube is filled with saturated helium vapor in equilibrium with He II. The cryogenic part of the shock tube has a rectangular cross section ($34\,\text{mm} \times 19\,\text{mm}$). The temperature of He II in the test section is controlled through the vapor pressure control

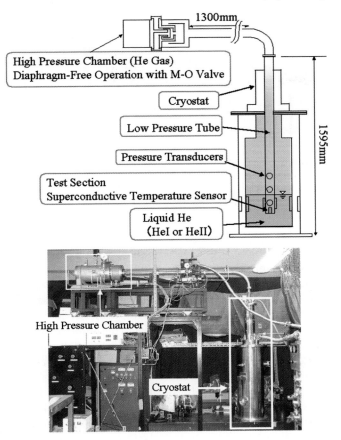

Fig. 4.15. Overview of the Superfluid Shock Tube Facility. (**a**) Schematic drawing. (**b**) Photograph of the overall view of the Facility

with an automatic control valve. Thus the pressure of the test fluid, He II, is equal to the saturated vapor pressure corresponding to the temperature. The He II vessel surrounding the test section works as a guard tank to maintain a cryogenic state in the test section.

In this experimental facility, a gas-dynamic shock wave is generated in helium vapor by using a conventional shock tube technique. The shock impinges on the free surface of He II in the same manner as in the cryogenic shock tube at Caltech. Some part of an impinging shock wave penetrates into the He II and the remainder is reflected back to the vapor as indicated in Fig. 4.9. Upon the impingement, a thermal shock wave is also generated and propagates through He II at lower velocity than the compression shock wave.

The measurements of the temperature and the pressure and optical visualization are adopted for the experiments. In addition, the temperature and the pressure of the quiescent He II are monitored for the purposes of house

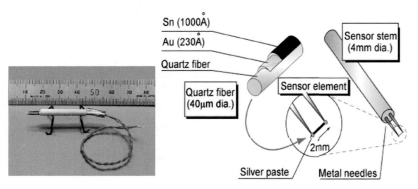

Fig. 4.16. Superconductive thin film temperature sensor

keeping and control of the experimental condition at cryogenic temperature. The pressure is measured by commercially available piezo-type pressure transducers (PCB Piezotronics, Model-PCB 102A) mounted on the shock tube wall at three locations including one or two in He II as illustrated in Fig. 4.15a. This pressure transducer can be somehow used at cryogenic temperatures but is not available in the calibrated model at the liquid helium temperature. It was, however, found that at the liquid helium temperature the deviation from the calibrated value at 80 K was relatively small.

Some advanced technique is needed for the detection of such very fast temperature variation as caused by shock waves in He II. The superconductive temperature sensor is one of the solutions as depicted in Fig. 4.16. It resembles a conventional hot wire anemometer in shape. The sensing element is made of gold–tin thin film vacuum-deposited on the surface of a fine quartz filament with a diameter of 40 μm. This type of sensor was originally developed by Laguna [29] and Cummings et al. [30]. The superconductive temperature sensor has a high temperature resolution as small as 1 mK and a short response time, about 1 μs, sufficient for the measurement of shock waves in He II. The working principle of the sensor is the abrupt change in the electrical resistance of the thin superconductive metal film due to the superconductive-normal transition. On the other hand, for monitoring the thermophysical state of the test fluid, He II, the vapor pressure that is also the pressure of the test fluid and the temperature are measured. The measurement of the vapor pressure of He II can be performed without any difficulty because a pressure transducer located outside of a cryostat at room temperature can be used. For monitoring the He II temperature, a number of commercially available thermometers, such as a germanium, composite carbon resistor and ruthenium oxide resistance thermometers, can be used in both calibrated and uncalibrated models.

A number of optical visualization methods can be utilized for the visualization study of He II shock wave phenomena. However, some special cares are required for the successful application of the methods. First of all, superleak-tight windows must be attached to He II vessels. He II extremely

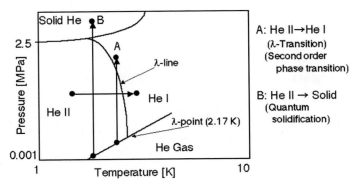

Fig. 4.17. A number of interesting shock compression processes in He II

easily leaks through the seal (usually indium metal seal) between the window glass plate and the metal window frame of a cryostat due to the superfluid superleak frictionless flow. Application of optical fibers would be an alternative to superleak-tight optical windows. The second difficulty results from the use of multiwindow glass plates, that is two outside of the vacuum shell, two at the intermediate temperature stage usually at liquid nitrogen temperature to absorb thermal radiation from the room temperature environment, and two superleak-tight windows. Totally six glass plates are used, though three may be a possible design if a mirror or a prism is located inside the cryostat. The third difficulty is that the variation in the refractive index induced by the density variation of He II is very small. Therefore, at most one interferogram fringe appears in the infinite fringe mode.

There are a number of interesting shock compression processes in He II as schematically shown in Fig. 4.17. In most cases where the magnitude of shock compression is small and the temperature is not so close to the lambda line, the final state after shock compression still remains He II. For moderately strong shock compressions so as to cross the λ-line, the final state turns to He I. This process is indicated by process A in Fig. 4.17, which involves a shock induced λ-phase transition. The λ-phase transition is usually induced by a temperature rise or drop as indicated by a horizontal line connecting He II and He I in Fig. 4.17. If the pressure after the shock compression exceeds the solidification pressure, about 25 bar, solid helium may be produced in quite a nonequilibrium manner.

A typical example of some time records of a shock wave experiment are shown in Fig. 4.18. The initial temperature of the test gas in a quiescent state in the low-pressure section varies over a wide range from 80 K at the top down to about 2 K at the free surface. The shock wave signal is first detected by the pressure transducer B in the vapor as indicated by the trace (A) in this diagram. A pressure jump at the wave front of the incident gasdynamic shock is clearly recorded. The trace (B) is the transmitted compression wave propagating through He II recorded by the pressure transducer C located in He II.

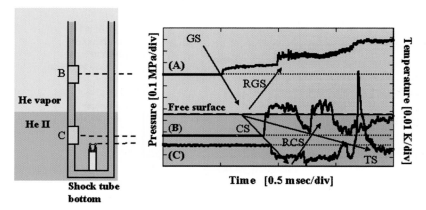

Fig. 4.18. Typical example of time records of detected signals. Driver gas $p_4 =$ 704 kPa (300 K); test gas: saturated He II vapor $p_1 = 2.20$ kPa ($T_1 = 2.00$ K); p_{41} (= p_4/p_1) = 320. Incident shock speed = 307 m s^{-1}; shock Mach number $M_{SV} = 4.5$. Transmitted compression shock speed in He II = 252 m s^{-1}; transmitted shock Mach number in He II $M_{SL} = 1.15$. Detected signals are: (**A**) measured by pressure transducer B; (**B**) measured by pressure transducer C; (**C**) measured by superconductive temperature sensor. GS: gasdynamic shock wave, RGS: reflected gasdynamic shock wave; CS: transmitted compression shock wave; RCS: reflected compression shock wave; TS: thermal shock wave

This pressure rise is very sharp unlike compression waves in ordinary liquids. The reason for this is that He II has a considerably higher compressibility than any other liquids resulting in a better shock impedance matching for a pressure wave transmission through a gas–liquid interface. This means that a gasdynamic shock wave is transmitted more efficiently into He II than into any other liquids. This result indicates that the present method of impinging a gasdynamic shock wave onto a He II free surface is an effective means to generate a strong compression shock wave in He II. The arrival of the transmitted compression shock wave is registered even by the superconductive temperature sensor, as a negative temperature jump as small as about 20 mK, as seen in trace (C). The second temperature drop in trace (C) results from the compression shock wave reflected from the bottom of the shock tube. These temperature drops arising in shock compression process can be quite naturally understood if we know the fact that the thermal expansion coefficient of He II is negative in the present temperature range, which means the temperature of He II decreases by compression. This is a characteristic feature of compression shock waves in He II, and different from thermal shock waves originating from the second sound. It is, however, known that the magnitude of the temperature jump induced by a compression wave process is of higher order smallness. Some other wave signals are also registered on the traces of Fig. 4.18. The second positive pressure jump of trace (A) results from the reflected gasdynamic shock wave from the He II free surface. The second pressure jump of the trace (B) is a record of the arrival of the reflected compression shock wave from the shock tube bottom. It is interesting to note that the third pressure rise in trace (A) is caused by the penetration of a He II compression shock wave that is reflected from the shock tube bottom through the He II–vapor interface into the vapor phase. The large temperature spike, as high as 100 mK, seen on the latter half of trace (C) is a thermal shock wave that propagates much more slowly ($\sim$20 m s^{-1}) than a He II compression shock wave ($\sim$250 m s^{-1}).

Some Experimental Results

First, the transmitted compression shock wave is investigated. Shown in Fig. 4.19 is the plot of the pressure and the temperature values at the front of transmitted shock front plotted on the p–T diagram, together with the saturated vapor pressure line, the λ-line and the isentropes of liquid helium starting from each initial temperature. It is known that the isentrope almost coincides with the shock adiabatic curve in weak shock compression process as in our experiment. It is seen from this result that the experimental data well agree with the isentropes for both He II and He I. It is also clearly seen from the figure that the temperature drops for He II and rises for He I as a result of shock compression.

In the experiments, compression shocks in He II induce some degree of temperature drops as seen from Fig. 4.19, though the Khalatnikov approximate theory [14] predicts that the compression shock process in He II is regarded as an isothermal one. And of course, in He I the temperature rises as a result of shock compression as is evident in Fig. 4.19. This means that effects of higher order than described by the first order Khalatnikov approximation, could be observed in the experiments. The transient temperature variations

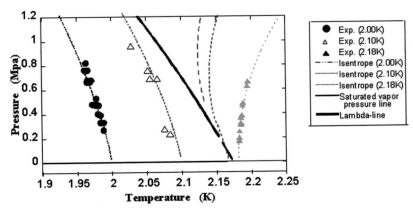

Fig. 4.19. Pressure and temperature values at the front of transmitted shock wave plotted on the p-T diagram, together with the saturated vapor pressure line, the λ-line and the isentropes of liquid helium starting from each initial temperature

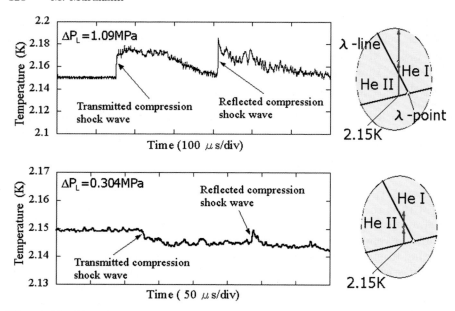

Fig. 4.20. Transient temperature variations under the condition of the shock induced λ-phase transition. The initial state of He II; the temperature at 2.15 K, the pressure differences Δp_L $(= p_4 - p_1)$ of 1.09 Mpa (upper chart) and 0.304 Mpa, (lower chart), respectively. Each shock compression process is also schematically shown

under the condition that a shock induced λ-phase transition takes place are shown in Fig. 4.20. In the upper chart two successive positive temperature jumps are recorded. This indicates that the initial liquid state, He II, changes to He I after the first compression by the transmitted compression shock and the state remains He I after the further compression by the reflected compression shock from the shock tube bottom. In the lower transient record, we recognize one negative and subsequent one positive temperature jumps. This result indicates that the liquid state still remains He II after the first compression by the transmitted compression shock and then the state changes to He I after the second compression by the reflected compression shock.

The impingement of a gasdynamic shock wave on the He II free surface induces a thermal shock wave as well as a transmitted compression shock wave. The thermal shock wave profiles measured by a superconductive temperature sensor are shown in Fig. 4.21. A thermal pulse is generated primarily by impulsive heating of the He II free surface by the impinging gasdynamic shock wave, though the compression of He II by a gasdynamic shock has a little negative effect on thermal pulse generation. It should be noted from these temperature profiles that the peak amplitudes are much larger than those generated by pulsed heating from a heater, and the effect of high-density quantized vortices is so large as to lead to a change of the profiles to triangular waveforms. It is interesting to note in the profiles (a) and (b) that the waveform varied

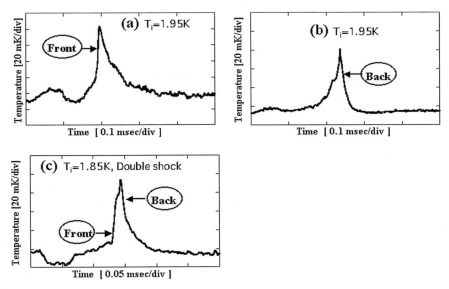

Fig. 4.21. Thermal shock wave profiles measured by a superconductive temperature sensor at two initial temperatures, 1.95 and 1.85 K. (**a**) and (**b**) are the profiles of a thermal shock wave measured at two locations; (**a**) for immediately below the free surface and (**b**) for 15 mm below the free surface under the condition of the initial temperature of 1.95 K. (**c**) is an example of a double shock wave, the temperature in the initial state of 1.85 K

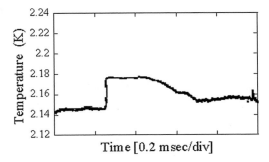

Fig. 4.22. Thermal shock wave profile measured by a superconductive temperature sensor for the case of the initial temperature very close to T_λ

during its propagation. The profile (a) measured at immediately below the free surface still maintains a steep front due to initial shock heating, but in profile (b) the feature of a back shock with steepened backside and a front that turns into gently sloping becomes pronounced while propagating, as the initial temperature is in the range of a negative steepening coefficient. The profile (c) is an example of a double shock that has discontinuities in temperature at both the wave front and the back. The profile shown in Fig. 4.22 is generated at the temperature very close to T_λ. It seems that the peak is truncated above T_λ and the pulse length is elongated. The thermal shock does not appear in He I above T_λ.

References

1. Liepmann, H.W., Cummings, J.C., Rupert, V.C.: Cryogenic shock tube. Phys. Fluids. **16**(2), 332–333 (1973)
2. Cummings, J.C.: Development of a high-performance cryogenic shock tube. J. Fluid Mech. **66**(1), 177–187 (1974)
3. Cummings, J.C.: Experimental investigation of shock waves in liquid helium I and II. J., Fluid, Mech. **75**(2), 373–383 (1976)
4. Liepmann, H.W., Laguna, G.A.: Nonlinear interactions in the fluid mechanics of helium II. Ann. Rev. Fluid Mech., **16**, 139–177 (1984)
5. Liepmann, H.W., Toczynski, J.R.: Shock waves in helium at low temperatures. In Proceedings of the 15th International Symposium on Shock Waves and Shock Tubes, 87–96 (1986)
6. Maeno, K., Orikasa, S.: Study on shock waves in low temperature gas by means of a non-diaphragm shock tube. In Proceedings of 15th International Symposium on Shock Waves and Shock Tubes, 69–78 (1986)
7. Reppy, J.D., Depatie, P.A.: Persistent currents in superfluid helium. Phys. Rev. Lett. **12**(8), 187–189 (1964)
8. Reppy, J.D., Depatie, P.A.: Phys. Rev. Lett. **14**, 733 (1965)
9. Daunt, J.G., Mendelssohn, K.: Proc. Roy. Soc. (London) A170 423–439 (1939)
10. Kapitza, P.: The study of heat transfer in helium II. J. Phys. USSR **5**, 59–88, (1941), or Phys. Rev. **60**, 354 (1941)
11. Tisza, L.: Transport phenomena in helium II. Nature **141**, 913 (1938)
12. Landau, L.D.: Zh. Eksp. Teor. Fiz., 11 592 (1941), J. Phys, USSR, **5**, 71 (1941)
13. Landau, L.D., Lifshiz, E.M.: *Fluid Mech.* Pergamon, London Chap. 16 (1987)
14. Khalatnikov, I.M.: Zh. Ekspp. Teor. Fiz., **23**, 253 (1952), Introduction to the Theory of Superfluidity Benjamin, New York, Chap. 13 (1965)
15. Gorter, C.J., Mellink, J.H.: On the irreversible process in liquid helium II. Physica **15**, 285 (1949)
16. Vinen, W.F.: Mutual friction in a heat current in liquid helium II. I. Experiments on steady heat currents. II. Experiments I n transient effects. Proc. R. Soc. London Ser. A **240**, 114–127, 128–143 (1957)
17. Vinen, W.F.: Mutual friction in a heat current in liquid helium II. III. Theory of mutual friction. Proc. R. Soc. London Ser. A **242**, 493–515 (1957)
18. Osborne, D.V.: Second sound in liquid helium II. Proc. Phys. Soc. London Ser. A **64**, 114–123 (1951)
19. Dessler, A.J., Fairbank, W.M.: Amplitude dependence of velocity of second sound. Phys. Rev. 104 (195) 6–10
20. Iida, T., Shimazaki, T., Murakami, M.: Laser holographic interferometer visualization of a thermal shock wave in he II. Adv. Cryo. Eng., 39B 1851–1857 (1993)
21. Borner, H., Schmeling, T., Schmidt, D.W.: Experimental investigations of fast gold-tin metal film second –sound detectors and their applications. J. Low Temp. Phys. **50**, 405–435 (1983)
22. Turner, T.N.: Using second-sound shock waves to probe the intrinsic critical velocity of liquid helium II. Phys. Fluids. **26**(11), 3227–3241 (1983)

23. Shimazaki, T., Murakami, M., Iida, T.: Second sound wave heat transfer, thermal boundary layer formation and boiling: highly transient heat transport phenomena in He II. Cryogenics, **35**, 645–651 (1995)
24. Gulyaev, A.I.: Schlieren photography of thermal pulses in liquid He 4. Sov. Phys., JETP **30**, 34–43 (1960)
25. Torczynski, J.R., Gerthsen, D., Roesgen, T.: Schlieren photography of second sound shock waves in superfluid helium. Phys. Fluids **27**(10), 2418–2423 (1984)
26. Shimazaki, T., Murakami, M., Kanari, T.: Measurement of characteristic time for quantized vortex tangle development in He II. Cryogenics, **38**, 601–606 (1998)
27. Nagai, H., Yang, H.S., Ueta, Y., Murakami, M., Takano, N., Yanaka, K.: Application of superfluid shock tube facility to experiments of highly transient low-temperature thermo-fluid dynamic phenomena. Cryogenics, **41**, 421–428 (2001)
28. Ueta, Y., Yanaka, K., Murakami, M., Nagai, H., Yang, H.S.: Experimental study of λ-phase transition induced by shock compression. Cryogenics, **42**, 645–651 (2002)
29. Laguna, G.A.: Photolithographic fabrication of high frequency second sound detectors. Cryogenics **16**, 241–243 (1976)
30. Cummings, J.C., Schmidt, D.W., Wagner, W.J.: Experiments on second-sound shock waves in superfluid helium. Phys. Fluids **21**(5), 713–717 (1978)

Part II

Shock Waves and Phase Transition

Shock Waves in Fluids with Interphase Transport of Mass, Momentum and Energy (Vapour–Droplet Mixtures and Solid-Particle-Laden Gases)

Abhijit Guha

5.1 Introduction

In this part of the book we discuss the structure of shock waves in vapour–droplet mixtures and solid-particle-laden gases. Both media exhibit relaxation effects. Hence, we first discuss the effect of one general internal variable in a gas, whose departure from equilibrium results in a typical relaxation phenomenon, and explain how this relaxation process affects the speed of sound in the gas and the qualitative structure of shock waves in it. We then show how a complex set of relaxation phenomena develops in a vapour–droplet mixture as a result of interphase transport of mass, momentum and energy. We first tackle a pure vapour–droplet mixture in which both the vapour and liquid are of the same chemical species. Later we generalize the situation with the introduction of an inert carrier gas. We analyze the fluid dynamics and shock structure in vapour–droplet flow both with differential equations, which determine the detailed structure within a shock wave, and the integral approach, which typically relates the end states of a shock wave. We use both analytical results and numerical computations (Computational Fluid Dynamics, CFD).

Having understood the structure of shock waves in vapour–droplet mixtures, we then turn our attention to a solid-particle-laden gas. In the absence of any mass transfer between the solid particles and the gas, the analysis is simpler and fewer relaxation times are necessary to describe a solid-particle-laden gas than it is necessary for a vapour–droplet mixture. The integral relations or jump conditions across a shock wave in a solid-particle-laden gas are also correspondingly simpler. One particular extra complexity has been highlighted for the case of vapour–droplet mixtures in that complete evaporation of the liquid phase may take place across a shock wave: in this case the fluid may be a two-phase mixture upstream of the shock wave but is a single-phase fluid downstream of the shock wave. Careful use of appropriate equations is therefore necessary.

Along with the analysis of the structure of shock waves and of speeds of sound, we have also presented some related aspects, e.g. the issue of thermal choking and the interpretation of total pressure and temperature in multiphase mixtures. The decelerating field in front of a stagnation point gives rise to a relaxing flow, sometimes with an in-built frozen aerodynamic shock wave. Although we have provided the following exposure in relation to normal shock waves, similar considerations are valid for oblique shock waves also.

Any abrupt and appreciable change in flow-variables such that the wave-front remains perpendicular to the flow direction is termed as a normal shock. The analysis of such stationary shock waves in a perfect gas is standard and may be found in any textbook on gas dynamics. It consists of writing the one-dimensional "inviscid" equations for conservation of mass, momentum and energy together with an equation of state and arriving at the jump-conditions, collectively known as the Rankine-Hugoniot relations. If the upstream flow variables are specified, the downstream values (w.r.t. the shock) can be determined completely (including entropy) and the apparently "inviscid" equations of motion predict a fixed entropy generation across the shock wave.

It is clear that there must be some real physical mechanism in the flow to produce the dissipation which the equations predict and that no real discontinuities can exist. In fact across such high gradients of velocity and temperature, viscous effects and heat conduction, however small they might be, come into play and give rise to a transition-layer of finite thickness over which the changes in flow properties take place. The Rankine-Hugoniot relations apply quite correctly to the flow ahead of and behind this transition region. It is instructive to point out clearly that although viscosity and thermal conductivity constitute the sources of dissipation, they do not determine the total magnitude of dissipation. Rather, together they determine the distance over which the total dissipation, fixed by Rankine-Hugoniot relations, is distributed.

If one includes the axial viscous forces in the momentum equation, in addition to the inertia and pressure forces, and also the terms for work done by axial viscous stresses and for axial heat conduction in the energy equation (collectively known as the Navier-Stokes equations), it is possible to analyze the flow structure within the narrow transition layer mentioned earlier, at least for sufficiently weak shocks. Although different flow-variables approach their far upstream and downstream values asymptotically, one can define a finite thickness of the shock wave, over which most of the changes in flow properties occur. Such calculations show that the thickness of the shock wave is of the order of a few mean free paths of the gas molecules. Objections have therefore been raised against treating a fluid as a continuum there, and suggestions have been made rather to apply directly the Boltzman Equation from kinetic theory. For our purpose it is sufficient to note that a normal shock wave in a perfect gas is confined to a very narrow zone and for many practical purposes it may be treated as a mathematical discontinuity. That is, however, not the case in a dispersive, relaxing medium such as a vapour–droplet mixture or a solid-particle-laden gas.

References [1] and [2] provide an excellent introduction to general relaxation effects, Relevant material on relaxation effects and shock waves in vapour–droplet flow may be found in [3–20]. Solid-particle-laden gases are discussed in [21–27]. Effects due to non-equilibrium condensation may be found in [3–8, 14–17, 28]. CFD techniques are discussed in [9, 14, 15, 29, 30].

5.2 Relaxation Gas Dynamics for Pure Vapour–Droplet Mixtures and Detailed Structure of Shock Waves

5.2.1 Relaxation Phenomena

Introduction

If a property of a medium is perturbed from its equilibrium state and the restoration to equilibrium occurs at a *finite rate*, the medium is called a relaxing medium and the process of restoration is termed relaxation. In certain occasions relaxation effects play an important role in gas dynamics. For example, consider a body moving through a gas at high speed so that the gas near the body is heated considerably. At sufficiently high temperature, the gas may dissociate, chemical reactions may take place or the internal degrees of freedom (e.g. the vibrational degree of freedom in the case of a diatomic or polyatomic gas) of the molecules may become excited. Since all these processes, e.g. molecular vibration, dissociation, and chemical reaction, occur at a finite rate, the gas is not able to respond instantaneously to a very rapid change in its external conditions. For example, if the temperature and pressure of a mixture of chemically reacting species are changed very rapidly, the equilibrium concentrations of different species, as prescribed by the law of mass action, corresponding to the new pressure and temperature are not attained instantaneously. The medium needs some finite time to attain the new thermodynamic equilibrium by completing its internal rate processes: in other words, it needs some time to "relax" towards its new equilibrium state.

Simple relaxing media follow the archetypal rule of the restoration process:

$$\frac{\mathrm{d}(\Delta q)}{\mathrm{d}t} = -\frac{\Delta q}{\tau},\tag{5.1}$$

where, q is an internal state variable, Δq is the departure from equilibrium and τ is the relaxation time. The relaxation time τ is a measure of the order of the time required for the system to relax to its new equilibrium state, if the external conditions are kept fixed at their new changed values during the process of relaxation. Equation 5.1 is the archetypal relaxation equation.

In a real flow process the external conditions will, however, change continuously, for instance, as the gas flows past an aerofoil, or passes through a nozzle or blade passage. Obviously a competition takes place between these

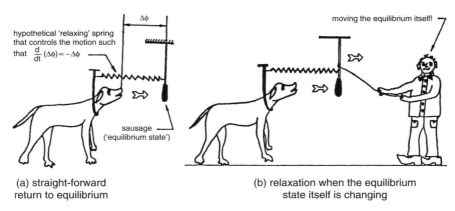

(a) straight-forward return to equilibrium

(b) relaxation when the equilibrium state itself is changing

Fig. 5.1. Illustration of a general relaxation phenomenon

two opposite phenomena: the rapid change of external conditions that tend to create further departure from equilibrium and the relaxation process that tries to catch up with the current equilibrium state. [Figure 5.1 gives the author's artistic expression of this "competition". The author initially termed it "The Dog and the Jogger Model": the dog trying to reach its equilibrium position, the sausage, while the jogger pulling the sausage away from the dog thereby creating further non-equilibrium. In a graduate environment, the illustration, however, became very popular as "The Professor and the PhD Model"!] Which one of these two effects dominates, depends on the relative magnitude of the time scales associated with them. For a body with characteristic linear dimension l moving with a velocity V through the gas, the time scale of the change of external conditions is of the order of $\tau_{ext} = l/V$. The ratio of the two time scales $D = \tau/\tau_{ext}$ is a measure of the relative importance of the internal rate process and is usually called the "Damköhler's parameter".

If $D \to 0$, the internal rate processes are very fast and their lag behind the changing external conditions can be neglected. The gas is in thermodynamic equilibrium everywhere in the flow field and is governed by the well-established rules of classical gas dynamics. In the other limiting case as $D \to \infty$, the external conditions change so rapidly in terms of the internal time scale τ that the internal rate processes are essentially "frozen", i.e. they do not take place at all. Although the gas is not under thermodynamic equilibrium under these circumstances, classical gas dynamics is again valid! (with different equation of state and isentropic exponent). It is when $\tau \sim \tau_{ext}$ (i.e. $D \sim 1$) that pronounced relaxation effects take place and relaxation gas dynamics is needed.

Relaxation Processes in a Pure Vapour–Droplet Medium

A vapour–droplet medium is assumed to be a homogeneous two-phase mixture of a large number of fine droplets (denoted here by subscript l) dispersed in the continuous vapour phase (subscript g). The droplet cloud may exhibit an arbitrary level of polydispersity with a spectrum of different sizes of droplets. Numerical illustrations in the present analysis are presented for a mono-dispersed, spherical droplet population for the simplicity of description. For a polydispersed droplet population, one has to discretize into a suitable number of size groups and then apply the droplet equations for each such group. Pure substances imply that the phase change is heat transfer rather than diffusion controlled.

In order to identify the various relaxation processes that may occur in a vapour–droplet mixture, we examine what determines the state of equilibrium between the two phases. At equilibrium, the velocity of the vapour phase, V_g, is equal to that of the liquid phase, V_l, (or the difference between them remains constant) and also the temperatures of both the vapour phase, T_g, and the droplets, T_l, are equal to the saturation temperature, T_s, corresponding to the prevailing pressure (neglecting the capillary supercooling for established droplet sizes). Thus obviously the system may go out of equilibrium in three different ways:

(i) The droplet temperature is not equal to the saturation temperature ($T_l \neq T_s$). $\Delta T_l = T_s - T_l$ is the relevant non-equilibrium variable and the corresponding relaxation time is τ_D.

(ii) The velocity of the two phases may not be equal; in other words there may be slip between the two phases ($V_g \neq V_l$). $\Delta V = V_g - V_l$ is the relevant non-equilibrium variable and the corresponding inertial relaxation time is τ_I.

(iii) The temperature of the vapour phase may be different from the saturation temperature ($T_g \neq T_s$). $\Delta T = T_s - T_g$ is the relevant non-equilibrium variable (called vapour subcooling or supercooling) and the corresponding relaxation time is τ_T. [Note that negative ΔT means that the vapour is superheated.]

It is clear that under non-equilibrium conditions mass, momentum and energy transfers would take place between the two phases. It is through these interphase transfer processes that the system would return to equilibrium following a perturbation. The equations representing the different interphase transfer processes have been derived in [10]. Here we mention the salient points only. If the velocity slip between droplets and vapour is small, the mass and energy transfer equations can be decoupled from the momentum transfer equation. Relaxation associated with the velocity slip then becomes the most straight forward and is given by,

$$\frac{dV_l}{dt_l} = \frac{\Delta V}{\tau_I}, \tag{5.2}$$

where $d/dt_l = \partial/\partial t + V_l \partial/\partial x$ is the substantive derivative following the droplets. In steady flow, (5.2) can be equivalently expressed as

$$\frac{d(\Delta V)}{dx} + \frac{\Delta V}{V_l \tau_I} = \frac{dV_g}{dx}. \tag{5.2a}$$

The inertial relaxation time τ_I is given by

$$\tau_I = \frac{2r^2\rho_l[\phi(Re) + 4.5Kn]}{9\mu_g},$$ (5.3)

where r is the droplet radius, ρ_l is the density of the liquid phase and μ_g is the dynamic viscosity of the vapour. $\phi(Re)$ is an empirical correction factor based on the Reynolds number Re corresponding to the slip velocity ($Re = 2\rho_g r|\Delta V|/\mu_g$) and is given by $\phi(Re) = [1 + 0.15\,Re^{0.687}]^{-1}$. For small slip Reynolds number and continuum flow ($Re \ll 1, Kn \ll 1$) the quantity within the brackets in (5.3) becomes unity and (5.2) reduces to the Stokes drag formula for a sphere. For free molecule flow ($Kn \gg 1$) an expression derivable from kinetic theory is obtained. The expression within the brackets in (5.3) provides a simple interpolation formula for intermediate Knudsen numbers. The inclusion of the correction term based on the slip Reynolds number in the expression for the inertial relaxation time means that (5.2) remains no longer restricted just to linear relaxation processes but can also be applied to practical non-linear processes; such non-linearity can easily be accommodated in a numerical calculation scheme.

The mass and energy transfer equations are, however, strongly coupled. They can be written in the abbreviated form as,

$$\frac{dm}{dt_l} = -\dot{M}$$ (5.4)

and,

$$m\frac{dh_l}{dt_l} = -(h_g - h_l)\dot{M} - \dot{Q},$$ (5.5)

where $\dot{M}$ is the mass transfer rate and $\dot{Q}$ is the heat transfer rate from a droplet. m is the mass of a single droplet, h_g and h_l are specific enthalpies of the vapour and liquid phase, respectively. The reversion to thermal equilibrium following a disturbance usually takes place in two, almost independent stages. (Mass transfer takes place continuously during both stages.) In the first (normally extremely fast) stage of relaxation the droplet temperature approaches exponentially to the saturation temperature. Negligible heat is transferred between the phases and the droplet radius changes only slightly (as the latent heat is very high compared to the thermal capacity of the droplets). In the second (much longer) stage of relaxation, the vapour temperature rises (or falls) to the saturation value due to the interphase heat transfer as the droplets grow (or evaporate). Thus in the first stage $\dot{Q}$ is negligible in (5.5) and the latent heat released as a result of phase-change alters the droplet temperature. On the contrary, in the second stage the L.H.S. of (5.5) is negligible and the latent heat liberated during phase change is conducted back to the vapour and changes the vapour phase temperature. The droplet temperature relaxation is given by [obtained after linearizing the mass transfer equation (5.4) and combining with the energy transfer equation (5.5), [10]

$$\frac{dT_l}{dt_l} = \frac{\Delta T_l}{\tau_D}.$$ (5.6)

In steady flow, (5.6) can be equivalently expressed as,

$$\frac{d(\Delta T_l)}{dx} + \frac{\Delta T_l}{V_l \tau_D} = \frac{dT_s}{dx}.$$ (5.6a)

The droplet temperature relaxation time τ_D is given by [10]

$$\tau_D = \frac{2 - q_c}{2q_c} \left(\frac{RT_s}{h_{fg}} \right)^2 \left(\frac{r\rho_l c_l}{3R} \right) \frac{\sqrt{2\pi R T_s}}{p},$$ (5.7)

where c_l is the specific heat capacity for the liquid phase, q_c is the condensation coefficient, R is the gas constant, h_{fg} is the specific enthalpy of evaporation and p is the pressure.

Substitution of an expression for the heat transfer rate $\dot{Q}$ and the droplet temperature relaxation (5.6) into the energy transfer equation (5.5) gives [10],

$$(h_g - h_l)n\frac{dm}{dt_l} = \frac{(1 - y)c_{pg}(T_l - T_g)}{\tau_T} + \frac{yc_l\Delta T_l}{\tau_D},$$ (5.8)

where n is the number of droplets per unit mass of the mixture and y is the wetness fraction and c_{pg} is the specific heat capacity at constant pressure for the vapour phase. The vapour thermal relaxation time τ_T is given by [10],

$$\tau_T = \frac{(1 - y)c_{pg} \, r^2\rho_l}{3\lambda_g y}(1 + 4.5Kn/Pr),$$ (5.9)

where λ_g is the thermal conductivity of the vapour. The first term in the R.H.S. of (5.8) corresponds to the heat transfer rate $\dot{Q}$. For small Kn, this reduces to the continuum expression for steady-state heat transfer from a sphere. For large Kn the kinetic theory (free molecule) result is regained.

Both velocities slip relaxation, (5.2), and droplet temperature relaxation, (5.6), follow the archetypal form, (5.1). Assuming $\tau_D \ll \tau_T$ (which will be shown to be true in the next section) in (5.8) and making use of the different conservation equations for the vapour–droplet mixtures (Sect. 5.2.2) it can be shown that the variation of the vapour supercooling ΔT (with equilibrium droplet temperature and velocity slip) is given by

$$\frac{d(\Delta T)}{dx} + \frac{\Delta T}{V_g \tau_T} = \frac{RT_s}{h_{fg}} \left(1 - \frac{h_{fg}}{c_{pg}T_s} \right) \frac{T_s}{p} \frac{dp}{dx}.$$ (5.10)

Equation (5.10) shows clearly that ΔT is the non-equilibrium variable for this relaxation process and τ_T is the conjugate relaxation time.

Comparative Magnitudes of Different Relaxation Times

One of the simplifications in the analysis and the understanding of the relaxation processes in pure vapour–droplet flows arises from the fact that different

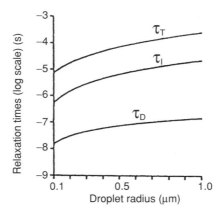

Fig. 5.2. Relaxation times for water droplets in pure steam; $p = 0.5$ bar, $y = 0.1$

relaxation times are of quite different magnitudes and hence each relaxation process may be studied almost independently of the others. Taking the ratio of (5.7) and (5.9), it can be shown that

$$\frac{\tau_D}{\tau_T} = \frac{2 - q_c}{2q_c} \frac{y}{1 - y} \frac{c_l}{R} \left(\frac{RT_s}{h_{fg}}\right)^2 \left(\frac{Kn/Pr}{1 + 4.5\,Kn/Pr}\right). \qquad (5.11)$$

Detailed calculations [13] with (5.11) shows that, in pure vapour–droplet flow, the droplet temperature relaxation time, τ_D, is, in general, several orders of magnitude lower than the vapour thermal relaxation time, τ_T, the exception to this general rule being in the unlikely event that $q_c \ll 1$.

Now consider the inertial relaxation time. Comparison of (5.3) and (5.9) shows that the ratio τ_I/τ_T is of the order of y. [For $Pr = 1$ and $Re \to 0$, $\tau_I/\tau_T = 2y/3(1 - y)$.] Since the analysis is anyway restricted to low wetness fractions ($y < 0.2$), the inertial relaxation time is at least one order of magnitude less than the vapour thermal relaxation time. Figure 5.2 shows the variation of the three relaxation times in pure steam as a function of droplet radius.

Thus, in general, $\tau_D \ll \tau_I \ll \tau_T$. Therefore following a disturbance, on a very short time scale the droplet temperature reaches equilibrium, then the velocity slip and finally the vapour temperature relaxes to the equilibrium value. *While considering any particular relaxation process, therefore, relaxation processes with smaller relaxation times may be assumed to have been equilibrated whereas relaxation processes with larger time scales may be assumed to be frozen.* Mathematically, for example, *equilibrium* droplet temperature means that ΔT_l and τ_D simultaneously tend to zero, the ratio $\Delta T_l/\tau_D$ remaining finite such that $\Delta T_l/\tau_D \to V_l(dT_s/dx)$, and, for instance, *frozen* heat transfer implies $\tau_T \to \infty$.

5.2.2 Gas Dynamics

Conservation Equations

The droplet cloud is assumed to be uniformly distributed throughout the vapour and of sufficient number density that the interaction with the vapour can be described by a continuous variation. We therefore adopt what is commonly called the "two-fluid" model and view the droplets as providing sources or sinks of mass, momentum and energy for the vapour.

The analysis is valid for low wetness fractions ($y < 0.2$). The volume occupied by the liquid phase and the partial pressure of the droplet cloud are negligibly small. Coagulation or fragmentation of the droplets are neglected. Thus each droplet is assumed to retain its individual identity and its radius changes solely by pure evaporation or condensation. All these assumptions apply to most wet vapour flows formed initially by homogeneous nucleation. For example, the radius of water droplets in wet steam, after the nucleation process is complete, is expected to lie in the range 0.05–$2.0\,\mu\text{m}$. For a wetness fraction of 0.2, this corresponds to a number concentration varying from $1.5 \times 10^{11}\,\text{cm}^{-3}$ to $2.4 \times 10^{6}\,\text{cm}^{-3}$. The average distance between droplets is about 23 droplet diameters and the volume fraction occupied by the liquid phase is only 8×10^{-5}. Even for an interphase slip velocity as high as $200\,\text{m\,s}^{-1}$, the droplets satisfy the Weber number criterion for stability against fragmentation. Under these circumstances, the conservation equations for quasi-one-dimensional unsteady flow in a duct of flow cross-sectional area A are [8]

Droplet number density
$$\frac{\partial}{\partial t}\left(\frac{An\rho_\text{g}}{1-y}\right) + \frac{\partial}{\partial x}\left(\frac{An\rho_\text{g}V_\text{l}}{1-y}\right) = 0. \qquad (5.12)$$

Mass
$$\frac{\partial}{\partial t}\left(A\rho_\text{g} + \frac{Amn\rho_\text{g}}{1-y}\right) + \frac{\partial}{\partial x}\left(A\rho_\text{g}V_\text{g} + \frac{Amn\rho_\text{g}V_\text{l}}{1-y}\right) = 0. \qquad (5.13)$$

Momentum
$$\frac{\partial}{\partial t}\left(A\rho_\text{g}V_\text{g} + \frac{Amn\rho_\text{g}V_\text{l}}{1-y}\right) + \frac{\partial}{\partial x}\left(A\rho_\text{g}V_\text{g}^2 + \frac{Amn\rho_\text{g}V_\text{l}^2}{1-y}\right)$$
$$+ A\frac{\partial p}{\partial x} = 0. \qquad (5.14)$$

Energy
$$\frac{\partial}{\partial t}\left[A\rho_\text{g}\left(h_\text{g} + \frac{V_\text{g}^2}{2}\right) + \frac{Amn\rho_\text{g}}{1-y}\left(h_\text{l} + \frac{V_\text{l}^2}{2}\right)\right] - A\frac{\partial p}{\partial t}$$
$$+ \frac{\partial}{\partial x}\left[A\rho_\text{g}V_\text{g}\left(h_\text{g} + \frac{V_\text{g}^2}{2}\right) + \frac{Amn\rho_\text{g}V_\text{l}}{1-y}\left(h_\text{l} + \frac{V_\text{l}^2}{2}\right)\right] = 0. \quad (5.15)$$

Equation of state $p = \rho_\text{g}RT_\text{g}.$ \qquad (5.16)

The equation set (5.12)–(5.16) is incomplete and must be supplemented by the three interphase transport laws, (5.2), (5.6) and (5.8).

Numerical Solution of the Equations

For the investigation of stationary shock waves in steady flow, all partial derivatives with respect to time are removed from (5.12)–(5.16) and (5.2)–(5.8), and, A is assumed constant. The continuity, momentum and energy equations (5.13)–(5.15) may then be recast as a set of three simultaneous equations for dV_g/dx, dT_g/dx and dp/dx. Equations (5.2), (5.6), (5.8) and (5.12) furnish expressions for dV_l/dx, dT_l/dx, dm/dx and dn/dx. The resulting set of seven simultaneous first-order differential equations can then be integrated numerically using a fourth order Runge-Kutta procedure. In the case of polydispersed droplet size distribution, the droplet population is discretized into a suitable number of groups (say, N). Equations (5.2), (5.6), (5.8) and (5.12) can then be written for each droplet group. In this case a set of $(3 + 4N)$ simultaneous differential equations need to be solved. Details of the numerical solution procedure are given in [9, 11].

A computational procedure that marches forward in space must necessarily start from an initial condition that represents a deviation from equilibrium. For a partly dispersed shock wave (See "Structure of Stationary Shock Waves"), the difference in the vapour and liquid phase flow variables just downstream of the frozen shock discontinuity constitute the required initial departure from equilibrium. For a fully dispersed shock wave an initial, arbitrary perturbation of the flow must be specified. Step-by-step integration of the conservation equations then automatically generates the wave profile. Provided the initial perturbation is sufficiently small, the numerical results closely approach the exact solution. Thus, if two calculations are performed for the same upstream flow conditions but with different initial small perturbations, it is possible to superpose the results by a relative shift in the x-direction. The physical and mathematical reasons why such a computational procedure can successfully generate the dispersed wave profiles have been explained in Sect. 5.2.3.

For calculating moving shock waves, the unsteady forms of the basic equations are retained and solved by a mixed Eulerian–Lagrangian time-marching technique. For this purpose, the continuity, momentum and energy equations (5.13)–(5.16) are solved by the Denton, finite volume, timemarching method [29]. The interphase transfer processes, however, are more easily describable in a Lagrangian framework. Equations (5.2), (5.6) and (5.8) are therefore integrated along the droplet path lines and are coupled to the unsteady Euler solver by introducing their effects as source terms in the continuity, momentum and energy equations. For simplicity, droplet temperature relaxation is neglected in the unsteady flow analysis and the droplet temperature is assumed to equal the local saturation temperature at all times. Details of the numerical procedure for the calculation of the unsteady evolution of a moving shock wave in vapour–droplet mixture have been given in [9]; an efficient numerical procedure for the computation of oscillating condensation waves is given in [15].

Speeds of Sound

In a Simple Relaxing Medium with one Internal Variable

In contrast to the case of an ideal gas as covered in classical gas dynamics, the speed of an harmonic sound wave in a relaxing medium is not a local thermodynamic property but is a function of the frequency of the sound wave itself (called "frequency dispersion"). Harmonic sound waves of high frequency (representing fast-changing external conditions) can pass through the medium before any significant change in the internal state variable can take place: the speed of such a sound wave is termed the frozen speed of sound, a_f. On the other hand, harmonic sound waves of very low frequency (representing slow-changing external conditions when the internal state variable has enough time to "relax" completely to its equilibrium value) might pass through the medium with equilibrium speed of sound, a_e such that at every moment the system passes through states that are in thermodynamic equilibrium. For intermediate frequencies, the speed of sound varies continuously between these two extreme limits. Maximum dispersion in the sound speed is obtained when $\omega\tau \sim 1$, ω being the circular frequency of the wave and τ the relaxation time of the medium ($\omega\tau$ is the appropriate Damköhler parameter in this case). It can be shown that, in general, $a_f > a_e$ [1].

In a Pure Vapour–Droplet Mixture

As a result of three relaxation processes in vapour–droplet flow, four different sound speeds (a_f, a_{e1}, a_{e2}, a_e) may be defined subject to different mechanical and thermodynamic constraints (Fig. 5.3). The full frozen speed of sound, a_f, corresponds to the speed of an harmonic acoustic wave of such high frequency that the response of the droplets is negligible (i.e. zero mass, momentum and energy transfer). The full equilibrium speed of sound, a_e, corresponds to the speed of an harmonic acoustic wave of such low frequency that liquid–vapour equilibrium is maintained at all times. The two intermediate speeds

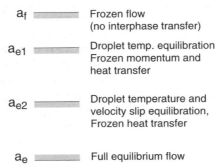

Fig. 5.3. Various speeds of sound in a pure vapour–droplet mixture

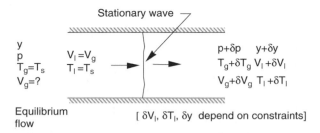

Fig. 5.4. Flow configuration for calculation of the various sound speeds

correspond to (a) the case of equilibrium droplet temperature relaxation but frozen momentum and heat transfer (a_{e1}), and, (b) the case of equilibrium droplet temperature and velocity slip relaxation but frozen heat transfer (a_{e2}).

The derivation of the sound speeds parallels the simple analysis, to be found in most elementary textbooks, for the speed of sound in single-phase flow. Figure 5.4 shows a stationary plane wave in a one-dimensional duct and the conservation equations (5.13)–(5.15) are applied across the wave and solved to give the vapour flow velocity relative to the wave under the specific constraints of interest. This results in the following four expressions [10]:

$$a_f^2 = \gamma\, RT_g, \tag{5.17}$$

$$a_{e1}^2 = \frac{\gamma\, RT_g}{[1 + (y/(1-y))(\gamma c_l/R)(RT_s/h_{fg})^2]}, \tag{5.18}$$

$$a_{e2}^2 = \frac{(1-y)\gamma\, RT_g}{[1 + (y/(1-y))(\gamma c_l/R)(RT_s/h_{fg})^2]}, \tag{5.19}$$

$$a_e^2 = \frac{(1-y)\gamma\, RT_g}{\gamma[1 - (RT_s/h_{fg})(2 - cT_s/h_{fg})]}, \tag{5.20}$$

where $c = c_{pg} + yc_l/(1-y)$. Equation (5.20) shows that the equilibrium speed of sound can be calculated in terms of the equilibrium isentropic index γ_e in pure vapour–droplet flow, given by [12]

$$\gamma_e = \left(1 - \frac{2RT_s}{h_{fg}} + \frac{cT_s}{h_{fg}}\frac{RT_s}{h_{fg}}\right)^{-1}. \tag{5.21}$$

The value of γ_e is, in general, less than γ. For low-pressure steam, $\gamma \sim 1.32$ and $\gamma_e \sim 1.12$. The relationship of the four sound speeds to each other is of great importance. As a typical example, the ratios of different sound speeds for steam at 1 bar and 0.1 wetness fraction are given by, a_f: a_{e1}: a_{e2}: $a_e \equiv 1 : 0.997 : 0.945 : 0.878$.

Qualitative Aspects of Shock Structure

In a Simple Relaxing Medium with One Internal Variable

One can define at least two Mach numbers corresponding to the limiting speeds of sound: a frozen Mach number $M_f = V/a_f$, and an equilibrium Mach

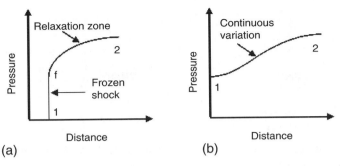

Fig. 5.5. Schematic structure of shock waves in a relaxing medium: (**a**) Partly dispersed, (**b**) Fully dispersed shock wave

number $M_e = V/a_e$, where V is the flow velocity. In general, $M_e > M_f$. As a result of the frequency dispersion, therefore, two distinct types of shock waves might form in a relaxing medium [1, 2], the far upstream and far downstream conditions being at equilibrium (Fig. 5.5):

(a) A partly dispersed shock wave, where a discontinuous jump in flow variables (dominated by viscosity and thermal conductivity as in Rankine–Hugoniot shocks in classical gas dynamics) is followed by a continuous relaxation zone in which the system returns to equilibrium by performing the relaxation process. The internal state variable remains frozen at its upstream value as a fluid particle passes through the (almost) discontinuous wave front and hence the system goes out of equilibrium downstream of the frozen shock. Such a case arises if the upstream velocity is greater than the frozen speed of sound, i.e. if $V_{go} > a_{fo}$ or, $M_{fo} > 1$. The subscript o denotes upstream condition.

(b) A fully dispersed shock wave, where a *continuous* change of flow properties takes place from the upstream to the downstream equilibrium state. This corresponds to the case when $a_{fo} \le V_{go} < a_{eo}$; i.e., when $M_{fo} \le 1$, but $M_{eo} > 1$.

In a Pure Vapour–Droplet Mixture

With four limiting speeds of sound, four types of stationary shock wave structures may arise (Fig. 5.6).

Type I waves	corresponding to	$a_{eo} < V_{go} < a_{e2o}$
Type II waves	corresponding to	$a_{e2o} < V_{go} < a_{e1o}$
Type III waves	corresponding to	$a_{e1o} < V_{go} < a_{fo}$
Partly dispersed waves	corresponding to	$V_{go} > a_{fo}$

Type I, II and III waves are sub-categories of the well-known fully dispersed waves, where the steepening effect of the non-linear terms in the equations of motion is just balanced by the dispersive effect of the relaxation processes. Type I waves are dominated by vapour thermal relaxation, Type II waves by

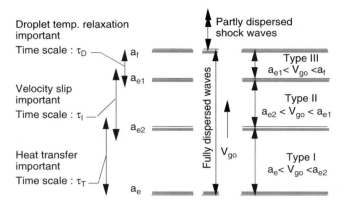

Fig. 5.6. Various shock structures in pure vapour–droplet flows

both velocity and vapour thermal relaxation and Type III waves by all three relaxation processes.

5.2.3 Detailed Structue of Shock Waves

Structure of Stationary Shock Waves

Figure 5.7 shows numerical solutions for a typical Type I and a typical Type II wave in pure wet steam. In "Numerical Solution of the Equations" on numerical algorithm we stated that, for a fully dispersed shock wave, an initial, arbitrary perturbation of the flow must be specified. Step-by-step integration of the conservation equations then automatically generates the wave profile. The following analysis gives physical insight into the detailed structure of various types of shock waves, which also clarifies why such a numerical procedure can generate the wave profiles.

Type I Fully Dispersed Wave

In this case the upstream velocity lies in the range $a_{\mathrm{eo}} < V_{\mathrm{go}} < a_{\mathrm{e2o}}$. Assuming *droplet temperature and velocity slip equilibration*, the continuity, momentum and energy conservation equations for steady, constant area duct flow of the two-phase mixture may be solved [10] to give the variation of vapour supercooling as

$$\frac{1}{T_{\mathrm{g}}}\frac{\mathrm{d}(\Delta T)}{\mathrm{d}x} + \left[\frac{1 - M_{\mathrm{e}}^2}{1 - M_{\mathrm{e2}}^2}\right]\frac{\Delta T}{V_{\mathrm{g}}\tau_{\mathrm{T}}T_{\mathrm{g}}} = 0, \tag{5.22}$$

where M_{e} and M_{e2} are Mach numbers based on a_{e} and a_{e2}, respectively. The quantity within the bracket is negative for the range $a_{\mathrm{e}} < V_{\mathrm{g}} < a_{\mathrm{e2}}$, otherwise it is positive:

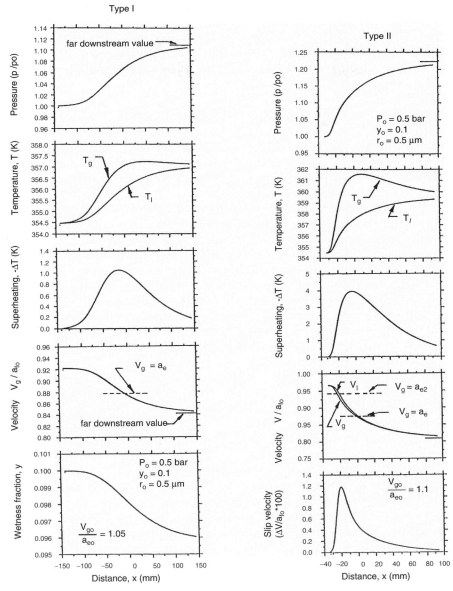

Fig. 5.7. Numerical solution for a Type I fully dispersed stationary wave in steady flow of wet steam ($p_o = 0.5$ bar, $y_o = 0.1$, $r_o = 0.5\,\mu$m, $V_{go}/a_{eo} = 1.05$) Numerical solution for a Type II fully dispersed stationary wave in steady flow of wet steam ($p_o = 0.5$ bar, $y_o = 0.1$, $r_o = 0.5\,\mu$m, $V_{go}/a_{eo} = 1.1$)

(i) If $V_g < a_e$ or $V_g > a_{e2}$

then $\Delta T > 0$ implies $\frac{\mathrm{d}(\Delta T)}{\mathrm{d}x} < 0$ and $\Delta T < 0$ implies $\frac{\mathrm{d}(\Delta T)}{\mathrm{d}x} > 0$; i.e. in these ranges of velocity, ΔT shows stable behaviour, it always tends to zero. If the flow is disturbed from its equilibrium state, it returns to equilibrium.

(ii) If $a_e < V_g < a_{e2}$

then $\Delta T > 0$ implies $\mathrm{d}(\Delta T)/\mathrm{d}x > 0$ and $\Delta T < 0$ implies $\mathrm{d}(\Delta T)/\mathrm{d}x < 0$.

Within this range of velocities, ΔT shows unstable behaviour: it always tends away from zero and consequently a small departure from thermal equilibrium tends to grow. It is this characteristic which is responsible for the existence of Type I fully dispersed shock waves.

Far upstream and downstream of the wave the flow is in equilibrium and $|\Delta T|$ attains the maximum value where the vapour phase velocity equals the local equilibrium speed of sound ($V_g = a_e$). The far upstream (V_{go}) and far downstream ($V_{g\infty}$) velocities are related by Prandtl's relation $V_{go}V_{g\infty} = a_e^2$. In [10], an analytical theory has been formulated, considering vapour thermal relaxation only. The theory predicts the variation of velocity and various thermodynamic properties through the wave in close agreement with full numerical calculations. An example of numerically calculated pressure, velocity and superheat profiles for a typical Type I fully dispersed wave is shown in Fig. 5.7. It can be seen that the magnitude of the departures from equilibrium is quite small everywhere but this still drives the system from one equilibrium state to another. Also the shock waves in vapour–droplet flows are of considerable thickness depending on the relaxation times (especially τ_T) and the upstream Mach number. In exact analogy with the role of viscosity in a perfect gas, the relaxation times here determine the shock thickness and influence the wave profile for any of the thermodynamic properties through scaling in the x-direction.

As the upstream velocity approaches a_{e2} from below, the forward part of the profiles of different flow variables (like pressure, vapour-phase velocity etc.) becomes increasingly steeper and the wave becomes more asymmetric with respect to the position that corresponds to $V_g = a_e$. The validity of the assumption of equilibrium velocity slip also deteriorates, especially near the wave tip where the velocity gradient is particularly high.

Type II Fully Dispersed Wave

If $V_{go} > a_{e2o}$, no continuous solution of the gas dynamic equations is possible if the assumption about equilibrium velocity slip is still maintained. There are two ways of tackling the problem as shown in Fig. 5.8.

Option 1: Introduce a discontinuity (approximately centred around a_{e2} and specified by Prandtl's relation) that would bring the vapour phase velocity within a_{e2} and a_e. Then the analysis for thermal non-equilibrium alone can be performed as was done in the case of Type I fully dispersed shock

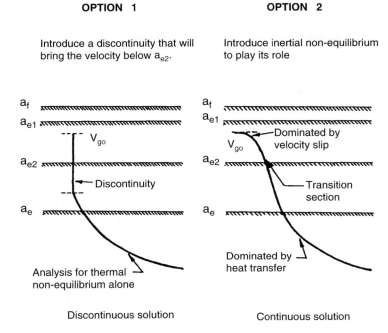

Fig. 5.8. Schematic structure of a Type II wave: velocity profile

wave. The assumption is acceptable in situations where the extent of the zone over which velocity relaxation takes place is short compared to the physical dimensions of the flow domain. The discontinuity appears here in the same way as a normal Rankine–Hugoniot shock in an ideal gas is treated as a mathematical discontinuity when the effects of viscosity and thermal conductivity are neglected. a_{e2} here plays the same role as that of the frozen speed of sound in an ideal gas.

Option 2: Relax the assumption about equilibrium velocity slip and let the inertial non-equilibrium play its role. Assuming *equilibrium* droplet temperature and *frozen* heat transfer, the continuity, momentum and energy equations for the two-phase mixture can be solved to show that the variation of the velocity slip through the shock wave is given by [10],

$$\frac{1}{V_g}\frac{d(\Delta V)}{dx} + \left[\frac{1 - M_{e2}^2}{1 - M_{e1}^2}\right]\frac{\Delta V}{V_g \tau_I V_g} = 0, \tag{5.23}$$

where $M_{e1} = V_g/a_{e1}$. In analogy with the analysis following (5.22), it can be shown from (5.23) that ΔV becomes unstable in the range $a_{e2} < V_g < a_{e1}$ (Fig. 5.9) and it is this characteristics which allows Type II fully dispersed waves to exist stably in steady flow. The front part (upstream end) of the wave is steep and is governed by velocity relaxation (unless V_{go} is just above a_{e2}, see Fig. 5.6). The tail of the wave (downstream end) extends over a much

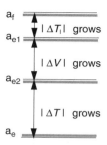

Fig. 5.9. Reasons for the existence of fully dispersed waves in vapour–droplet flow

longer distance and its structure is governed (just as in a Type I wave) by vapour thermal relaxation. Connecting the two regions is a transition section where the effects of velocity slip and vapour thermal relaxation are of comparable magnitude (Fig. 5.8). The solution for the transition section may be constructed geometrically, given the analytical solution for the front and the rear part of the wave. In exact analogy with the occurrence of maximum $|\Delta T|$ at $V_g = a_e$, (5.23) shows that the maximum of $|\Delta V|$ occurs at the position where $V_g = a_{e2}$. A numerical solution for Type II wave is presented in Fig. 5.7.

Type III Fully Dispersed Wave

When the upstream velocity is in the range $a_{elo} < V_{go} < a_{fo}$ a continuous solution of the equations of motion is impossible without relaxing the constraint on the droplet temperature equilibration. However, if this is done, a fully dispersed wave can exist, the leading tip of the wave being dominated by the effects of droplet temperature relaxation. Assuming *frozen* momentum *and* heat transfer, it can be shown that the variation of droplet temperature supercooling is given by,

$$\frac{1}{T_s}\frac{\mathrm{d}(\Delta T_l)}{\mathrm{d}x} + \left[\frac{1 - M_{el}^2}{1 - M_f^2}\right]\frac{\Delta T_l}{V_g \tau_D T_s} = 0. \tag{5.24}$$

Equation (5.24) shows that when the vapour velocity is in the range $a_{el} < V_g < a_f$, ΔT_l becomes unstable and is responsible for the existence of Type III waves (Fig. 5.9).

In practice, the difference between a_{el} and a_f is very small. The leading part of the wave is very steep indeed as the relaxation time τ_D is very small and under certain circumstances the thickness may even approach a value where the neglect of viscosity and thermal conductivity is not a good approximation. Inserting numerical values in the expression for τ_D (5.7) shows that the approximate length of the droplet temperature relaxation zone in low pressure wet steam is about (500 r) mean free paths where r is the droplet radius in micron.

Partly Dispersed Shock Waves

For $V_{\mathrm{go}} > a_{\mathrm{fo}}$ no continuous solution of the conservation equations (5.12)–(5.15) is possible and the dispersed part of the wave is preceded by a near discontinuity dominated by viscosity and thermal conductivity. Detailed analysis of this type of waves is presented in [11].

The usual model of a partly dispersed shock wave assumes that the interphase transfer processes are frozen during the passage through the discontinuity and the vapour properties just downstream of the discontinuity can be calculated using a standard Rankine–Hugoniot analysis. The liquid droplets therefore pass through the shock without change in radius, temperature and velocity. A comparison of the relaxation times τ_{D}, τ_{I} and τ_{T} with the usual estimates of shock wave thickness (a few vapour mean free paths) shows this assumption to be quite acceptable with the possible exception of droplet temperature relaxation.

The conditions downstream of the discontinuity provide the initial values for integrating the conservation equations through the relaxation zones. The droplet temperature relaxes very quickly, followed by velocity slip and finally the vapour temperature. The lengths of the relaxation zones are in the approximate ratios $\tau_{\mathrm{D}} : \tau_{\mathrm{I}} : \tau_{\mathrm{T}}$. Although linearized analyses are often presented for the relaxation zone downstream of the frozen shock [2], they are of limited applicability to vapour–droplet flows [11].

The results of a typical numerical calculation for low pressure steam are shown in Fig. 5.10. Upstream of the discontinuity, the flow is in equilibrium at a pressure of 0.35 bar and a wetness fraction of 0.1. The frozen Mach number, M_{fo}, is 1.5 and the droplets are monodispersed with radius 0.1 μm. The numerical results show that, downstream of the discontinuity, the droplet temperature relaxes very quickly. The velocity slip, which is initially very large, then also relaxes and is finally followed by the vapour superheat. The initial increase in vapour temperature just after the discontinuity is due to the effect of velocity slip. Details may be found in [11].

Role of Coupled Relaxation Processes

To gain further insight into the relaxation processes in the *transonic* regime we note from Fig. 5.6 that when the velocity V_g is close to a_{e1} or a_{e2}, more than one relaxation processes are active [13]. For example, if V_g is in the vicinity of a_{e2}, the inertial and vapour thermal relaxations become coupled. Assuming equilibrium droplet temperature and neglecting second-order terms of small quantities, the coupled differential equations for the variation of ΔV and ΔT in a constant-area-duct can be obtained from (5.2)–(5.16), and are given by

$$\left(1 - M_{\mathrm{e1}}^2\right) \frac{1}{V_{\mathrm{g}}} \frac{\mathrm{d}(\Delta V)}{\mathrm{d}x} + \left(1 - M_{\mathrm{e2}}^2\right) \frac{\Delta V}{V_{\mathrm{g}} \tau_{\mathrm{I}} V_{\mathrm{g}}} - \left(1 - \frac{c_{\mathrm{pg}} T_{\mathrm{g}}}{h_{\mathrm{fg}}}\right) \frac{\Delta T}{V_{\mathrm{g}} \tau_{\mathrm{T}} T_{\mathrm{g}}} = 0, \quad (5.25)$$

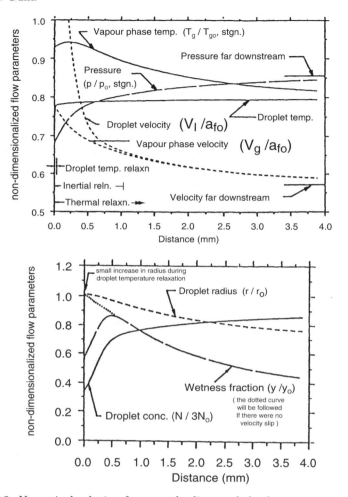

Fig. 5.10. Numerical solution for a partly dispersed shock wave in steady flow

$$\left(1 - M_{e1}^2\right) \frac{1}{T_g} \frac{\mathrm{d}(\Delta T)}{\mathrm{d}x} + \left(1 - M_e^2\right) \frac{\Delta T}{V_g \tau_T T_g} - \frac{V_g^2}{c_{pg} T_g} \frac{y}{1-y} \left(1 - \frac{c_{pg} T_g}{h_{fg}}\right) \frac{\Delta V}{V_g \tau_I V_g} = 0.$$

(5.26)

Note that when V_g is close to a_{e2}, $(1-M_{e2}^2) \to 0$. Hence the third term in (5.25) may become significant in relation to the second term, despite $\tau_T \gg \tau_I$. In a compressive wave both ΔV and ΔT are negative. Thus if $M_{e2} < 1$, the second term in the L.H.S. of (5.25) tends to maintain equilibrium whereas the third term tends to increase the slip velocity. [Compare (5.25) with (5.23) which was derived assuming frozen heat transfer, and (5.26) with (5.22) which was derived assuming equilibrium velocity slip.] Equation (5.25), therefore, implies that the slip velocity $-\Delta V$ may sometimes grow even when $V_g < a_{e2}$, but it does so because of the positive feedback from the unstable amplification of $-\Delta T$ in the regime $a_e < V_g < a_{e2}$. (Equation (5.23) shows that ΔV itself becomes unstable in the regime $a_{e2} < V_g < a_{e1}$.) If V_g is slightly above a_{e2}, both the second and third terms in (5.25) tend to increase the slip velocity. Thus the net effect of the heat transfer would be to delay the position of maximum velocity slip slightly (i.e. the maximum velocity slip occurs slightly downstream of the point where $V_g = a_{e2}$). Similar arguments concerning (5.26) show that the superheat $-\Delta T$ grows also in the regime $a_{e2} < V_g < a_{e1}$ as a result of the unstable amplification of $-\Delta V$. This would be so even if the heat transfer is frozen, which can be verified by letting $\tau_T \to \infty$ in (5.25) and (5.26). Numerical solution (Fig. 5.7) of the superheat profile in a Type II wave illustrates the point.

Although a linearized solution of the simultaneous equations (5.25) and (5.26) is possible (for example, inside the relaxation zone of a partly dispersed shock wave), the wave profile is best determined numerically. If the frozen shock is substantial, second-order non-equilibrium effects involving $(\Delta V/V_g)^2$ etc. dominate the initial stage of relaxation and, when, the shock is weak, the change of sign of the terms such as $(1 - M_e^2)$ and $(1 - M_{e2}^2)$ in (5.25) and (5.26) within the relaxation zone precludes the use of linearized theory.

Time Evolution of Shock Waves

The physical significance of the various wave profiles discussed earlier can be appreciated more readily by considering their development under unsteady flow conditions. As a typical example, we now discuss wave generation in one-dimensional flow by an instantaneously accelerated piston in a frictionless pipe initially containing stationary wet steam. Unsteady flows with shock waves are difficult to compute even in single-phase flow. To demonstrate the accuracy of the numerical scheme [9], we therefore present results for the piston-in-a-pipe problem using a perfect gas rather than wet steam. Figure 5.11 shows the theoretical $(t - x)$ characteristics diagram which consists, quite simply, of a shock wave of constant strength propagating at constant velocity into the stationary gas. The gas velocity behind the wave is constant and equal to the speed of the piston. The numerical solution for the pressure distribution is also shown in Fig. 5.11 and it is evident that the computed shock profile and wave velocity are extremely accurate and remain remarkably constant as the wave propagates along the pipe.

Figure 5.12 shows the rather different behaviour when the wave propagates into wet steam (of pressure 0.35 bar, wetness fraction 0.1 and droplet radius 0.1μm). The $(t - x)$ diagram was constructed from the results of the unsteady time-marching calculation.

At the instant of initiation, all the interphase transfer processes are frozen and the shock velocity corresponds to the propagation velocity into a single-phase vapour at the same temperature. Behind the shock, the mixture relaxes to equilibrium along the particle pathlines. The droplet temperature relaxes

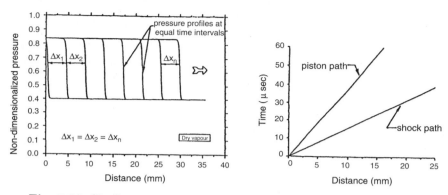

Fig. 5.11. Shock propagation in a pipe: single-phase flow of a perfect gas

first on the very short timescale τ_D and is followed by the velocity slip and vapour temperature on timescales τ_I and τ_T, respectively. Changes along the particle paths are propagated upstream and downstream along the left and right running Mach lines (based on the frozen speed of sound). The right running Mach lines overtake the shock wave, weakening it and causing it to slow down. The shock path therefore curves in the $(t - x)$ diagram until it reaches a constant equilibrium speed. When this occurs, the dispersive effects of the relaxation processes are just balanced by the steepening effects of the non-linear terms and the wave structure is identical to that of the stationary waves in steady flow described earlier. Whether the final equilibrium structure is partly or fully dispersed depends on the piston velocity (Sect. 5.3.3).

The variation of the wave pressure profile with time is also shown in Fig. 5.12 and the deceleration and weakening of the wave front are clearly visible. The behaviour of the superheat, vapour temperature, droplet radius and wetness fraction is shown by the curves in Fig. 5.13 which are self-explanatory. As with stationary partly dispersed shock waves, the increase in wetness fraction downstream of the frozen shock wave is due to the effects of velocity slip.

5.3 Integral Analysis: Jump Conditions for Pure Vapour–Droplet Flows

5.3.1 Stationary Shock Waves

A detailed integral, control-volume, analysis relating the end states of a condensation zone as well as aerodynamic shock waves, showing similarities and differences between the two, is presented in [8]. Detailed study on the jump conditions across shock waves in pure vapour–droplet flow has been made in [12]. Here, we can only mention the bare minimum, and the control volume

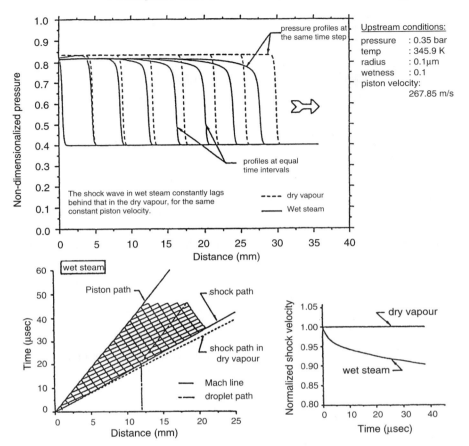

Fig. 5.12. Shock propagation in a pipe: two-phase flow of wet steam

analysis is depicted in Fig. 5.14. The subscripts 1 and 2 refer to the far upstream and far downstream conditions, respectively.

In a simple relaxing medium, e.g. a solid particle laden gas, the form of the jump conditions (R in Fig. 5.15) are identical to the Rankine–Hugoniot relations for an ideal gas, if one uses the equilibrium Mach number, $M_e = V/a_e$, and the equilibrium isentropic index, γ_e. [Expression for γ_e in pure vapour–droplet mixtures is given by (5.21), that in the presence of a carrier gas is later given by (5.55), that in a solid-particle-laden gas is given in Sect. 5.6] However, it has been shown in [12] that although jump conditions of the same form as Rankine–Hugoniot relations can be formulated for vapour–droplet mixtures, they are *approximate* and hold only *conditionally*.

A solid-particle-laden-gas can be treated as a modified gas. Equations governing vapour–droplet flow are much more complex. An additional complexity

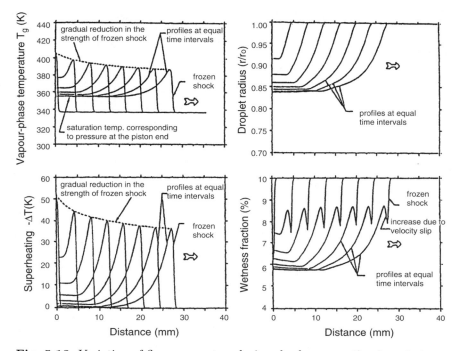

Fig. 5.13. Variation of flow parameters during shock propagation in wet steam

Two phase equilibrium	Aerodynamic shock wave	Two phase equilibrium or superheated vapour
(1)	(Evaporation of liquid phase)	(2)

Fig. 5.14. Control volume analysis of aerodynamic shock wave

(and *novelty*) occurs due to interphase mass transfer. As depicted in Fig. 5.14, liquid phase evaporates inside a dispersed wave. If the strength of the wave is sufficient, complete evaporation may result, thereby rendering a two-phase medium into a single-phase one!

The integral analysis of [12] reveals that, depending on the upstream wetness fraction and the pressure ratio across the wave, four types of shock structures may result in vapour–droplet flow. They are: I Equilibrium fully dispersed, II Equilibrium partly dispersed, III Fully dispersed with complete evaporation, IV Partly dispersed with complete evaporation. Figure 5.16 shows the boundaries of the four regimes in low-pressure wet steam.

Jump conditions across all types of aerodynamic waves are derived in [8] and [12]. Figure 5.17 shows the predictions of the integral analysis compared

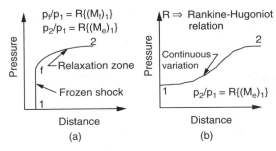

Fig. 5.15. Schematic diagram showing the meaning of jump conditions (**a**) Partly dispersed wave $(M_e)_1 > 1$, $(M_f)_1 > 1$, (**b**) Fully dispersed wave $(M_e)_1 > 1$, $(M_f)_1 \leq 1$

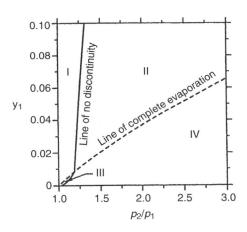

Fig. 5.16. Phase diagram of different shock structures in low-pressure steam

with numerical solutions of the wave profile as discussed in Sect. 5.2. The dotted lines in Fig. 5.17 represent the jump conditions. The three solid lines represent numerical solutions for three categories of shock waves: I, partly dispersed with complete evaporation ($p_1 = 0.35$ bar, $M_{f1} = 1.6$, $y_1 = 0.05$, $r_1 = 0.1\,\mu m$); II, equilibrium partly dispersed ($p_1 = 0.35$ bar, $M_{f1} = 1.2$, $y_1 = 0.05$, $r_1 = 0.1\,\mu m$); III, equilibrium fully dispersed ($p_1 = 0.35$ bar, $M_{f1} = 0.97$, $y_1 = 0.1$, $r_1 = 0.1\,\mu m$). Results from integral analyses agree remarkably well with solutions of the differential equations of motion, thereby confirming independent theoretical consistency.

5.3.2 Jump Relations for Pure Vapour–Droplet Flow

These relations have been fully derived in [12], which should be referred to for extended physical insight. Here we quote a few results.

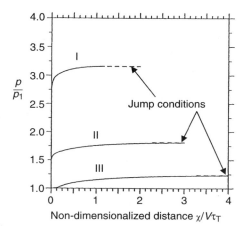

Fig. 5.17. Comparison of asymptotic pressure profiles with generalized Rankine–Hugoniot equations

Approximate Rankine–Hugoniot Relations for Specified p_1, V_1 and y_1

The fluid is an equilibrium two-phase vapour–droplet mixture both upstream and downstream of the shock wave. The jump relations for pressure and velocity are:

$$\frac{p_2}{p_1} = \frac{2\gamma_e}{\gamma_e + 1}(M_e)_1^2 - \frac{\gamma_e - 1}{\gamma_e + 1}, \tag{5.27}$$

$$\frac{V_2}{V_1} = \frac{(\gamma_e - 1)(M_e)_1^2 + 2}{(\gamma_e + 1)(M_e)_1^2}. \tag{5.28}$$

The temperature ratio across the shock, however, is not given by the perfect gas type Rankine–Hugoniot relation, but must be calculated by integrating the Clausius-Clapeyron equation. Neglecting the specific volume of the liquid and introducing the perfect gas equation of state for vapour phase, the Clausius-Clapeyron equation can be written as $dT_s/T_s = (RT_s/h_{fg})(dp/p)$. Assuming that the value of RT_s/h_{fg} does not change appreciably across a weak shock wave, (which is in keeping with the assumption of constant γ_e), the temperature ratio is approximately given as,

$$\frac{T_{s2}}{T_{s1}} = \left[\frac{p_2}{p_1}\right]^{\frac{RT_s}{h_{fg}}}. \tag{5.29}$$

Although not exact, these equations have two advantages (1) all shock relations could be explicitly written in terms of the upstream parameters only, and (2) the form of equations is similar to the well-known Rankine–Hugoniot relations for ideal gas.

It should be noted that (5.27)–(5.29) are valid for both partly and fully dispersed shock waves. Partly dispersed shock waves are characterized by an almost discontinuous wave-front followed by a relaxation zone and occur when the upstream frozen Mach number is greater than unity ($M_{f1} > 1$, which also implies that $(M_e)_1 > 1$). The overall changes across such a shock wave are given by the earlier equations while the changes just across the discontinuous wave-front are given by the classical Rankine–Hugoniot equation based on M_{f1}. If $M_{f1} < 1$ but $(M_{e_1}) > 1$, then a fully-dispersed shock wave, which does not have a discontinuous wave-front, will appear in the flow-field. The earlier equations again give the overall changes across such waves. Figure 5.15 schematically depicts the point.

Exact Jump Relations for Specified p_1, p_2 and y_1

The fluid is an equilibrium two-phase vapour–droplet mixture both upstream and downstream of the shock wave. For a prescribed pressure ratio (rather than specifying the upstream velocity V_1) an accurate closed form analytical solution of the equations is possible and is given by [12] :

$$y_2 = \frac{2c_{\mathrm{pg}}(T_{s2} - T_{s1}) + 2y_1 h_{\mathrm{fg1}} - \frac{p_2 - p_1}{\rho_1} - RT_{s2}\frac{p_2 - p_1}{p_2}}{2h_{\mathrm{fg2}} - RT_{s2}\frac{p_2 - p_1}{p_2}}, \tag{5.30}$$

$$V_1^2 = \frac{p_2 - p_1}{\rho_1}\left(1 - \frac{\rho_1(1 - y_2)RT_{s2}}{p_2}\right)^{-1}. \tag{5.31}$$

Exact Jump Relations for Complete Evaporation

If the imposed pressure ratio is too high, complete evaporation of the droplets may take place inside the relaxation zone. In such a case, upstream of the shock wave, the fluid is an equilibrium two-phase mixture whereas downstream of the shock, it is single-phase vapour alone. (This problem does not arise in a solid particle-laden gas.) The downstream vapour-phase temperature is no longer the saturation temperature corresponding to the prevailing pressure. The jump relations are given by [12] :

$$T_2 = \frac{T_{s1}\left(1 + \frac{p_2 - p_1}{p_1}\frac{\gamma - 1}{2\gamma}\right) - \left(\frac{h_{\mathrm{fg1}}}{c_{\mathrm{pg}}} + \frac{p_2 - p_1}{p_1}\frac{\gamma - 1}{2\gamma}T_{s1}\right)y_1}{\left(1 - \frac{p_2 - p_1}{p_2}\frac{\gamma - 1}{2\gamma}\right)}, \tag{5.32}$$

$$V_1^2 = \frac{p_2 - p_1}{\rho_1}\left(1 - \rho_1\frac{RT_2}{p_2}\right)^{-1}. \tag{5.33}$$

There is a limiting value for the upstream wetness fraction ($y_{1,\mathrm{lim}}$), such that below this limit complete evaporation takes place inside the shock wave.

$y_{1,\text{lim}}$ predominantly depends on the pressure ratio p_2/p_1, and can be shown to be given by [12]:

$$y_{1,\text{lim}} = \frac{T_{s1}\left(\frac{p_2-p_1}{p_1}R + 2c_{\text{pg}}\right) - T_{s2}\left(2c_{\text{pg}} - \frac{p_2-p_1}{p_2}R\right)}{\left(2h_{\text{fg1}} + \frac{p_2-p_1}{p_1}RT_{s1}\right)}. \tag{5.34}$$

Thus if $y_1 = 0$, the vapour is dry throughout, and the velocity of the shock is given by the frozen shock relation. If y_1 is greater than or equal to the limiting value $y_{1,\text{lim}}$, for which the vapour is wet at downstream of the shock, the velocity of the shock is given by the equilibrium relation, (5.31). For intermediate values of y_1, i.e. when $0 \leq y_1 \leq y_{1,\text{lim}}$, complete evaporation takes place inside the dispersed shock and the earlier theory (5.33) shows that there is a continuous transition in the shock velocity from the frozen to the equilibrium value. Thus (5.33) becomes identical with the frozen shock relation (classical gas dynamic result) in the limit $y_1 = 0$, and becomes identical with the equilibrium shock relation (5.31) in the limit $y_1 = y_{1,\text{lim}}$.

Generation of Entropy Inside a Shock Wave

In addition to the effects of viscosity and thermal conduction, entropy is created due to the relaxation processes inside a dispersed shock wave [12]. Figure 5.18 shows the rise in entropy across shock waves as a function of upstream frozen Mach number. Two cases are considered: dry steam, and, wet steam with upstream wetness fraction 0.1. Shock waves may occur in wet steam even when $M_{\text{f1}} < 1$, due to the existence of fully dispersed waves. In a partly dispersed wave, entropy rises across the frozen shock as well within the relaxation zone that follows. The figure shows that the contribution of the relaxation processes is extremely significant in the overall creation of entropy.

In an ideal gas, the total rise in entropy across a shock wave is fixed by Rankine–Hugoniot relations, the magnitudes of viscosity and thermal conductivity only determine the thickness of the shock wave. It has been shown in [12] that the total rise in entropy across a dispersed shock wave in vapour–droplet mixtures is similarly fixed by the jump conditions. In exact analogy with the role of viscosity, various relaxation processes and their timescales determine the thickness of the shock wave. Overall jump conditions in any property, including entropy, can be determined without any explicit reference to the processes which make the "jump" happen.

5.3.3 Unsteady Development of Shock Waves

An integral analysis can again provide much insight [13]. In order to appreciate how the different shock structures originate through an unsteady process, we consider the classical piston and cylinder problem. Initially the

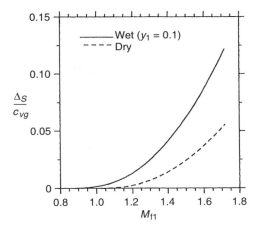

Fig. 5.18. Entropy rise across shock waves in low-pressure steam

cylinder is filled with a vapour–droplet mixture at rest. As soon as the piston is given an impulsive constant velocity V_p, a frozen shock wave starts
moving into the undisturbed vapour–droplet mixture with a velocity equal
to V_s^i. At the moment of initiation, all interphase transfer processes are
frozen and the shock velocity corresponds to the propagation velocity into
a single-phase vapour at the same temperature. If it were an ideal gas, the
shock wave would continue to move undiminished in strength. However, in
a vapour–droplet mixture, as the droplets start interacting with the vapour
phase because of the various relaxation processes, the discontinuous wave
front weakens, the shock velocity decreases and a relaxation zone develops behind the discontinuity. This process continues until an equilibrium
structure is obtained when the non-linear steepening terms are exactly balanced by the linear dispersion terms. The wave profile no longer changes but
convects downstream with a velocity V_s^f. Whether this final, stable shock
wave profile is partly dispersed or fully dispersed, depends on the piston
velocity V_p.

If we fix the reference frame with the moving shock wave, then it is possible
to write the Rankine–Hugoniot relations across the frozen shock wave at the
moment of its initiation. The initial frozen Mach number M_{fo}^i, can then be expressed [13] as a function of the non-dimensionalized piston velocity (V_p/a_{fo}).
Thus

$$M_{fo}^i = \frac{(\gamma+1)V_p}{4\,a_{fo}} + \sqrt{\frac{V_p^2(\gamma+1)^2}{16\,a_{fo}^2} + 1}, \tag{5.35}$$

where the subscripts o and f refer to the upstream and frozen conditions,
respectively, and the superscript i refers to the moment of initiation. Equation
(5.35) correctly points out that as $V_p \to 0$, $M_{fo}^i \to 1$; in other words, the wave
becomes an acoustic one.

The frozen Mach number of the wave when it has attained its final stable structure (denoted by superscript f), is given by [13]

$$M_{\text{fo}}^{\text{f}} = \frac{V_{\text{s}}^{\text{f}}}{a_{\text{fo}}} = \frac{(\gamma_{\text{e}} + 1)V_{\text{p}}}{4\,a_{\text{fo}}} + \left[\sqrt{\frac{V_{\text{p}}^2(\gamma_{\text{e}} + 1)^2}{16\,a_{\text{eo}}^2} + 1}\right]\frac{a_{\text{eo}}}{a_{\text{fo}}} \tag{5.36}$$

The final shock structure would be partly dispersed if $M_{\text{fo}}^{\text{f}} > 1$ and fully dispersed if $M_{\text{fo}}^{\text{f}} \leq 1$. Thus the limiting piston velocity V_{pl}, corresponding to $M_{\text{fo}}^{\text{f}} = 1$, can be found from (5.36),

$$\frac{V_{\text{pl}}}{a_{\text{fo}}} = \frac{2}{\gamma_{\text{e}} + 1}\left[1 - \left(\frac{a_{\text{eo}}}{a_{\text{fo}}}\right)^2\right]. \tag{5.37}$$

For steam at $p = 0.35$ bar, $y = 0.1$, $\gamma = 1.32$, $\gamma_{\text{e}} = 1.13$; (5.37) shows $V_{\text{pl}} = 0.22\,a_{\text{fo}}$. Thus if $V_{\text{p}} \leq 0.22\,a_{\text{fo}}$, the final structure would be fully dispersed, whereas if $V_{\text{p}} > 0.22\,a_{\text{fo}}$, a partly dispersed shock structure would arise (for the steam conditions mentioned earlier). It is evident that V_{pl} would be a function of p and y. Figure 5.19 shows the variation of V_{pl} as a function of wetness fraction.

As explained in [10], a_{e} does not tend to a_{f} in the limit y tends to zero and hence $(V_{\text{pl}}/a_{\text{fo}})$ in (5.37) does not tend to zero as $y \to 0$, which it should do in the frozen limit. It is assumed while deriving (5.36) that the fluid far upstream and downstream of the shock wave is in two-phase, vapour–droplet equilibrium. This may be violated if the piston velocity is too high for the upstream wetness level resulting in complete evaporation of the liquid phase. This aspect has been elaborated in [12]. Integral analysis can explain many other flow features [13], e.g. why the pressure at the piston end ($x = 0$ in Fig. 5.12) in wet vapour is lower than that in dry vapour for the same piston velocity.

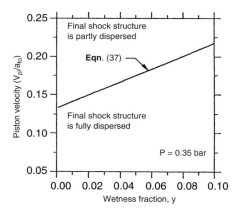

Fig. 5.19. Limiting piston velocity

5.4 Coupled Relaxation Processes and Thermal Choking

5.4.1 Differential Approach

Equations (5.2), (5.6), (5.8) and (5.12)–(5.16) may be combined (after neglecting second-order small terms involving the products of any two of $\Delta V/V_{\mathrm{g}}$, $\Delta T/T_{\mathrm{g}}$ and $\Delta T_{\mathrm{l}}/T_{\mathrm{s}}$) to give explicitly the variation in different vapour properties in a quasi-one-dimensional nozzle.

$$\left(1 - M_{\mathrm{f}}^2\right)\frac{DV_{\mathrm{g}}}{V_{\mathrm{g}}} = -\frac{DA}{A} + \left(1 - \frac{c_{\mathrm{pg}}T_{\mathrm{s}}}{h_{\mathrm{fg}}}\right)\Theta + M_{\mathrm{f}}^2\Pi - \sigma, \tag{5.38}$$

$$\left(1 - M_{\mathrm{f}}^2\right)\frac{Dp}{p} = \gamma M_{\mathrm{f}}^2\left[\frac{DA}{A} - \left(1 - \frac{c_{\mathrm{pg}}T_{\mathrm{s}}}{h_{\mathrm{fg}}}\right)\Theta - \Pi + \sigma\right], \tag{5.39}$$

$$\left(1 - M_{\mathrm{f}}^2\right)\frac{DT_{\mathrm{g}}}{T_{\mathrm{g}}} = (\gamma - 1)M_{\mathrm{f}}^2\frac{DA}{A} + \left[1 - \gamma M_{\mathrm{f}}^2\left(1 - \frac{RT_{\mathrm{s}}}{h_{\mathrm{fg}}}\right)\right]\Theta$$
$$- (\gamma - 1)M_{\mathrm{f}}^2\left(\Pi - \sigma\right), \tag{5.40}$$

where the shorthand notation $D = \mathrm{d}/\mathrm{d}x$ is used for convenience. The variables Θ, Π and σ represent the contributions from three relaxation phenomena and are given by

$$\Theta = \frac{\Delta T}{V_{\mathrm{g}}\tau_{\mathrm{T}}T_{\mathrm{g}}}, \tag{5.41}$$

$$\Pi = \frac{y}{1 - y}\frac{\Delta V}{V_{\mathrm{g}}\tau_{\mathrm{I}}V_{\mathrm{g}}}, \tag{5.42}$$

$$\sigma = \frac{y}{1 - y}\frac{c_{\mathrm{l}}T_{\mathrm{s}}}{h_{\mathrm{fg}}}\frac{\Delta T_{\mathrm{l}}}{V_{\mathrm{g}}\tau_{\mathrm{D}}T_{\mathrm{s}}}. \tag{5.43}$$

Θ is associated with vapour thermal relaxation, Π with inertial (or velocity slip) relaxation and σ with droplet temperature relaxation. It is easy to draw from (5.38)–(5.40) a table of influence coefficients for non-equilibrium condensing flow [8].

Comparison of (5.38)–(5.40) with similar equations for a perfect gas with external heat addition [8] shows the similarities and differences between the characteristics of a condensing flow and the flow of a perfect gas with external heat addition. In particular, note that the mass depletion due to condensation and the use of the Clausius–Clapeyron equation (to obtain the vapour temperature) result in terms like $(1 - c_{\mathrm{pg}}T_{\mathrm{s}}/h_{\mathrm{fg}})$ and $(1 - RT_{\mathrm{s}}/h_{\mathrm{fg}})$ in the flow equations. (These could give unusual results in fluids with $c_{\mathrm{pg}}T_{\mathrm{s}}/h_{\mathrm{fg}} > 1$, or, $RT_{\mathrm{s}}/h_{\mathrm{fg}} > 1$.)

The general expressions for the variation of the subcooling, ΔT, and the frozen Mach number, M_{f}, can be derived from (5.38)–(5.40).

Sign of Influence Coefficients in Condensing Flow

In a continually expanding flow, in general, $\Theta > 0$, Π 0 and $\sigma < 0$.

	$M_f < 1$				$M_f > 1$			
	Contribution from				Contribution from			
	DA/A	Θ	Π	σ	DA/A	Θ	Π	σ
DV_g/V_g	$\mp$ [†]	$+$	$+$	$+$	$\pm$ [†]	$-$	$-$	$-$
Dp/p	$\pm$ [†]	$-$	$-$	$-$	$\mp$ [†]	$+$	$+$	$+$
DT_g/T_g	$\pm$ [†]	$+$ if $M_f < M_*^{\ddagger}$ $-$ if $M_f > M_*$	$-$	$-$	$\mp$ [†]	$+$	$+$	$+$

[a]Upper sign if $DA > 0$, lower sign if $DA < 0$.

[b]In an ideal gas $M_* = 1/\sqrt{\gamma}$ Equation (5.40) shows that in a condensing flow.

$M_* = 1/\sqrt{\gamma(1 - RT_s/h_{fg})}$.

$$\left(1 - M_f^2\right) \frac{1}{T_g} \frac{d(\Delta T)}{dx} + \left(1 - \frac{M_f^2}{M_+^2}\right)\Theta$$
$$= M_f^2(\gamma - 1)\left(1 - \frac{c_{pg}T_s}{h_{fg}}\right)\left[-\frac{1}{A}\frac{dA}{dx} + \Pi - \sigma\right], \qquad (5.44)$$

$$\left(1 - M_f^2\right)\frac{1}{M_f}\frac{dM_f}{dx} = \left(1 + \frac{\gamma - 1}{2}M_f^2\right)\left[-\frac{1}{A}\frac{dA}{dx} - \sigma\right] + \frac{\gamma + 1}{2}M_f^2\Pi$$
$$+ \left[\frac{1}{2}\left\{1 + \gamma M_f^2\left(1 - \frac{RT_s}{h_{fg}}\right)\right\} - \frac{c_{pg}T_s}{h_{fg}}\right]\Theta, \quad (5.45)$$

where M_f is the frozen Mach number. M_+ in (5.44) is given by

$$M_+ = a_+/a_f, \qquad (5.46)$$

$$\left(\frac{a_+}{a_f}\right)^2 = \left[\gamma\left(1 - 2\frac{RT_s}{h_{fg}} + \frac{RT_s}{h_{fg}}\frac{c_{pg}T_s}{h_{fg}}\right)\right]^{-1}. \qquad (5.47)$$

Equation (5.47) shows that the ratio a_+/a_f is a weak function of pressure and does not depend on the wetness fraction. For wet steam, $a_+/a_f \approx 0.92$ over a rather wide range of conditions (the ratio varies from 0.919 at 0.1 bar to 0.934 at 2 bar). Note that

$$a_+ \approx \frac{a_e}{\sqrt{1 - y}}, \qquad (5.48)$$

where the equilibrium speed of sound, a_e, is given by (5.20).

Equations (5.44) and (5.45) show that, depending on the shape of the flow passage and the flow conditions, quite complicated variations in ΔT and M_f are possible. Equation (5.45) also shows that, when all non-equilibrium

processes are properly considered, the physical condition of thermal choking is obtained when the *frozen* Mach number is unity. Simplifications of
the flow equations alter the mathematical shape of the differential equations and the location of the singularity. For example, if droplet temperature equilibration (with frozen interphase momentum and heat transfer) is
assumed and then the differential equations are re-derived systematically, the
common factor $(1 - M_f^2)$ appearing in the LHS of equations (5.38)–(5.40)
and (5.44), (5.45) is replaced by another common factor $(1 - M_{e1}^2)$, where
the characteristic speed of sound a_{e1} is given by (5.18). Similarly, the assumption of equilibrium droplet temperature and velocity slip (with frozen
heat transfer) results in a common factor $(1 - M_{e2}^2)$, where the characteristic speed of sound a_{e2} is given by (5.19). Finally, the assumption of full
equilibrium results in a common factor $(1 - M_e^2)$, where a_e is the equilibrium speed of sound given by equation (5.20). Clearly, these simplified models
give mathematical singularities when the flow velocity equals a_{e1}, a_{e2} and a_e,
respectively.

5.4.2 Integral Approach

At a particular flow Mach number, a fluid can absorb a maximum quantity
of heat before the local Mach number equals unity and the flow becomes
thermally choked. Textbooks on classical gas dynamics show that the value of
the critical quantity of heat, $q^*_{\text{classical}}$, for simple heat addition (external heat
addition without any change in flow cross-sectional area) to an ideal gas is
given by,

$$\frac{q^*_{\text{classical}}}{c_{\text{pg}}T_{o1}} = \left[M_1^2 - 1\right]^2 \left(2(\gamma + 1)M_1^2 \left[1 + \frac{\gamma - 1}{2}M_1^2\right]\right)^{-1}, \tag{5.49}$$

where M_1 is the Mach number at which the heat addition begins and T_{o1} is
the stagnation temperature before heat addition.

Similar to the case of external heat addition, the Mach number decreases
in the condensation zone (the flow being supersonic). Therefore, for particular combinations of nozzle geometry, supply conditions and the working fluid,
the liberation of latent heat could be such that the minimum Mach number
becomes unity and the flow is thermally choked. If the inlet total temperature, T_{o1}, is reduced any further, keeping the inlet total pressure, p_{o1}, fixed, a
continuous variation of the flow variables is no longer possible and an aerodynamic shock wave appears inside the condensation zone.

Although widely referred, it has been shown in [16] that (5.49) is not appropriate for a condensing flow in a pure vapour primarily for two reasons (1)
in case of a condensation shock, the heat is added as a result of condensation
of a part of the fluid itself. Therefore, the mass flow rate of the condensable vapour changes as the vapour is continually transformed into the liquid
phase. Equation (5.49), which is derived for external heat addition to an ideal

gas, does not take into account this mass depletion. (2) The droplets formed through homogeneous nucleation grow at a finite rate by exchanging mass and energy with the surrounding vapour. Therefore, the energy addition due to condensation is not instantaneous and takes place over a short but finite zone. Since condensation shocks normally occur in the diverging section (with dry vapour at inlet), this means the flow area increases between the upstream and downstream of the condensation zone. Equation (5.49), on the other hand, is derived by assuming heat addition in a constant area duct.

Reference [16] gives details about the relative importance of the above two effects, and derives an appropriate expression for the critical quantity of heat in condensing two-phase flow. Although the flow is overall adiabatic, we define the critical quantity of heat (for comparison with previous cases) as

$$q^*_{\text{integral}} \equiv y^*_2 h_{\text{fg}}, \tag{5.50}$$

where y^*_2 is the critical amount of condensation, and is given by the analysis of [16],

$$\left(1 + \frac{y^*_2 h_{\text{fg}}}{c_{\text{pg}} T_{\text{o1}}}\right) \left[\gamma + \frac{\bar{A}}{A_2}(1 - y^*_2)\right]^2 = \frac{\gamma + 1}{2} \frac{\left[\gamma M_1^2 + \frac{\bar{A}}{A_1}\right]^2}{M_1^2 \left[1 + \frac{\gamma - 1}{2} M_1^2\right]}. \tag{5.51}$$

Equation (5.51) can be solved as a cubic equation in y^*_2 (or q^*_{integral}).

Detailed calculations [16] with equation (5.51) show that the effects of mass depletion and (even a small) area variation are quite dramatic, especially when the Mach number is close to unity. Figure 5.20 shows a full numerical, time-marching, solution of the differential equations of motion giving the detailed structure of a condensation shock wave leading to thermal choking. Reference [16] shows that the time-marching results agree with the integral analysis, (5.51). The usually quoted (5.49) underestimates the critical heat by a factor of three in the example calculation presented.

5.5 Shock Waves in Vapour–Droplet Flow with a Carrier Gas

5.5.1 Introduction

A comprehensive treatment of condensation and aerodynamic shock waves when a carrier gas is present is given in [8]. Here, we just quote a few final results from [8].

Consider a gas–liquid droplet mixture. The gas phase is the continuous phase and, in general, consists of a mixture of an inert gas and a condensable vapour. The liquid phase is the discontinuous phase and consists of a poly-dispersed population of spherical droplets of the same chemical species as the

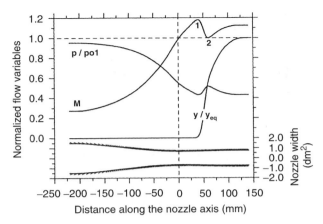

Fig. 5.20. Time-marching solution of a limiting condensation shock in a quasi-one-dimensional convergent–divergent nozzle. (Continuous variation of flow variables leading to thermal choking at point 2)

condensable vapour. The inert gas, vapour and liquid droplets are referred to by subscripts g, v and l, respectively.

If the mass fraction of the inert gas per unit mass of the mixture is denoted by g, the mass of vapour plus liquid per unit mass of mixture is $(1-g)$. The wetness fraction y is then defined as the mass of liquid per unit mass of vapour plus liquid. Therefore, the mass of vapour per unit mass of mixture is $(1-g)(1-y)$ and the total mass of liquid per unit mass of mixture is $(1-g)y$.

The earlier definitions of g and y are very convenient. In the absence of velocity slip between the gas and liquid phases, elemental fluid particles retain their identity in that the total mass of vapour plus liquid contained in the particle remains constant even though condensation or evaporation may occur. Under these conditions, it follows that g remains constant along a particle path line in unsteady flow or along a streamline in steady flow. Furthermore, any generalized set of equations can be converted to the special case by letting the appropriate quantity to its limiting value. For example, for a pure vapour $g = 0$, and y is synonymous with the conventional definition of wetness fraction. The condition $g = 1$ renders the medium to consist of the inert gas only, and $y = 0$ renders the medium to be a mixture of the inert gas and superheated vapour. In passing, note that the specific humidity (defined for mixtures of low vapour content as the ratio of the mass of vapour to the mass of inert gas) is given by $(1-g)(1-y)/g$.

Weighted average values of a few properties are useful in the analysis:

$$R = \frac{gR_g + (1-g)(1-y)R_v}{g + (1-g)(1-y)},$$

$$\bar{R} = gR_g + (1-g)(1-y)R_v,$$

$$c_{\mathrm{p}} = \frac{gc_{\mathrm{pg}} + (1-g)(1-y)c_{\mathrm{pv}}}{g + (1-g)(1-y)},$$

$$\tilde{c}_{\mathrm{p}} = gc_{\mathrm{pg}} + (1-g)c_{\mathrm{pv}},$$

$$\bar{c}_{\mathrm{p}} = gc_{\mathrm{pg}} + (1-g)(1-y)c_{\mathrm{pv}} + (1-g)yc_{\mathrm{l}}.$$

The isentropic exponent for frozen flow, γ_{f}, is given by [8],

$$\gamma_{\mathrm{f}} = \frac{gc_{\mathrm{pg}} + (1-g)(1-y)c_{\mathrm{pv}}}{g(c_{\mathrm{pg}} - R_{\mathrm{g}}) + (1-g)(1-y)(c_{\mathrm{pv}} - R_{\mathrm{v}})}. \qquad (5.52)$$

Accordingly the frozen speed of sound, a_{f}, can be defined by

$$a_{\mathrm{f}} = \sqrt{\gamma_{\mathrm{f}} RT}. \qquad (5.53)$$

If the flow velocity is V, then a frozen Mach number may be defined as $M_{\mathrm{f}} = V/a_{\mathrm{f}}$.

γ_{f}, given by equation (5.52), can be interpreted as usual as $\gamma_{\mathrm{f}} = c_{\mathrm{p}}/(c_{\mathrm{p}}-R)$. The relevant properties are weighted averages according to the composition of the gas phase only (not the mixture). It is important to note that the liquid phase does not take part in the dynamics of the flow process under the assumed conditions and the liquid flow properties do not change. But, with the present notations, different values of y for a fixed g changes the ratio of the mass fractions of the inert gas and the vapour. Hence the wetness fraction y appears in the expression for the frozen exponent γ_{f} in (5.52). In the limits $g = 0$ and $g = 1$, γ_{f} given by (5.52) reduces to the isentropic exponents of the vapour phase, γ_{v}, and of the inert gas, γ_{g}, respectively.

When the liquid droplets are always in complete equilibrium with their own vapour, the laws of equilibrium thermodynamics hold good and any entropy production due to relaxation mechanism vanishes. Conditions of equilibrium are that the velocity and temperature (neglecting surface effects for established droplet sizes) of the gas and liquid phases are equal and also that the partial pressure due to the vapour is equal to the saturation pressure corresponding to its temperature. Hence the vapour pressure changes according to the Clausius-Clapeyron equation,

$$\frac{\mathrm{d}p_{\mathrm{v}}}{p_{\mathrm{v}}} = \frac{h_{\mathrm{fg}}}{R_{\mathrm{v}}T}\frac{\mathrm{d}T}{T}, \qquad (5.54)$$

where, h_{fg}, is the specific enthalpy of evaporation which is a known function of temperature. For particular values of p, T and g, the equilibrium wetness fraction y is fixed. Any arbitrary combination of p, T and g may not satisfy the requirements of two-phase equilibrium. For example, p cannot be less than the saturation pressure at T. Also note that to maintain the equilibrium partial pressure there must be some vapour in the mixture. In other words, at equilibrium, y can never be equal to 1. On the other hand, a mixture of inert gas and superheated vapour (without any liquid) can be at equilibrium.

It is derived in [8] that the equilibrium isentropic index, γ_e, is given by

$$\gamma_e = \frac{1 + \left(\dfrac{R_g}{\bar{R}}\right)\left[\dfrac{g}{(1-g)(1-y)}\right]\left(\dfrac{\bar{c}_p T}{h_{fg}}\right)\left(\dfrac{R_v T}{h_{fg}}\right)}{1 - 2\dfrac{R_v T}{h_{fg}} + \left[\dfrac{1}{(1-g)(1-y)}\right]\left(\dfrac{R_v T}{h_{fg}}\right)\left(\dfrac{\bar{c}_p T}{h_{fg}} - \dfrac{g R_g T}{h_{fg}}\right)} \tag{5.55}$$

In case of pure vapour–droplet flow, $g \to 0$ and γ_e given by (5.55) reduces to the same expression given earlier for pure vapour–droplet flow (5.21). The equilibrium sonic speed in gas–vapour–droplet mixture is simply given by

$$a_e = \sqrt{\gamma_e \bar{R} T}. \tag{5.56}$$

If the flow velocity is V then an equilibrium Mach number, M_e, may be defined such that $M_e = V/a_e$. The equilibrium speed of sound is always less than the frozen speed of sound in any relaxing medium ($a_f > a_e$). Therefore, it follows, that the local equilibrium Mach number at any point in the flow field is higher than the local frozen Mach number ($M_e > M_f$).

If ρ is the density of the gas phase (inert gas + vapour) and ρ_l is the material density of the liquid phase, then the mixture density $\bar{\rho}$ is calculated from

$$\frac{1}{\bar{\rho}} = \frac{g + (1-g)(1-y)}{\rho} + \frac{(1-g)y}{\rho_l}. \tag{5.57}$$

The second term in RHS of (5.57) is usually negligible. By Dalton's law of partial pressure one can further show the following two relations at any state point

$$p = \rho RT = \bar{\rho}\bar{R}T, \tag{5.58}$$

$$\frac{p_v}{p} = \frac{(1-g)(1-y)R_v}{gR_g + (1-g)(1-y)R_v}. \tag{5.59}$$

The specific enthalpy of the gas–vapour–droplet mixture, $\bar{h}$, comprises of the contributions from individual components and is given by

$$\bar{h} = g h_g + (1-g)(1-y)h_v + (1-g)y h_l. \tag{5.60}$$

Equation (5.60) is a general expression valid under non-equilibrium conditions as well, in which case h_l would be calculated at the liquid phase temperature (different from gas phase temperature) and also y would not be the equilibrium wetness fraction.

5.5.2 Jump Conditions Across Normal Shock Waves when both Boundary Conditions are Equilibrium States

We now consider the structure of stationary, finite-amplitude waves in one-dimensional steady flow of a gas–vapour–droplet mixture. Far upstream of the wave (denoted by subscript 1) the flow is assumed to be in the thermodynamic and inertial equilibrium with a prescribed pressure, temperature and mass fraction g. Far downstream of the wave (denoted by subscript 2) a new two-phase equilibrium condition is re-established. The continuity, momentum and energy equations for the two-phase mixture connecting the two end equilibrium states across a normal shock wave take the form:

$$\text{continuity} \quad \bar{\rho}_1 V_1 = \bar{\rho}_2 V_2 \,, \tag{5.61}$$

$$\text{momentum} \quad p_1 + \bar{\rho}_1 V_1^2 = p_2 + \bar{\rho}_2 V_2^2 \tag{5.62}$$

$$\text{energy} \quad \bar{h}_1 + \frac{1}{2} V_1^2 = \bar{h}_2 + \frac{1}{2} V_2^2 \,. \tag{5.63}$$

Equation (5.63) can be written as

$$\tilde{c}_{\mathrm{p}} T_1 - (1-g) y_1 h_{\mathrm{fg}1} + V_1^2/2 = \tilde{c}_{\mathrm{p}} T_2 - (1-g) y_2 h_{\mathrm{fg}2} + V_2^2/2 \,. \tag{5.64}$$

Note that the same value of g applies to the far upstream and far downstream ends and $\tilde{c}_{\mathrm{p}}$ remains constant. Equations (5.54), (5.58) and (5.59) are applicable to both ends of the shock wave as well. The equation set (5.54), (5.58), (5.59), (5.61), (5.62) and (5.64) therefore furnish altogether nine relations between 13 variables and hence can be solved if any four of the variables are prescribed. However, no general analytical solution is possible if all four are specified at the upstream and the equations have to be solved by an iterative numerical scheme. For moderate strengths of shock waves, it is, however, possible to obtain an approximate analytical solution and this has been derived later. Later it is also shown that if instead of specifying the upstream velocity V_1, the temperature ratio across the shock is prescribed a completely general analytical solution of the earlier-mentioned equation set is possible.

Approximate Rankine–Hugoniot Relations (for Specified p_1, T_1, V_1 and g)

The ratio of the different flow variables between the two end states of a normal shock wave in an ideal gas can be expressed as functions of the upstream Mach number. They are generally referred to as the Rankine–Hugoniot relations. In the case of a simpler relaxing medium (e.g. solid-particle-laden gas), it can be shown that these relations remain identical if the upstream equilibrium Mach number $(M_{\mathrm{e}})_1$ is used instead of the frozen Mach number $(M_{\mathrm{f}})_1$. These relations are exact and hold unconditionally. Derivation of similar relations in gas–vapour–droplet flow involves approximations and the derived relations are of conditional applicability. Difficulties arise mainly because of the mass transfer between the vapour and the liquid phase and also because the partial pressure and temperature of the vapour at equilibrium with liquid droplets are not independent of each other but are connected via the Clausius-Clapeyron equation $[p_{\mathrm{v}} = p_{\mathrm{s}}(T)]$.

If the shock wave is weak so that the entropy change is small, one can show [8] that the approximate Rankine–Hugoniot relations are given by (5.27) and (5.28). The equilibrium Mach number is calculated by $(M_e)_1 = V_1/(a_e)_1 = V_1/\sqrt{\gamma_e p_1/\bar{\rho}_1}$, where γ_e is given by (5.55). Although not exact, (5.27) and (5.28) have two advantages (1) all shock relations could be explicitly written in terms of the upstream parameters only, and (2) the form of the equations is similar to the well-known Rankine–Hugoniot relations for ideal gas.

It should be noted that (5.27) and (5.28) are valid for both partly and fully dispersed shock waves. Partly dispersed shock waves arise when the upstream frozen Mach number is greater than unity $[(M_f)_1 > 1$ implies $(M_e)_1 > 1]$. They are characterized by an almost discontinuous wave front, dominated by viscous dissipation and thermal conduction, followed by a continuous relaxation zone. The jump conditions across the discontinuous wave front (termed frozen shock) is given by the classical Rankine–Hugoniot relations based on M_{f1}. Equations (5.27) and (5.28), on the other hand give the overall changes in flow properties across the entire shock wave (frozen shock + relaxation zone). Fully dispersed waves do not involve any frozen discontinuity and give rise to continuous variation of flow properties from one equilibrium state to another. They may appear in the flow field if the upstream velocity is such that $(M_f)_1 \leq 1$, but $(M_e)_1 > 1$. Equations (5.27) and (5.28) specify the jump conditions across fully dispersed waves as well. This point is schematically presented in Fig. 5.15.

An Exact Jump Condition (For Specified p_1, T_1, T_2 and g)

If instead of specifying the upstream velocity V_1, the temperature ratio T_2/T_1 across the shock wave is treated as an independent variable, an exact solution of (5.61), (5.62) and (5.64) can be formulated (without requiring any approximate energy equation such as that used in "Approximate Rankine–Hugohiot Relations (For Specified p_1, T_1, V_1 and g")). The solution is

$$1 - y_2 = \frac{B + \sqrt{B^2 + 4AC}}{2A} \tag{5.65}$$

$$\text{where,} \quad A = 2(1-g)h_{fg2} - \frac{p_{v2} - p_1}{p_{v2}}T_2(1-g)R_v,$$

$$B = 2(1-g)(h_{fg2} - y_1 h_{fg1}) + 2\tilde{c}_p(T_1 - T_2) + gR_gT_2 + \frac{p_{v2} - p_1}{\bar{\rho}_1},$$

$$C = \frac{gR_g}{(1-g)R_v}\frac{p_{v2}}{\bar{\rho}_1}.$$

The vapour pressure at 2 is known from the relation $p_{v2} = p_s(T_2)$. The dimensions of A, B and C in (5.65) are that of energy per unit mass. A and C are positive. Therefore the requirement that $0 \leq y_2 \leq 1$ eliminates the other root in (5.65). Equation (5.65) reduces to (5.30) in the special case of pure vapour–droplet mixture ($g = 0$, $C = 0$).

Once y_2 is found, p_2 may be calculated from (5.59) and $\bar{\rho}_2$ from (5.58). The flow velocities V_1 and V_2 may then be determined from (5.61) and (5.62).

It has been shown in [8] that the pressure ratios calculated by the approximate relation, (5.27), compare reasonably well with the values calculated from the exact formulation, (5.65) and (5.59), until the mixture is close to complete evaporation.

5.5.3 Limiting Wetness Fraction

In the previous section it was assumed that downstream of the shock wave the medium is an equilibrium mixture of the gas phase and the liquid droplets. However, the liquid droplets evaporate in a fully dispersed wave or in the relaxation zone of a partly dispersed wave. Hence if the strength of the shock wave is substantial, the whole of the liquid phase may evaporate. This would render the medium at the downstream end as a single-phase mixture of the inert gas and superheated vapour. The vapour pressure is no longer restrained to the saturation pressure corresponding to the mixture temperature and none of the equations derived in "Approximate Rankine–Hugoniot Relations (For Specified P_1, T_1, V_1 and g)" and "An Exact Solution (For specified P_1, T_1, V_1 and g)" would be valid. Note, however, that the value of g does not change as a result of the complete evaporation of the liquid phase.

The limiting wetness fraction, $y_{1,\mathrm{lim}}$, can be determined by letting y_2 to zero in (5.65). If the upstream wetness fraction is less than this limiting value corresponding to a shock of particular strength, complete evaporation takes place in the dispersed wave. Thus

$$
y_{1,\mathrm{lim}} =
$$

$$
\frac{\frac{T_2}{T_1}\left(\frac{p_{v2}-p_1}{p_{v2}}+\frac{gR_\mathrm{g}-2\bar{c}_\mathrm{p}}{(1-g)R_\mathrm{v}}\right)+\left(1+\frac{g}{1-g}\frac{R_\mathrm{g}}{R_\mathrm{v}}\right)\left(\frac{p_{v2}-p_1}{p_1}+\frac{g}{1-g}\frac{R_\mathrm{g}}{R_\mathrm{v}}\frac{p_{v2}}{p_1}\right)+\frac{2\bar{c}_\mathrm{p}}{(1-g)R_\mathrm{v}}}{\frac{2h_\mathrm{fg1}}{R_\mathrm{v}T_1}+\frac{p_{v2}-p_1}{p_1}+\frac{g}{1-g}\frac{R_\mathrm{g}}{R_\mathrm{v}}\frac{p_{v2}}{p_1}}. \tag{5.66}
$$

p_{v2} in (5.66) is still given by the saturation pressure at T_2. Equation (5.66) is an implicit relation because, for $y_1 = y_{1,\mathrm{lim}}$, only two parameters out of p_1, T_1 and g can be specified independently. Equation (5.66) reduces to (5.34) in the special case of pure vapour–droplet mixture ($g = 0$).

5.5.4 Jump Conditions Across Shock Waves with Complete Evaporation of Liquid Phase ($y_2 = 0$)

If the amount of liquid phase in the mixture at the upstream of the shock wave is less than that given by (5.66) for a shock wave of a particular strength, then complete evaporation takes place. Upstream of the dispersed wave the medium is an equilibrium mixture of the gas, vapour and liquid, whereas downstream it is a mixture of the gas and superheated vapour. The vapour pressure and the temperature at point 2 are then independent variables (i.e. they are not connected by the Clausius-Clapeyron equation) and, therefore, neither the approximate jump relations (5.27) and (5.28) nor the exact solution (5.65) are applicable. Mathematically, the condition $y_2 = 0$ implies $p_{v2} < p_s(T_2)$. The conservation equations (5.61)–(5.63) are general and applicable in this case as well. Care must be exercised, however, in determining the mixture enthalpy, density and pressure.

An Exact Jump Condition (For Specified p_1, T_1, V_1 and g; $y_2 = 0$)

This is the conventional case when all upstream parameters are known and all the downstream parameters are to be predicted. In this case we can formulate an exact analytical solution of (5.61), (5.62) and (5.64). The solution [8] is

$$p_2 = \frac{E + \sqrt{E^2 - 4DF}}{2D}, \qquad (5.67)$$

where:

$$D = 2\tilde{c}_p/\tilde{R} - 1,$$

$$E = 2F_1(\tilde{c}_p/\tilde{R} - 1),$$

$$F = 2H_1\bar{\rho}_1^2V_1^2 - F_1^2,$$

$$F_1 = p_1 + \bar{\rho}_1V_1^2,$$

$$H_1 = \tilde{c}_pT_1 - (1-g)y_1h_{fg1} + V_1^2/2.$$

The pressure increases across an aerodynamic shock wave and, hence, $p_2 > p_1$. This condition eliminates the other root in (5.67). Once p_2 is known, V_2 and T_2 can be calculated. Note that (5.67) should reduce to the classical Rankine–Hugoniot solution if $y_1 = 0$ (in other words, if the medium just consists of the inert gas and the superheated vapour - a mixture of ideal gases under the assumptions made).

The pressure ratio p_2/p_1 calculated for an air–water mixture by (5.67) is plotted as a function of upstream frozen Mach number (V_1/a_{f1}) as the chain-dotted line in Fig. 5.21. Also included in the figure is the prediction of (5.65) and (5.59) giving results when two-phase equilibrium exists at both ends of a shock wave. The solutions of the two equations, (5.65) and (5.67), coincide at the point of limiting upstream wetness fraction (5.66). The dotted line in the same figure gives the pressure ratio across shock waves in a dry mixture (with the same value of g as in the wet mixture). The rise in pressure in the wet mixture is more than that in the dry mixture because further increase in pressure takes place in the relaxation zone following the frozen shock. It is to be noted from Fig. 5.21 that steady shock waves exist in a wet mixture even for $M_{f1} \leq 1$. These waves are termed fully dispersed waves. Both (5.27) and (5.65) apply across such waves.

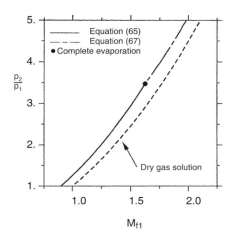

Fig. 5.21. Analytical solutions for pressure ratios across aerodynamic shock waves in air–water mixture as a function of upstream frozen Mach number ($p_1 = 1\,\mathrm{bar}$, $T = 293\,\mathrm{K}$, $g = 0.95$)

An Exact Jump Condition (For Specified p_1, T_1, p_2 and g; $y_2 = 0$)

If, instead of the upstream velocity V_1, p_2 is specified, then also an exact solution of (5.61), (5.62) and (5.64) can be obtained. The solution is [8]

$$T_2 = \frac{2\tilde{c}_\mathrm{p} T_1 - 2(1-g)y_1 h_{\mathrm{fg}1} + \frac{p_2 - p_1}{\bar{\rho}_1}}{2\tilde{c}_\mathrm{p} - \tilde{R}\frac{p_2 - p_1}{p_2}}, \tag{5.68}$$

$$V_1^2 = \frac{p_2 - p_1}{\bar{\rho}_1}\left[1 - \bar{\rho}_1\frac{\tilde{R}T_2}{p_2}\right]^{-1}. \tag{5.69}$$

5.6 Shock Waves in a Solid Particle-Laden Gas

5.6.1 Equilibrium Gas Dynamics for Gas-Particle Mixture

If there is no interphase mass transfer, the size of the individual particles does not change. A corollary of this would be that in steady, equilibrium flow (no slip) the mass fraction y remains constant along streamlines. At equilibrium, the velocity and temperature of the two phases are the same. We assume that the volume fraction and partial pressure of the particles are negligible ([23] shows how the equations are modified if the particle volume is not insignificant). The gas phase is assumed to be governed by perfect gas laws. The mixture enthalpy, $\bar{h}$, can be calculated from

$$\mathrm{d}\bar{h} = [(1-y)c_\mathrm{pg} + yc_\mathrm{s}]\mathrm{d}T. \tag{5.70}$$

The subscript s denotes variables for the solid particles. c_s is the specific heat capacity of the solid particles and T is the mixture temperature. Thus a solid-particle-laden gas at equilibrium can be described by perfect gas laws, if mixture specific heat, $\bar{c}$ $[\bar{c} = (1-y)c_{pg} + yc_s]$ and mixture specific gas constant, $\bar{R}$ $[\bar{R} = (1-y)R_g]$ are used. Mixture density is given by $\bar{\rho} = \rho_g/(1-y)$. Simple manipulation of equations gives

$$d\left(\frac{p}{\bar{\rho}}\right) = \frac{\gamma_e - 1}{\gamma_e}\bar{c}\,dT,\tag{5.71}$$

$$\text{where}\quad \gamma_e = \frac{(1-y)c_{pg} + yc_s}{(1-y)(c_{pg} - R_g) + yc_s}.\tag{5.72}$$

For a particular mass fraction, γ_e remains *constant* and represents the equilibrium isentropic exponent of the gas–particle mixture. Hence, for example, $p/\bar{\rho}^{\gamma_e}$ remains constant in an isentropic flow. Moreover, (5.71) is a general identity and hence it may be substituted in the energy equation and then integrated to give

$$\frac{\gamma_e}{\gamma_e - 1}\frac{p_1}{\bar{\rho}_1} + \frac{V_1^2}{2} = \frac{\gamma_e}{\gamma_e - 1}\frac{p_2}{\bar{\rho}_2} + \frac{V_2^2}{2}.\tag{5.73}$$

Equation (5.73) is *exact,* generally valid for *any* steady, adiabatic flow as long as state points 1 and 2 are equilibrium states, and similar in form to its counterpart in single-phase, ideal gas flow. It is therefore applicable even across a shock wave. Thus the familiar Rankine–Hugoniot equations for ideal gas also hold unconditionally for solid-particle-laden gas provided γ_e and equilibrium speed of sound are used. The equilibrium sound speed is calculated from

$$a_e = \sqrt{\frac{\gamma_e p}{\bar{\rho}}}.\tag{5.74}$$

Thus, in essence, the equilibrium flow and shock wave phenomena in a mixture of a perfect gas and solid particles are governed by perfect gas laws. Contrast this with the situation for pure vapour–droplet flow in which mass transfer is present, (5.71) and (5.73) are only approximately valid (when the change in entropy is small), and γ_e given by (5.21) varies slightly within the flow field.

5.6.2 Relaxation Phenomena

The relaxation phenomena are simpler in gas–particle mixtures without mass transfer, as compared to vapour–droplet flow, and only interphase momentum and heat transfer are considered. Under the assumptions, (5.2) for velocity slip equilibration and (5.3) for the inertial relaxation time are also applicable to gas–particle mixture (solid properties are used in the expressions instead of the corresponding liquid properties).

The thermal relaxation time can be determined as follows. In the case of a solid-particle-laden gas, the heat transfer to a single particle of mass m and radius r can be written as:

$$\left(\frac{4}{3}\pi r^3 \rho_s\right) c_s \frac{dT_{sp}}{dt} = 4\pi r^2 \frac{\lambda_g}{r}(T_g - T_{sp}), \tag{5.75}$$

$$\text{or,}\quad \frac{d(T_g - T_{sp})}{dt} + \frac{3\lambda_g}{r^2 \rho_s c_s}T_g - T_{sp} = \frac{dT_g}{dt}. \tag{5.76}$$

T_{sp} is used to denote the temperature of solid particles, instead of T_s, in order to avoid any potential confusion with the symbol for saturation temperature used earlier in the context of vapour–droplet flow. Equation (5.76) is deduced by using a lumped heat capacity model for the solid particle and by assuming steady state heat conduction to a sphere from a surrounding gas. Now if we assume for simplicity that the gas temperature T_g does not change during the relaxation of the solid particles then (5.76) can be expressed in the archetypal form of a general relaxation process given by (5.1) with time constant τ_{Ts}:

$$\tau_{Ts} = \frac{r^2 \rho_s c_s}{3\lambda_g}. \tag{5.77}$$

Thus (5.77) becomes physically meaningful (not just a convenient, shorthand form) in the case of constant gas temperature and denotes the time scale with which any difference in the temperatures of the two phases ($\Delta T = T_g - T_{sp}$) is equilibrated.

Equations (5.5), and (5.12)–(5.16) are equally applicable to a solid particle laden gas. The main difference between the equations for vapour–droplet and gas–particle mixtures lies in the facts that there is no interphase mass transfer in the latter case and that the pressure and temperature of a gas–particle mixture, even at equilibrium, are independent of each other. (The pressure and temperature of an equilibrium, wet vapour are, on the other hand, related by the Clausius-Clapeyron equation.) A similar analysis would lead to $\tau_{Ts} = \left(r^2 \rho_s c_s\right)\left(1 + 4.5Kn/Pr\right)/(3\lambda_g)$, which reduces to (5.77) in the continuum regime (large particles, $Kn \to 0$).

Equations (5.75)–(5.77) were written assuming the only mode of heat transfer between the particles and the gas is steady state conduction. However, the heat transfer coefficient may depend on the particle Reynolds number Re, particularly when the slip velocity is high. Such dependence is normally modelled by empirical correlations. For example, the widely used Knudsen and Katz [25] model gives the correlation for the Nusselt number Nu by $Nu = 2 + 0.6\sqrt[3]{Pr}\sqrt[2]{Re}$. $Nu = 2$ corresponds to the steady state conduction, as assumed in (5.75)–(5.77). Calculations by Rudinger [22] showed that the Reynolds number correction had little effect on the computed structure of a partly dispersed wave even when the upstream Mach number was 1.7. Similar conclusions were made by Kriebel [24].

Various correlations for the drag coefficient and Nusselt number are available in the literature (e.g. [25–27]). These are dependent on various parameters such as the slip Reynolds number, Mach number, etc. Any such appropriate expressions can be used in practical numerical calculations, being mindful of their range of validity. It is useful to note that τ_{Ts} and Nu are related by,

$$\tau_{\mathrm{Ts}} = \frac{2}{3} \frac{r^2 \rho_{\mathrm{s}} c_{\mathrm{s}}}{\lambda_{\mathrm{g}} Nu}. \qquad (5.78)$$

5.6.3 Comparative Magnitudes of Relaxation Times and Structure of Shock Waves

We have seen that two relaxation times τ_{Is} and τ_{Ts} characterize a solid particle-laden gas and they are given by (5.3) and (5.78), respectively. Taking the ratio of (5.3) and (5.78) for small Reynolds number and small Knudsen number,

$$\frac{\tau_{\mathrm{Ts}}}{\tau_{\mathrm{Is}}} = 1.5 \, Pr \frac{c_{\mathrm{s}}}{c_{\mathrm{pg}}}, \qquad (5.79)$$

where Pr is the Prandtl number of the vapour phase, c_{s} and c_{pg} are the specific heat capacities of the solid and the gas, respectively. Equation (5.79) shows that the two relaxation times are of comparable magnitude unless the ratio of the specific heats is very high (or low). Some typical examples may be found from the different cases discussed by Rudinger [22]. For glass spheres suspended in air, $Pr = 0.7$, $c_{\mathrm{s}}/c_{\mathrm{pg}} = 1.125$ and hence $\tau_{\mathrm{Ts}}/\tau_{\mathrm{Is}} = 1.2$. For water droplets suspended in argon, $Pr = 0.67$, $c_{\mathrm{s}}/c_{\mathrm{pg}} = 8$ and hence $\tau_{\mathrm{Ts}}/\tau_{\mathrm{Is}} = 8$. On the other hand, Kriebel [24] gives the property values typical of a mixture in a solid-propellant rocket motor as $Pr = 0.57$, $c_{\mathrm{s}}/c_{\mathrm{pg}} = 0.57$. Substituting in (5.79) gives, $\tau_{\mathrm{Ts}}/\tau_{\mathrm{Is}} = 0.5$. Thus it may be expected in this case that temperature relaxation will be over before velocity slip relaxes completely (as is borne out by Kriebel's numerical calculations of shock wave structures.) It may be recalled from "Comparative Magnitudes of Different Relaxation Times" that in pure vapour droplet flow, for $Pr \approx 1$ and $Re \to 0$, $\tau_{\mathrm{I}}/\tau_{\mathrm{T}} = 2y/3(1-y)$ and hence usually $\tau_{\mathrm{I}} \ll \tau_{\mathrm{T}}$. Hence, the inertial relaxation zone is much shorter than vapour thermal relaxation zone in pure vapour–droplet flow, which can clearly be seen in Fig. 5.10.

5.6.4 Jump Conditions

It has been shown in Sect. 5.6.1 that a gas–particle mixture at equilibrium may be modelled as a modified gas. Hence the Rankine–Hugoniot relations for a single-phase perfect gas are applicable to a mixture of solid particles and a perfect gas. The fluid is an equilibrium two-phase gas–particle mixture both at upstream and at downstream of the shock wave. The relations are

$$\frac{p_2}{p_1} = \frac{2\gamma_{\mathrm{e}}}{\gamma_{\mathrm{e}}+1} (M_{\mathrm{e}})_1^2 - \frac{\gamma_{\mathrm{e}}-1}{\gamma_{\mathrm{e}}+1}, \qquad (5.80)$$

$$\frac{V_2}{V_1} = \frac{(\gamma_e - 1)(M_e)_1^2 + 2}{(\gamma_e + 1)(M_e)_1^2}, \tag{5.81}$$

where, γ_e is given by (5.72), and the equilibrium speed of sound is calculated from (5.74). It should be noted that (5.80) and (5.81) are valid for both partly and fully dispersed shock waves. Partly dispersed shock waves appear when $M_{f1} > 1$ (which also implies that $(M_e)_1 > 1$). The overall changes across such a shock wave are given by the earlier equations while the changes just across the discontinuous wave-front are given by the classical Rankine–Hugoniot equation based on M_{f1}. If $M_{f1} < 1$ but $(M_e)_1 > 1$, then a fully-dispersed shock wave may appear in the flow-field. The earlier equations again give the overall changes across fully dispersed waves. Figure 5.15 schematically depicts the point.

For a perfect gas–particle mixture, (5.80) and (5.81) are exact. On the other hand, analogous relations for vapour–droplet flows are approximate (See "Approximate Rankine–Hugoniot Relations for Specified p_1, v_1 and y_1" and "Approximate Rankine–Hugoniot Relations (for Specified p_1, T_1, V_1 and g")).

5.7 Interpretation of Total Pressure and Total Temperature

In this section we discuss briefly some interesting effects of the non-equilibrium, interphase transfer mechanisms in a stagnation process in multiphase flow. This analysis gives insight into a physically appropriate construction of the concepts of total pressure and total temperature in multiphase mixtures.

Pitot measurements are often used for inferring velocity or loss (entropy generation) in multiphase mixtures. In single phase fluids, the fluid is assumed to be brought to rest at the mouth of the Pitot tube isentropically. Hence flow Mach number and entropy generation (in steady, adiabatic flow) are uniquely determined by the total pressure measured by a Pitot tube, together with an independent measurement of the static pressure. (In supersonic flow in an ideal gas, application of Rankine–Hugoniot equations across the detached shock wave in front of a Pitot tube retains the utility of Pitot measurements for deducing flow Mach number and entropy generation). The interpretation of total pressure in a multiphase mixture, however, requires careful considerations [18, 19]. Similar non-equilibrium effects on the interpretation of total temperatures are discussed in [18].

The solid particles or the liquid droplets respond to changes in temperature, velocity, etc. of the gas phase through interphase exchanges of mass, momentum and energy. These are essentially rate processes and hence significant departures from equilibrium can take place if the rate of change of external conditions, imposed by the deceleration in the stagnating flow, is comparable to the internal time scales. Thus, for example, if the size of the liquid droplets or the solid particles is very small, then inertial and thermodynamic equilibrium between the two phases are maintained always, and a Pitot tube would measure the equilibrium total pressure, p_{oe}. On the other hand, if the size of the droplets or the particles is very large, all interphase transfer processes remain essentially frozen. The Pitot tube records the pressure which it would have recorded if the vapour phase alone was brought to rest from the same velocity. The total pressure in this case is termed the frozen total pressure, p_{of}.

Analytical expressions for p_{oe} and p_{of}, both in vapour–droplet and gas–particle flow, are derived in [18]. In subsonic flow p_{of} is given by

$$\frac{p_{of}}{p_\infty} = \left[1 + \frac{\gamma - 1}{2} M_{f\infty}^2\right]^{\gamma/(\gamma-1)} , \qquad (5.82)$$

where $M_{f\infty}$ is the unperturbed frozen Mach number far upstream of the measuring point and p_∞ is the static pressure. p_{oe} is given by

$$\frac{p_{oe}}{p_\infty} = \left[1 + \frac{\gamma_e - 1}{2} M_{e\infty}^2\right]^{\gamma_e/(\gamma_e-1)} , \qquad (5.83)$$

where $M_{e\infty}$ is the unperturbed equilibrium Mach number far upstream of the measuring point. The equilibrium isentropic exponent, γ_e, in pure vapour–droplet is given by equation (5.21), and that in solid-particle-laden gas is given by (5.72). Equation (5.83) is exact for a gas-particle mixture but is only *approximate* for a vapour–droplet mixture, and applies only if the stagnation process does not result in complete evaporation of the liquid phase. If a frozen shock wave is present (as in supersonic flow) and/or complete evaporation takes place, different equations are to be used [18].

As an example, consider low-pressure wet steam with a typical wetness fraction of 10% and at a Mach number 1.5. Calculations show that $p_{of}/p_\infty = 3.3$ and $p_{oe}/p_\infty = 3.79$, where p_∞ is the static pressure. Therefore, in this particular example, the equilibrium total pressure is about 15% higher than the frozen total pressure.

It is expected that for intermediate sizes of the droplets or particles, the pressure recorded by the probe would neither be the equilibrium nor the frozen value. The imposed deceleration in front of the Pitot tube would cause the two-phase mixture to deviate from equilibrium conditions, both inertially and thermodynamically. The deceleration process consequently ceases to be isentropic, as non-equilibrium exchanges of mass, momentum and energy between the two phases create entropy.

For a proper solution of the real flow field around the mouth of a Pitot tube, one therefore has to solve multidimensional (at least two-dimensional axisymmetric) conservation equations with viscous, and thermal and inertial non-equilibrium effects. It is important to incorporate the inertial non-equilibrium effects, i.e. to allow a velocity slip between the two phases. It is so, not only because the inertial effects are themselves significant but also because restraining the two phases to travel at the same velocity has serious implications for the thermal equilibration process. For example, a fluid particle moving along the stagnation streamline takes infinite time to reach the stagnation point. Therefore, in the absence of any velocity slip, the vapour–droplet mixture would have long enough time to come to thermal equilibrium, irrespective of the magnitude of the thermal relaxation time. In reality, however, inertia may cause large droplets to arrive at the mouth of the Pitot tube without significant interphase mass and energy transfer (the droplets would finally hit the back of the probe without influencing the pressure).

Most available computational schemes for wet steam, however, neglect any velocity slip between the phases. It is also quite likely that the numerical entropy generation by a CFD code would distort the entropy generation due to relaxation processes. Hence Guha [18] presented a simple, quasi-one-dimensional model in which some plausible assumptions about the variation of the gas phase velocity are made. Step-by-step numerical integration of the conservation equations for a multiphase mixture, between a far-upstream point and the stagnation point, then determines the pressure that would have been recorded by a measuring device. The same calculation procedure *simultaneously* gives the temperature attained at the measuring point, giving the total temperature. The principles of calculation are not very far from the numerical solution of shock waves described in Sect. 5.2.

Guha [18] considered a large number of two-phase mixtures, both gas–particle and vapour–droplet, at subsonic as well as supersonic velocities for many different sizes of the droplets (or particles). In the supersonic case a detached frozen shock wave stands in front of the Pitot tube. The relaxation mechanisms in a gas–particle mixture are different from those in a vapour–droplet flow. Despite all these complexities and differences, it was possible with proper non-dimensionalization of flow parameters to adopt a universal plot, within acceptable tolerance, of non-dimensional total pressure, R_p, versus Stokes number, St (which is a non-dimensional representation of particle inertia). St and R_p are defined by [18]

$$St \equiv \frac{\tau_I V_\infty}{L}, \tag{5.84}$$

$$R_p \equiv \frac{p_o - p_{of}}{p_{oe} - p_{of}}, \tag{5.85}$$

where, V_∞ is the unperturbed velocity of the two-phase mixture far upstream of the measuring device, p_o is the pressure attained at the measuring point under non-equilibrium conditions (the total pressure which is measured) and L is a characteristic length (in subsonic flow L is related to the Pitot diameter, in supersonic flow L is related to the distance between the frozen shock wave and the Pitot mouth). Larger droplets or particles correspond to higher St.

Figure 5.22 shows the variation of R_p with St, which may be adopted as the Pitot correction curve usable at a wide range of subsonic and supersonic

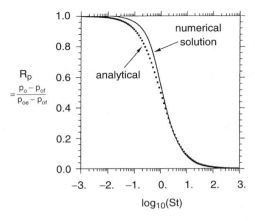

Fig. 5.22. Near-universal plot of R_p versus St: comparison of a simple analytical theory with numerical solutions

Mach numbers and for any two-phase flow (vapour–droplet or gas–particle mixtures). The variation is monotonic. *It should be noted that the denominator in the expression for R_p (5.85) is calculated using the equilibrium gasdynamic equations, whereas the numerator is calculated using non-equilibrium equations. That the value of R_p, shown in Fig. 5.22, indeed tends to unity and zero in the appropriate limits of St, demonstrates independent theoretical consistency of the calculation schemes.*

The calculations in [18] are, however, numerical in nature. In [19], an analytical theory for determining total pressure under non-equilibrium conditions is presented. The analytical theory is simple and is amenable to direct physical interpretation. The theory shows that

$$R_\mathrm{p} = \frac{1}{1 + St}. \tag{5.86}$$

The predicted total pressure correctly reduces to the frozen total pressure in the limit of large Stokes number (large particles) and to the equilibrium total pressure in the limit of small Stokes number (small particles). Maximum dependence of the total pressure on Stokes number is observed when the Stokes number is of the order unity. Equation (5.86) is also plotted in Fig. 5.22 for comparison. Under non-equilibrium conditions for intermediate St, the prediction of this equation compares well with results from full numerical solution of the gas dynamic equations for multiphase mixtures.

The associated problem of interpreting total temperature, and the relation between measured total pressure and entropy production in multiphase flow are discussed at length in [18]. It is shown there that the rate of entropy production in a multiphase mixture is maximum when the Stokes number is of the order unity (in accordance with "Introduction" in "Relaxation phenomena"), and a reduction in measured total pressure is not unequivocally related to a rise in entropy (as it is in steady, adiabatic flow of *single-phase* fluids).

It is shown earlier that the equilibrium total pressure is always higher than the frozen total pressure ($p_{oe} > p_{of}$). The same conclusions do not apply universally in the case of total temperature. The relation between frozen total temperature (T_{of}) and the equilibrium total temperature (T_{oe}) in a gas–particle mixture can be found from the energy equation [18]

$$\frac{T_{of} - T_\infty}{T_{oe} - T_\infty} = 1 + y_\infty(\delta - 1), \tag{5.87}$$

where, $\delta \equiv c_s/c_{pg}$. Equation (5.87) clearly shows that $T_{oe} > T_{of}$ if $\delta < 1$, whereas $T_{oe} < T_{of}$ if $\delta > 1$.

The equilibrium total temperature, T_{oe}, in vapour–droplet flow is always (whether or not complete evaporation takes place) less than the frozen total temperature, T_{of}. This may be physically explained by considering the fact that evaporation of the liquid phase takes place in a compression process. The latent enthalpy of evaporation is supplied by a reduction in the internal energy of the vapour phase, thereby reducing the temperature. Details may be found in [18].

Figure 5.23 plots the rise in mixture entropy, as mixtures of air and solid particles are decelerated by a measuring probe from their far upstream velocity to rest. Four different solid particles (hypothetical) with $\delta \equiv c_s/c_{pg} = 0.1, 0.8, 1.2$ and 4 are considered and the calculations are done for two Mach numbers. For the subsonic case ($M_{f\infty} = 0.8$), Fig. 5.23 shows that the rise in entropy is indeed maximum when $St \sim 1$, and is almost zero in the frozen and equilibrium limits. (Recall from Fig. 5.22 that the total pressures are different

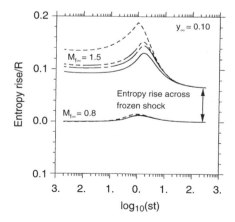

Fig. 5.23. Entropy rise versus Stokes number in air with different solid particles (———— $\delta = 0.1$, _ ———— _ $\delta = 0.8$, _ _ _ ———— $\delta = 1.2$, _ _ _ _ $\delta = 4.0$)

in these limits.) At $M_{f\infty} = 1.5$, the entropy rise is again maximum close to $St \sim 1$, but it has a finite value both at $St \to 0$ and at $St \to \infty$. The rise in entropy in the limit $St \to \infty$ is simply that across the frozen shock. (Since the same frozen shock is involved in all cases because the same $M_{f\infty}$ is used, this increase in entropy is the same for all four mixtures considered.) The rise in entropy in the limit $St \to 0$ is, however, different for different mixtures (it depends on the isentropic index of the *mixture* and hence on δ). However, it is shown in [12] that if the particles come to equilibrium downstream of a frozen shock wave, then the entropy rise (across the shock plus the relaxation zone) is not dependent on the particle size (and hence on the relaxation times) but is determined completely by Rankine–Hugoniot equations for two-phase flow. This fact is reflected in the straight, horizontal portions of the curves (at $M_{f\infty} = 1.5$) in Fig. 5.23 in the limit $St \to 0$.

The fact that the total pressure decreases monotonically from p_{oe} to p_{of} as St changes form 0 to ∞, whereas the entropy rise is zero (or low) at both limits and has a maxima when $St \sim 1$, demands care while interpreting total pressure in multiphase flow.

References

1. Becker, E.: Relaxation effects in gas dynamics. Aeronaut. J. **74**, 736–748 (1970)
2. Vincenti, W.G., Kruger, C.H.: Introduction to Physical Gas dynamics, Wiley, New York Chapter 8, (1965)
3. Guha, A.: Two-phase flows with phase transition. In: VKI Lecture Series 1995-06 ISSN 0377-8312, Belgium, pp. 1–110, von Karman Institute for Fluid Dynamics, Belgium, (1995)
4. Marble, F.E.: Some gas dynamic problems in the flow of condensing vapours. Astronautica Acta, **14**, 585–614 (1969)
5. Gyarmathy, G.: The spherical droplet in gaseous carrier streams: review and synthesis. Handbook of Chemistry and Physics, 62nd edition, 1981–1982. Multiphase Science and Technology 1, pp. 99–279, McGraw-Hill, New York (1982)
6. van Dongen, M.E.H.: Shock wave interaction with liquid gas suspensions. In: Ben-dor, Igra and Elperin (eds.). Handbook of Shock Waves [ISBN 0120864304], Academic, New York pp. 747–781 (2000)
7. Fuchs, N.A.: Evaporation and Droplet Growth in Gaseous Media. Translated by Pratt, J.M., Pergamon, New York (1959)
8. Guha, A.: A unified theory of aerodynamic and condensation shock waves in vapour-droplet flows with or without a carrier gas. Phys. Fluids, **6**(5), 1893–1913 (1994)
9. Guha, A., Young, J.B.: Stationary and moving normal shock waves in wet steam. In: Meier, G.E.A., Thompson, P.A. (eds.), Adiabatic Waves in Liquid-Vapour Systems [ISBN 3540502033], Springer, Berlin Heidelberg New York pp. 159–170 (1990)
10. Young, J.B., Guha, A.: Normal shock wave structure in two-phase vapour-droplet flows. J. Fluid Mech. **228**, 243–274 (1991)
11. Guha, A.: Structure of partly dispersed normal shock waves in vapour-droplet flows. Phys. Fluids A, **4**(7), 1566–1578 (1992)

12. Guha, A.: Jump conditions across normal shock waves in pure vapour-droplet flows. J. Fluid Mech. **241**, 349–369 (1992)

13. Guha, A.: The physics of relaxation processes and of stationary and non-stationary shock waves in vapour-droplet flows. In: Reizes, J.A. (ed.), *Transport Phenomena in Heat and Mass Transfer* [ISBN 0444898514], Elsevier, Amsterdam pp. 1404–1417 (1992)

14. Guha, A., Young, J.B.: The effect of flow unsteadiness on the homogeneous nucleation of water droplets in steam turbines. Philos. Trans. R. Soc. **349**, 445–472 (1994)

15. Guha, A., Young, J.B.: Time-marching prediction of unsteady condensation phenomena due to supercritical heat addition. In: *Turbomachinery: Latest Developments in a Changing Scene* [ISBN 0852987617], Inst. Mech. Engrs. London, pp. 167–177 (1991)

16. Guha, A.: Thermal choking due to non-equilibrium condensation. ASME J. Fluids Eng. **116**, 599–604 (1994)

17. Guha, A.: Computation, analysis and theory of two-phase flows. Aeronaut. J. **102** (1012), 71–82 (1998)

18. Guha, A.: A unified theory for the interpretation of total pressure and temperature in two-phase flows at subsonic and supersonic speeds, Proc. R. Soc. Ser. A, **454**, 671–695 (1998)

19. Guha, A.: A simple analytical theory for interpreting measured total pressure in multiphase flows. ASME J. Fluids Eng. **120**, 385–389 (1998)

20. Petr, V.: Variation of sound velocity in wet steam. Wet Steam 4, Conference, University of Warwick, Inst. Mech. Eng., London, pp. 17–20 (1973)

21. Marble, F.E.: Dynamics of dusty gases. Ann. Rev. Fluid Mech. **2**, 397–446 (1970)

22. Rudinger, G.: Some properties of shock relaxation in gas flows carrying small particles. Phys. Fluids, **7**(5), 658–663 (1964)

23. Rudinger, G.: Some effects of finite particle volume on the dynamics of gas-particle mixtures. AIAA J. **3**(7), pp. 1217–1222 (1965)

24. Kriebel, A.R.: Analysis of normal shock waves in particle-laden gas. J. Basic Eng. **86**, 655–665 (1964)

25. Knudsen, J.G., Katz, D.L.: Fluid Mechanics and Heat Transfer. McGraw Hill, New York (1958)

26. Klyachko, L.S.: Heat transfer between a gas and a spherical surface with the combined action of free and forced convection. ASME J. Heat Transfer, 355–357, (1963)

27. Clift, R., Grace, J.R., Weber, M.E.: Bubbles, Drops and Particles, Academic, New York (1978)

28. Schnerr, G.H., Dohrmann, U.: Transonic flow around airfoils with relaxation and energy supply by homogeneous condensation. AIAA J. **28**(7), 1187–1193 (1990)

29. Denton, J.D.: An improved time-marching method for turbomachinery flow calculation. ASME J. Eng. Power, **105**, 514–524 (1983)

30. Mei, Y., Guha, A.: Implicit numerical simulation of transonic flow through turbine cascades on unstructured grids. J. Power Energy, Proc. IMechE Part A, **219**(A1), 35–47 (2005)

6

Condensation Discontinuities and Condensation Induced Shock Waves

Can F. Delale, Günter H. Schnerr, and Marinus E.H. van Dongen

6.1 Introduction

In high speed expansion flows of a mixture of a condensable vapor and a carrier gas, the acceleration of the vapor component of the condensable vapor may change the state of the vapor from an unsaturated state to a supersaturated state in metastable equilibrium, as realized in the passage of the vapor through steam turbine blades and on the wings of a supersonic aircraft. In this case the vapor flow proceeds without condensation beyond the saturation state, being supercooled to a certain extent before the onset of condensation sets in. Such a phenomenon can be observed in expansion cloud chambers, in supersonic nozzles, in Prandtl–Meyer expansions and in the rarefaction waves of shock tubes, in which the thermodynamic state of change occurs in characteristic times of the order of milliseconds. The cooling rates, on the other hand, may vary in the range 10^{-3}–10^{0} K/μs and even cooling rates as high as 10^{2} K/μs can be achieved in free molecular jets with condensation. In this case a fog of tiny droplets can be observed following the onset of condensation with an increase in the temperature, the pressure and the density and with a decrease in flow speed due to the release of the latent heat of condensation. The two-phase nonequilibrium mixture then relaxes to a new saturated equilibrium state. In the case of heat release to the flow beyond a critical amount, condensation induced shock waves occur with different flow patterns (stationary normal shock waves, X-shocks, unsteady periodic flows with shocks in Laval nozzles, weak oblique shock waves in Prandtl–Meyer flows, unsteady shocks propagating in the opposite direction to the flow in the driver section of a shock tube, etc.). The phenomenon of condensation induced shock waves was first reported by Prandtl [1] in supersonic nozzle flow of moist air at the Volta Congress in Rome in 1936. The picture, reproduced in Fig. 6.1, shows a condensation induced X-shock. The phenomenon was later investigated both experimentally and theoretically by Hermann [2], Oswatitsch [3], Stever [4], Wegener and his collaborators [5–7] and Glass and his collaborators [8–11]. This investigation is devoted to fundamental aspects of condensation induced shock waves, as is

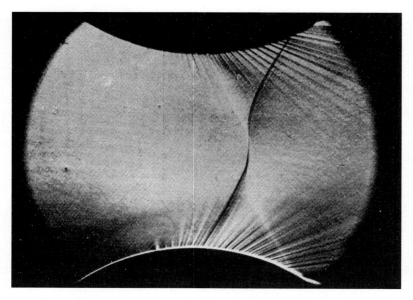

Fig. 6.1. Reproduction of schlieren photograph presented by Prandtl at the Volta Congress in 1936, showing a condensation induced X-shock in a circular arc nozzle

understood at present times. First, a brief account of the physics of nonequilibrium condensation is presented. A detailed account of high speed flows with heat addition from nonequilibrium condensation and/or from other external and internal sources, is presented. Condensation induced flow patterns with shock waves in supersonic nozzles, Prandtl–Meyer expansion flows and shock tube flows are then described in some depth.

6.2 Kinetics of Condensation and Condensation Models

The kinetics of condensation can be studied in two stages: The initial new phase formation in the form of condensation nuclei (nucleation) and the subsequent growth of these nuclei into droplets (droplet growth).

6.2.1 Nucleation

Nucleation may take place in the absence of any kind of impurities by thermal fluctuations in the condensing vapor itself. In this case condensation is said to take place by homogeneous nucleation and it may become much delayed with respect to the equilibrium state. Condensation may also take place on impurities such as ions, dust particles, salts present in the gaseous mixture, or on surfaces of the walls of a container. Under these conditions condensation is said to take place by heterogeneous nucleation near thermodynamic equilibrium with supersaturation ratios close to unity. The theory of new phase formation in the pure supersaturated vapor has been developed by Volmer and Weber [12], Farkas [13], Becker and Döring [14], Zel'dovich [15] and Frenkel [16] (today this theory is known as the classical theory of homogeneous nucleation). Theories employing methods of statistical mechanics were also considered by Lothe and Pound [17] and Feder et al. [18], but these theories did not prove to be successful in comparison with experiments. A comprehensive review of the theory of nucleation including recent developments can be found in Kashchiev [19]. Although the recent semiphenomenological theories of Dillmann and Meier [20], Delale and Meier [21], and Kalikmanov and van Dongen [22] have proved to be successful for some substances, the classical theory or a modified version is mostly used because of its simplicity and potential of applications in complex problems. Since the classical nucleation theory is based on a quasisteady state concept, it becomes inapplicable when the characteristic time required to reach steady-state becomes comparable to or larger than the characteristic time of change of state, as encountered in expansions of free molecular flows and cluster beams. Moreover, for very small clusters of formation, the bulk liquid surface tension expression has to be corrected for. Nevertheless, the classical steady-state isothermal nucleation rate expression given by

$$J = \left(\frac{2\,\sigma}{\pi\,m^3} \right)^{1/2} \frac{\rho_v^2}{\rho_l} \exp\left[-\frac{\Delta G^*}{k\,T} \right] \tag{6.1}$$

can be employed with some modification, especially for the condensation of water vapor. In (6.1), J is the rate of production of clusters of critical size, m is the mass of a single vapor molecule, ρ_v and ρ_l are, respectively, the actual vapor and condensate densities, $k = 1.38 \times 10^{-23}$ J/K is Boltzmann's constant, σ is the surface tension coefficient, T is the temperature, and ΔG^* is the Gibbs formation energy of a critical spherical cluster and is given by

$$\Delta G^* = \frac{4}{3}\,\pi\,(r^*)^2\,\sigma. \tag{6.2}$$

The critical cluster radius r^*, where the Gibbs formation energy is maximum, is given by the classical Thomson–Gibbs relation

$$r^* = \frac{2\,\sigma}{\rho_l\,R_v\,T\,(\ln S)}, \tag{6.3}$$

where R_v is the specific gas constant of the vapor and $S = p_v/p_s(T)$ is the supersaturation ratio with p_v denoting the partial vapor pressure of the vapor and $p_s(T)$ its saturation value at the local temperature T. The classical theory has further been modified by Courtney [23] and Girshick and Chiu [24]. This modified theory is called "internally consistent classical theory" (ICCT).

6.2.2 Droplet Growth

Once a cluster of critical size is formed, its further growth is determined by droplet growth considerations. In this case heat and mass transfer mechanisms, which assume no slip and infinite vapor surrounding, are described by the application of continuum or free molecular laws, depending on the Knudsen number Kn defined by

$$Kn = \frac{\ell}{2\,r}, \tag{6.4}$$

where ℓ is the mean free path of the vapor molecules and r is the radius of the droplet. When $Kn \gg 1$, the free molecular droplet growth law (Hertz–Knudsen) holds:

$$\frac{\mathrm{d}r}{\mathrm{d}t} = \alpha(T)\,\frac{p_{\mathrm{s}}(T)}{\rho_{\mathrm{l}}\,\sqrt{2\,\pi\,R_{\mathrm{v}}\,T}}\,(S-1), \tag{6.5}$$

where t denotes the time, $\alpha(T)$ is the condensation or accommodation coefficient characterizing the ratio of molecules sticking to the droplets to those impinging on it, and where the droplet temperature is assumed to be the same as the surrounding vapor temperature. When $Kn \ll 1$, the growth is controlled by the diffusion of molecules to the surface of the droplet and the so-called continuum droplet growth law

$$\frac{\mathrm{d}r}{\mathrm{d}t} = D\,\frac{p_{\mathrm{s}}(T)}{\rho_{\mathrm{l}}\,R_{\mathrm{v}}\,T\,r}\,(S-1) \tag{6.6}$$

holds where D is the binary diffusion coefficient of the condensing substance and where, once again, a one-temperature two-phase mixture of gas and droplets is assumed. A droplet growth model in the transitional region, where $Kn = O(1)$, is given by Gyarmathy [25], interpolating between the free molecular and continuum droplet growth laws. Alternatively, Young [26] has also proposed a droplet growth law valid in the transitional region. For expansion flows with nonequilibrium condensation, droplet growth initially obeys the Hertz-Knudsen law. As the droplets grow into larger size, the Hertz-Knudsen law becomes invalid. In this case the Gyarmathy law provides a valid description. Condensation models used in high speed flows differ by the expressions used for nucleation and droplet growth and for the poorly known thermodynamic functions such as surface tension and condensation coefficient. A critical comparison of some condensation models can be found in the recent work of Lamanna [27].

6.3 Flows with Heat Addition

In this section we briefly discuss flows with heat addition in relation to the heat released by nonequilibrium condensation. We recall the classical Rayleigh

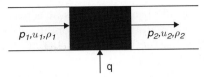

Fig. 6.2. Configuration of heat addition in one-dimensional channel flows with constant area

line relations for one-dimensional steady flows with heat addition as shown in Fig. 6.2.

$$\rho_1 \, u_1 \ = \ \rho_2 \, u_2, \tag{6.7}$$

$$p_1 \ + \ \rho_1 \, u_1^2 \ = \ p_2 \ + \ \rho_2 \, u_2^2 \tag{6.8}$$

and

$$h_1 \ + \ \frac{1}{2} \, u_1^2 \ + \ q \ = \ h_2 \ + \ \frac{1}{2} \, u_2^2, \tag{6.9}$$

where p, ρ, u, and h are the conventional symbols for the pressure, density, flow speed, and specific enthalpy of the gas, and q is the amount of heat per unit mass added to the flow between the states (1) and (2). In particular, for the case of a calorically perfect gas equations (6.7)–(6.9) can be cast into the explicit form [28–31]

$$\frac{p_2}{p_1} = \frac{1 \ + \ \gamma \, M_1^2}{1 \ + \ \gamma \, M_2^2}, \tag{6.10}$$

$$\frac{\rho_2}{\rho_1} = \frac{u_1}{u_2} = \left(\frac{1 \ + \ \gamma \, M_2^2}{1 \ + \ \gamma \, M_1^2} \right) \left(\frac{M_1}{M_2} \right)^2 \tag{6.11}$$

and

$$\frac{T_2}{T_1} = \left(\frac{1 \ + \ \gamma \, M_1^2}{1 \ + \ \gamma \, M_2^2} \right)^2 \left(\frac{M_2}{M_1} \right)^2, \tag{6.12}$$

where $\gamma = c_{\mathrm{p}}/c_{\mathrm{v}}$ is the ratio of specific heats (the adiabatic exponent) of the gas and M_1 and M_2 denote, respectively, the Mach numbers corresponding to states (1) and (2). The Mach number M_2 is given by the relation

$$M_2^2 = \left\{ 1 \ + \ \gamma \, M_1^4 \ - \ \kappa \gamma \, M_1^2 \left[2 \ + \ (\gamma - 1) \, M_1^2 \right] \right.$$
$$\pm \, (1 \ + \ \gamma \, M_1^2) \, \sqrt{ \left(M_1^2 \ - \ 1 \right)^2 \ - \ \kappa \, (\gamma + 1) \, M_1^2 \left[2 \ + \ (\gamma - 1) \, M_1^2 \right] } \, \right\}$$
$$\times \left\{ 2 \gamma \, M_1^2 \ - \ (\gamma - 1) \ + \ \kappa \gamma^2 \, M_1^2 \left[2 \ + \ (\gamma - 1) \, M_1^2 \right] \right\}^{-1} \tag{6.13}$$

with

$$\kappa \ = \ \frac{q}{c_{\mathrm{p}} \, T_{01}}, \tag{6.14}$$

where c_p denotes the specific heat of the gas at constant pressure and T_{01} is the total or stagnation temperature corresponding to state (1). When $\kappa = 0$ (no heat addition), it is instructive to note that (6.10)–(6.13) reduce to the normal shock relations. Consequently, when $\kappa = 0$, (6.13) yields the continuous solution $M_2 = M_1$ for the (+) sign and the classical normal shock relation between M_2 and M_1 for the (−) sign (physical solution exists only for $M_1 > 1$ which implies $M_2 < 1$). For flows with heat addition ($\kappa > 0$), in principle, there are also two branches of solution, corresponding to the (±) signs in (6.13). The one with the (+) sign should be chosen for subsonic ($M_1 < 1$) as well as supersonic ($M_1 > 1$) heat addition, as long as the argument of the square root in (6.13) remains positive (i.e. κ remains smaller than a critical value). In this case the effect of heat addition is to derive the downstream Mach number M_2 toward unity, decelerating a supersonic flow and accelerating a subsonic flow. We should also mention that the incoming flow Mach number M_1 can not be kept constant in the case of subsonic heat addition ($M_1 < 1$, $\kappa > 0$) so that direct application of (6.13) is not possible in this case. Solutions for finite κ with the (−) sign in (6.13) for which the flow jumps from supersonic to subsonic (strong compression) or from subsonic to supersonic (strong expansion) are excluded on the basis of physical grounds in this case [32]. The critical value κ_{max}, beyond which no steady heat addition is possible, is obtained by setting the square root argument in (6.13) equal to zero, and is given by

$$\kappa_{max} = \left(\frac{q}{c_p \, T_{01}} \right)_{max} = \frac{(M_1^2 - 1)^2}{2 \, (\gamma + 1) \, M_1^2 \, [\, 1 + (\gamma - 1) \, M_1^2 / 2 \,]}. \tag{6.15}$$

The variation of κ_{max} with M_1 is shown in Fig. 6.3. It can be shown that when $\kappa = \kappa_{max}$, the downstream Mach number becomes unity ($M_2 = 1$). The flow is then *thermally choked*. We now consider quasi-one-dimensional nozzle flows with heat addition. Without loss of generality we discuss nozzle flows with external heat addition as well as those cases where heat addition to the flow results from the latent heat released by nonequilibrium condensation (in the latter case $q = g \, L$, where g is the condensation mass fraction and L is the latent heat of condensation). The quasi-one-dimensional nozzle flow equations for a homogeneous two-phase mixture of a vapor and its droplets with a carrier gas can then be written as [3, 33–39]

$$\rho u \, A = \text{constant}, \tag{6.16}$$

$$\rho \, u \, \frac{du}{dx} = -\frac{dp}{dx} \tag{6.17}$$

and

$$h + \frac{1}{2} u^2 + q = h_o, \tag{6.18}$$

where p, ρ, u, and h now, respectively, refer to the pressure, the density, the flow speed, and the specific enthalpy of the mixture of condensable vapor and carrier gas with h_o denoting the total or stagnation specific enthalpy,

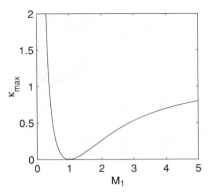

Fig. 6.3. The maximum amount of heat as a function of the Mach number M_1 for $\gamma = 1.4$

A is the cross-sectional area of the nozzle and x is the axial coordinate. The amount of heat per unit mass of the working fluid q in (6.18) can be thought to arise either from external heat addition or from the latent heat released by nonequilibrium condensation. For a calorically perfect vapor and carrier gas mixture, the specific enthalpy h is given by

$$h = c_{\mathrm{pm}}\, T + \text{constant}, \tag{6.19}$$

where $c_{\mathrm{pm}} = w_0\, c_{\mathrm{pv}} + (1 - w_0)\, c_{\mathrm{pg}}$ is the specific heat of the mixture at constant pressure, with c_{pv} and c_{pg}, respectively, denoting the corresponding vapor and carrier gas specific heats and w_0 denoting the specific initial humidity. The equation of state for this case assumes the form [35–39]

$$p = \rho \left(\frac{\Re}{\mu_{\mathrm{m}}} \right) T \left(1 - \frac{\mu_{\mathrm{m}}}{\mu_{\mathrm{v}}}\, g \right), \tag{6.20}$$

where g is the condensation mass fraction, μ_{m} is the mixture molecular mass defined by $\mu_{\mathrm{m}}^{-1} = w_0\, \mu_{\mathrm{v}}^{-1} + (1 - w_0)\, \mu_{\mathrm{g}}^{-1}$ with μ_{v} and μ_{g} denoting, respectively, the vapor and carrier gas molecular masses and $\Re$ is the universal gas constant. The governing differential equations of motion can then be written in a unified description as [40–42]

$$
\frac{1}{M} \frac{\mathrm{d}M}{\mathrm{d}x} = \frac{\left[1 + (\Gamma - 1)\, M^2/2 \right]}{(M^2 - 1)} \left\{ \frac{1}{A} \frac{\mathrm{d}A}{\mathrm{d}x} \right.
$$
$$
+ \frac{(\mu_{\mathrm{m}}/\mu_{\mathrm{v}}) \left[2 - \Gamma + (2\Gamma - 1)\, M^2 \right]}{2 \left[1 - (\mu_{\mathrm{m}}/\mu_{\mathrm{v}})\, g \right] \left[1 + (\Gamma - 1)\, M^2/2 \right]} \frac{\mathrm{d}g}{\mathrm{d}x}
$$
$$
\left. - \frac{\left(1 + \Gamma\, M^2 \right)}{2 \left[1 + q/(c_{\mathrm{pm}}\, T_{01}) \right]} \frac{\mathrm{d}}{\mathrm{d}x} \left(\frac{q}{c_{\mathrm{pm}}\, T_{01}} \right) \right\}, \tag{6.21}
$$

$$\frac{1}{\rho}\frac{d\rho}{dx} = -\frac{M^2}{(M^2-1)}\left\{\frac{1}{A}\frac{dA}{dx} + \frac{\mu_m/\mu_v}{M^2\left[1-(\mu_m/\mu_v)g\right]}\frac{dg}{dx}\right.$$
$$\left. - \frac{\left[1+(\Gamma-1)M^2/2\right]}{M^2\left[1+q/(c_{pm}T_{01})\right]}\frac{d}{dx}\left(\frac{q}{c_{pm}T_{01}}\right)\right\}, \tag{6.22}$$

$$\frac{1}{p}\frac{dp}{dx} = -\frac{\Gamma M^2}{(M^2-1)}\left\{\frac{1}{A}\frac{dA}{dx} + \frac{\mu_m/\mu_v}{\left[1-(\mu_m/\mu_v)g\right]}\frac{dg}{dx}\right.$$
$$\left. - \frac{\left[1+(\Gamma-1)M^2/2\right]}{\left[1+q/(c_{pm}T_{01})\right]}\frac{d}{dx}\left(\frac{q}{c_{pm}T_{01}}\right)\right\}, \tag{6.23}$$

$$\frac{1}{T}\frac{dT}{dx} = -\frac{(\Gamma-1)M^2}{(M^2-1)}\left\{\frac{1}{A}\frac{dA}{dx} + \frac{\mu_m/\mu_v}{\left[1-(\mu_m/\mu_v)g\right]}\frac{dg}{dx}\right.$$
$$\left. - \frac{\left(\Gamma M^2-1\right)\left[1+(\Gamma-1)M^2/2\right]}{\left(\Gamma-1\right)M^2\left[1+q/(c_{pm}T_{01})\right]}\frac{d}{dx}\left(\frac{q}{c_{pm}T_{01}}\right)\right\},$$
$$\tag{6.24}$$

where T_{01} is the stagnation or reservoir temperature and $M = u/a_f$ is the flow Mach number, based on the local frozen speed of sound a_f defined by

$$a_f^2 = \Gamma\left(\frac{\Re}{\mu_m}\right)T\left(1-\frac{\mu_m}{\mu_v}g\right) \tag{6.25}$$

with

$$\Gamma = \frac{\gamma}{\left[1+(\mu_m/\mu_v)(\gamma-1)g\right]}. \tag{6.26}$$

In (6.26) γ is the adiabatic exponent of the mixture and for most practical cases $\Gamma = \gamma$. The singularities of the differential system (6.21)–(6.24) for given heat addition distributions were considered by Möhring [41] (also discussed in detail in Zierep [30]) and its application to nonequilibrium condensation was given by Delale et al. [39, 40]. Here we discuss the latter case only. It is well known that the classification of the singularities of the differential system of (6.21)–(6.24) requires both the numerators and denominators to vanish. Applying l'Hospital's rule as $M \to 1$, say to (6.21), we obtain

$$4\psi^2 + 2\Gamma^\star\left(\frac{1}{A}\frac{dA}{dx}\right)^\star\psi - (1+\Gamma^\star)\left(\frac{dH}{dx}\right)^\star = 0, \tag{6.27}$$

where

$$\psi \equiv \left(\frac{\mathrm{d}M}{\mathrm{d}x} \right)^\star, \tag{6.28}$$

$$H(x) = r(x) - s(x) \tag{6.29}$$

with

$$r(x) = \frac{1}{A} \frac{\mathrm{d}A}{\mathrm{d}x} \tag{6.30}$$

and

$$s(x) = \frac{(1 + \Gamma)}{2 \left[1 + q/(c_{\mathrm{pm}} T_{01}) \right]} \frac{\mathrm{d}}{\mathrm{d}x} \left(\frac{q}{c_{\mathrm{pm}} T_{01}} \right) - \frac{\mu_{\mathrm{m}}/\mu_{\mathrm{v}}}{\left[1 - (\mu_{\mathrm{m}}/\mu_{\mathrm{v}}) g \right]} \frac{\mathrm{d}g}{\mathrm{d}x} \tag{6.31}$$

and where superscript $\star$ denotes evaluation at $M = 1$. In particular,

$$H^\star = H(x^\star) = r^\star - s^\star = 0. \tag{6.32}$$

Now, if we define G and D by

$$G = (1 + \Gamma^\star) \left(\frac{\mathrm{d}H}{\mathrm{d}x} \right)^\star \tag{6.33}$$

and

$$D = \left[\Gamma^\star \left(\frac{1}{A} \frac{\mathrm{d}A}{\mathrm{d}x} \right)^\star \right]^2 + 4G, \tag{6.34}$$

we obtain the solution of (6.27) for ψ as

$$\psi = \left(\frac{\mathrm{d}M}{\mathrm{d}x} \right)^\star = \frac{-\Gamma^\star \left(1/A \, \mathrm{d}A/\mathrm{d}x \right)^\star \pm D^{1/2}}{4}. \tag{6.35}$$

The classification of the singularities at $M = 1$ and $H^\star = 0$ of the differential system (6.21)–(6.24) then follows: The singular point is:

1. A turning point without heat addition if $G = D = 0$.
2. A saddle point with or without heat addition if $G \geq 0$ and $D > 0$.
3. A nodal point with heat addition if $G < 0$ and $D \geq 0$.
4. A spiral point with heat addition if $G < 0$ and $D < 0$.
5. A vortex point without heat addition if $G < 0$ and $D < 0$.

Possible global solutions of the system of (6.21)–(6.24) possessing one or three singularities in the $M - x$ phase plane are well-known (e.g., see [30, 39–42]). Figure 6.4 shows a global continuous solution with one singularity at $x_1 = 0$ (the classical saddle point singularity at the throat without heat addition). Figures 6.5a–6.5c show solutions with three singularities in the $M - x$ phase plane where $x_1 = 0$ is the classical saddle point at the throat, x_2 is a nodal point with heat addition, and x_3 is a saddle point with heat

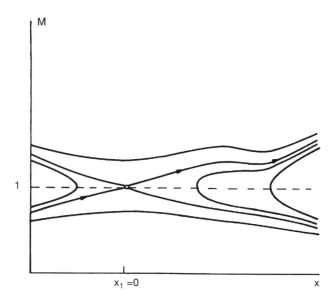

Fig. 6.4. The Mach number - axial coordinate phase plane corresponding to a global continuous solution (shown by arrows) of (6.21)–(6.24) with the classical throat singularity only (for a quantitative calculation, see Younis [43])

addition. Figures 6.6a–6.6c, on the other hand, show different $M - x$ phase portraits with three singularities: The classical saddle point at $x_1 = 0$ (at the throat), the spiral singularity at x_2, and the saddle point with heat addition at x_3. In cases where the flow Mach number M reaches unity due to considerable latent heat addition from condensation, it can be shown that the amount of heat q added to the flow becomes equal to the critical value q^* defined by (for details see [39–42])

$$\frac{q^*}{c_{\mathrm{pm}} T_{01}} = \frac{\left[(p + \rho u^2)/(\rho u) \right]^2}{4 \, \Theta(g) \, (\Re/\mu_{\mathrm{m}}) \, T_{01}} - 1, \tag{6.36}$$

where $\Theta(g)$ is given by

$$\Theta(g) = \frac{(\Gamma + 1)}{2 \, \Gamma} \left[1 - \frac{\mu_{\mathrm{m}}}{\mu_{\mathrm{v}}} g \right]. \tag{6.37}$$

The nonequilibrium flow with homogeneous condensation is then said to be *thermally choked*. In particular, (6.36) for q^* reduces to (6.15) for q_{max} for flows in constant area ducts. It should be mentioned that the critical amount of heat defined by (6.36) is not only valid for quasione-dimensional flows, but also for steady-state flows with heat addition along streamtubes in higher dimensions [42]. We can now classify the possible steady-state flow patterns,

a)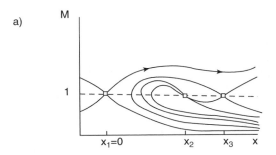

b)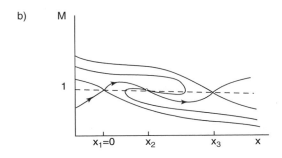

c)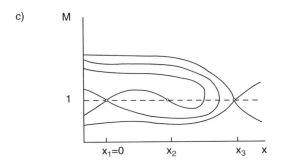

Fig. 6.5. The Mach number - axial coordinate phase planes corresponding to solutions of (6.21)–(6.24) with three singularities: the classical throat at $x_1 = 0$, a nodal point at x_2, and a saddle point with heat addition at x_3. (**a**) A continuous global solution (shown by arrows) for subcritical flows, (**b**) a continuous global solution (shown by arrows) for supercritical flows where both the nodal point at x_2 and the saddle point at x_3 are activated by supercritical heat addition (has never been observed), (**c**) only the nodal point at x_2 is activated by heat addition for which no continuous global solution is possible (details are given in Möhring [41])

a)

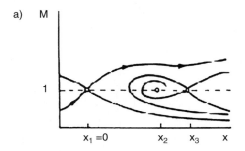

b)

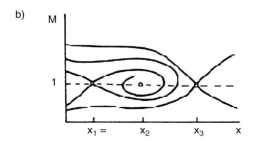

c)

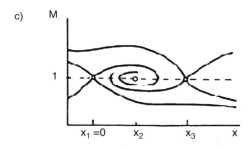

Fig. 6.6. The Mach number - axial coordinate phase planes corresponding to so-lutions of (6.21)–(6.24) with three singularities: the classical throat at $x_1 = 0$, a spiral point at x_2, and a saddle point with heat addition at x_3. (**a**) A continuous global solution (shown by arrows) for subcritical flows, (**b**) only the spiral point singularity at x_2 is activated by heat addition where no continuous global solution is possible, (**c**) both the spiral point at x_2 and the saddle point at x_3 are activated by heat addition where no continuous global solution is possible (details are given in Möhring [41])

that can be realized in condensing nozzle experiments, in connection with the above mentioned global solutions.

6.3.1 Subcritical Flows

For these flows the amount of heat added to the flow remains everywhere less than the critical amount given by (6.36). Consequently, the curve $s(x)$ lies completely below the curve $r(x)$ (see Fig. 6.7a), and the flow field remains everywhere continuous and supersonic in the diverging section of the nozzle. Such flows may assume any of the global solutions presented in Figs. 6.4 (with one singularity), 6.5a and 6.6a (with three singularities). All of these three types of global solutions, with different topological phase portraits, are possible and were calculated by Younis [43] for specified heat addition.

6.3.2 Supercritical Flows

These flows are defined as those flows for which the amount of latent heat released by nonequilibrium condensation exceeds the critical amount of heat given by (6.36). They correspond to global solutions with three singularities: the classical saddle at the throat, a nodal or spiral singularity, and a saddle point with heat addition (although they have been observed only with a spiral singularity). The underlying characteristics of supercritical flows is the existence of a localized subsonic region in the heat addition zones whether the transition occurs continuously with a nodal point singularity as shown in Fig. 6.5 or discontinuously with a spiral point singularity (with a normal shock) as shown in Fig. 6.6. Until now, no continuous supercritical solution with a nodal point singularity has been observed. In fact, only recently, have Delale et al. [40] proved that such flow solutions appear in the limit of the break-up of structural stability of the quasione-dimensional nozzle flow equations (6.16)–(6.20) and are, therefore, inherently unstable. Consequently, supercritical flows are realized with shock waves associated with a spiral singularity. Typical variations of the curves $r(x)$ and $s(x)$ for both the continuous and discontinuous cases (in the latter with a normal shock wave) are shown in Fig. 6.7.

So far we have discussed heat addition to steady flows. We now discuss in a simple model the effect of unsteady heat addition (for details see van Dongen et al. [44]). Let us consider a flow of a given Mach number M_1 with heat addition of constant relative strength $\kappa = q/h_1$, with h_1 denoting the specific enthalpy in state (1), starting at time $t = 0$. In this case unsteady compression waves will be generated. In subsonic flow, a shock wave with Mach number M_u will move upstream and a shock wave with Mach number M_d will move downstream. Figure 6.8 gives an overview of the different waves possible and of the nomenclature of the different uniform states. Heat is added to the flow at $x = 0$ in a compact manner so that relations (10)–(13) apply between states (2) and (3). However, the upstream condition is affected by the upstream

a)

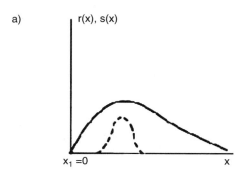

b)

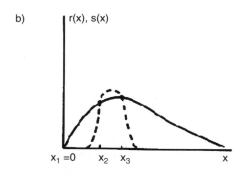

c)

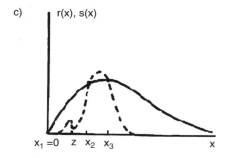

Fig. 6.7. Heat addition - area change diagrams for global solutions where the heat addition function $s(x)$ is represented by *dash lines* and the area change $r(x)$ is represented by *solid lines*. (**a**) Subcritical flows, (**b**) Continuous supercritical flows (not observed or calculated yet), (**c**) Supercritical flows with a stationary normal shock wave at $x = z$ [39, 40]

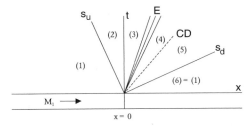

Fig. 6.8. Wave diagram resulting from the onset of heat release at $x = 0$. The occurrence of the upstream shock wave and the expansion fan depends on the flow Mach number and on the strength of heat release

moving shock (if present), so that in general $M_2 \leq M_1$. Downstream of the plane of heat addition, a contact discontinuity separates states of different entropy (4) and (5) with equal pressures and velocities. Three different regimes can be distinguished:

Regime A: Subsonic Flow with $M_3 < 1$ (No Thermal Choking)

If we have both $M_2 < 1$ and $M_3 < 1$ for given $M_1 < 1$, different states of Fig. 6.8 are connected in a straightforward manner. States (1) and (2), and states (5) and (6) are connected by the normal shock relations given by (6.10)–(6.13) for $q = 0$ with the $(-)$ sign chosen in (6.13). The Mach numbers in (6.10)–(6.13) in this case refer to the relative velocity between the gas and the wave. There is no expansion fan in this case so that states (3) and (4) of Fig. 6.8 are identical.

Regime B: Thermally Choked Flow with $M_2 < 1$ and $M_3 = 1$

For fixed κ and increasing M_1, there will be a point where the Mach number M_3 becomes unity. The unsteady flow is then *thermally choked*. In such a case the upstream shock Mach number and M_2 (in this case smaller than unity) are fully determined by the choking condition $M_3 = 1$. The flow is accelerated to state (4) by a right running expansion fan and decelerated back to the initial velocity by the downstream running shock wave. States (3) and (4) are connected by the Riemann invariants of the left running characteristics for isentropic flow as

$$u_3 + \frac{2\,a_3}{\gamma - 1} = u_4 + \frac{2\,a_4}{\gamma - 1} \tag{6.38}$$

with

$$\frac{a_3}{a_4} = \left(\frac{p_3}{p_4} \right)^{\frac{\gamma-1}{2\gamma}}, \tag{6.39}$$

where the symbol a stands for the speed of sound and γ is the adiabatic exponent. It should be mentioned that a thermally choked flow is also possible even when the incoming flow is supersonic, i.e., the Mach number $M_1 > 1$, provided that the strength of the upstream shock is sufficient enough to move upstream.

Regime C: Supersonic Flow Without Thermal Choking for $M_1 > 1$ and $M_3 > 1$

If both Mach numbers M_1 and M_3 exceed unity and M_1 is sufficiently large so that there is no upstream disturbance (with no upstream shock where states (1) and (2) in Fig. 6.8 are identical), the flow decelerates to a fixed downstream Mach number $M_3 < M_1$. In order to match the post shock conditions downstream an expansion fan remains essential. In this case the flow remains supersonic everywhere.

An important conclusion of this analysis is that the upstream moving shock, if it exists, is always much stronger than the downstream moving shock, which explains why usually only the upstream moving shock has been observed.

6.4 Condensation Induced Shock Waves in Nozzle Flows

The effects of homogeneous condensation in nozzle flows have been described in detail by Wegener and Mack [6] in a comprehensive article. In this context it is also instructive to mention the article by Hill [33]. The major effects of condensation on the flow is the depletion of the vapor phase and the heating of the remaining gas/vapor flow that absorbs the energy released by the latent heat of condensation. It has also been demonstrated that heating has a larger effect than vapor removal. Consequently, the effects of condensation are similar to those in flows with heat addition. The condensed phase is in the form of tiny droplets so that, except in a shock zone, the slip between the vapor and liquid phases is negligible. This suggests that the flow can be described by a homogeneous two-phase mixture of a vapor and droplets with or without a carrier gas. In what follows this flow description will be employed.

6.4.1 Quasi-One-Dimensional Nozzle Flows

The equations for quasi-one-dimensional nozzle flows with nonequilibrium condensation were formulated by Oswatitsch [3]. These equations consist of (6.16)–(6.20) for the homogeneous two-phase mixture of vapor and droplets with/without a carrier gas with $q = g\,L$, where g is the condensation mass fraction and L is the latent heat of condensation, supplemented by the nonequilibrium condensation rate equation. The nonequilibrium condensation rate equation can be constructed from a normalized homogeneous nucleation rate

$$J = \Sigma(p_{\mathrm{v}}, T) \exp\left[-K^{-1} B(p_{\mathrm{v}}, T)\right] \tag{6.40}$$

and a normalized Hertz-Knudsen droplet growth law

$$\frac{\mathrm{d}r}{\mathrm{d}x} = \lambda \, \Omega(p_{\mathrm{v}}, T, g). \tag{6.41}$$

Here p_{v} is the partial pressure of the condensable vapor, which can be related to the mixture pressure p, $\Sigma(p_{\mathrm{v}}, T)$ is a normalized pre-exponential factor of the nucleation rate equation, $B(p_{\mathrm{v}}, T)$ is the normalized activation function, K is the nucleation parameter (assumed small, $K \ll 1$), $\Omega(p_{\mathrm{v}}, T, g)$ is the normalized droplet growth function, λ is the droplet growth parameter (assumed large, $\lambda \gg 1$) and r is the normalized droplet radius (for normalization and relation of these functions to the nucleation rate equation and droplet growth law, see Delale et al. [37]). The nonequilibrium condensation rate equation can then be written, in normalized form, as [35, 36]

$$g(x) = \int_{x_{\mathrm{s}}}^{x} \left(r^{*}(\xi) + \lambda \int_{\xi}^{x} \Omega(\eta) \, \mathrm{d}\eta\right)^{3} \Sigma(\xi) \, A(\xi) \exp\left[-K^{-1} B(\xi)\right] \mathrm{d}\xi, \tag{6.42}$$

where r^{*} is the normalized critical nuclei radius and x_{s} is the distance along the nozzle where saturation conditions are reached. Preliminary calculations of (6.16)–(6.20) together with the condensation rate equation (6.42) were made by Oswatitsch himself [3]. Later the nonequilibrium condensation rate equation was cast into a system of four ordinary differential equations and were integrated together with (6.16)–(6.20) simultaneously by Hill [33] by adjusting the values of the accommodation coefficients. His theoretical predictions for subcritical (without shocks) flows using homogeneous nucleation and two different droplet growth laws against experimental data are shown in Fig. 6.9. As can clearly be seen from the figure, the pressure rise in the condensation zone (the droplet growth zone to be defined later) is not more than 10%. To obtain a larger pressure rise, the onset of condensation should shift considerably toward the throat, prior to which the flow becomes supercritical with the occurrence of a normal shock wave. Experiments and numerical calculations for the expansion of steam (as well as moist air) were also carried out by Barschdorff [34] for both subcritical and supercritical flows. An asymptotic solution to the condensation rate equation (6.42) with different ordering of the double limit as $K \to 0$ and $\lambda \to \infty$, signifying a large nucleation rate followed by a small droplet growth rate, was carried out by Blythe and Shih [35] and by Clarke and Delale [36] using Laplace's method [45]. In particular, Clarke and Delale identify the physically distinct condensation zones arising from the behavior of the normalized activation function B along the nozzle as shown in Fig. 6.10a. In the initial growth zone (IGZ) and further growth zone (FGZ), $\mathrm{d}B/\mathrm{d}x = O(1)$. The initial growth zone (IGZ) is defined as the zone where the number of condensation nuclei (clusters of critical size) created is not sufficient to influence the isentropic expansion. Thus the flow basically

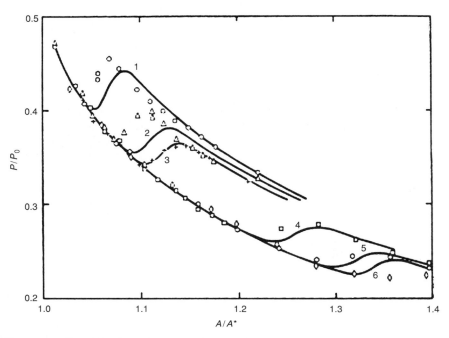

Fig. 6.9. Theoretical predictions of the pressure ratio as a function of the effective area ratio by Hill [33] against the experimental data of Binnie and Green for the expansion of steam (experimental points for reservoir conditions: upstream $P_{01} = 0.647$ atm. and $T_{01} = 101°C$; $P_{01} = 0.652$ atm. and $T_{01} = 108°C$; $+P_{01} = 1.022$ atm. and $T_{01} = 126°C$; $P_{01} = 0.646$ atm. and $T_{01} = 127°C$; downstream $P_{01} = 0.656$ atm. and $T_{01} = 136°C$; $P_{01} = 0.935$ atm. and $T_{01} = 153°C$; *solid lines* numbered by 1–6 denote the theoretical curves)

remains frozen (isentropic) throughout this zone and (6.16)–(6.20) with $g = 0$ reduce to the classical isentropic nozzle flow relations. In the further growth zone, B starts to deviate from its frozen value B_f (although negligibly small) where the derivatives dB/dx and dB_f/dx are now numerically distinct. In spite of the fact that the nucleation rate increases continuously throughout this zone, the rate is still not sufficient to produce enough nuclei for the onset of condensation. Although the two zones IGZ and FGZ are physically distinguished, they are not asymptotically distinct and the asymptotic solution of the rate equation (6.42) for $dB/dx = O(1)$ applies to both zones. As dB/dx diminishes to $O(K^{1/2})$ as the nucleation parameter $K \to 0$, the asymptotic solution in FGZ will no longer be valid and we are now in the rapid growth zone (RGZ) where the onset zone (OZ) is embedded. The onset zone (OZ) contains the onset point x_k (the so-called Wilson point), defined either empirically (e.g., by static pressure measurements) or numerically (e.g., by fixing a certain value of the condensation mass fraction g as an onset value for a

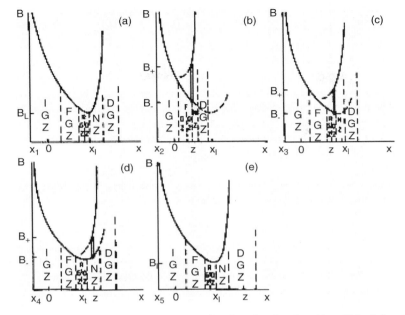

Fig. 6.10. Typical behavior of the normalized activation function B in (**a**) subcritical flows; (**b**) Regime I supercritical flows; (**c**) Regime II supercritical flows; (**d**) Regime III supercritical flows, and (**e**) Regime IV supercritical flows. (Here x_s is the saturation point, $x = 0$ is the throat location, z is the normal shock location, and x_l is the turning point of the activation function B) [38]

given working fluid under similar supply conditions and nozzle geometry). The onset point marks the beginning of the collapse of the supersaturated vapor. The nucleation rate peaks at a point x_1, defined as the point where the minimum of B is reached and called the relative onset point, that marks the end of the rapid growth and onset zones. Once again these two zones are not asymptotically distinct and the asymptotic expressions for the rate equation when $dB/dx = O(K^{1/2})$ apply to both zones. For $x \geq x_1$ we have two physically and asymptotically distinct condensation zones: The nucleation zone with growth (NZ) and the droplet growth zone (DGZ). The nucleation zone with growth (NZ) proceeds the onset zone OZ and ends as the nucleation rate diminishes. In this zone both nucleation and droplet growth are important and the nucleation rate decreases ($dB/dx > 0$) until it practically diminishes. Droplet growth on condensation nuclei exceeding the critical size results in latent heat release to the flow. As a result an increase in the pressure, temperature, and density and a decrease in the flow speed of the mixture are first observed in this zone. Downstream of NZ, droplet growth takes over and we are by definition in the droplet growth zone (DGZ). In this zone a remarkable amount of heat, but not exceeding the critical amount given by

(6.36) at any location within this zone, is set free. Consequently, the temperature, pressure, and density of the mixture increase over a relatively small thickness (this phenomenon is inappropriately referred to as a "condensation shock" in the literature). The two-phase mixture eventually approaches equilibrium in this zone. The asymptotic solution of the rate equation (6.42) and the complete solution in each zone for subcritical flows, where the amount of latent heat released to the flow does not exceed the critical amount given by (6.36) anywhere in the droplet growth zone, can be found in Delale et al. [37]. Figure 6.11 shows the predictions of the asymptotic theory for the transonic expansion of moist air through a circular arc nozzle (whose geometric configuration is shown) under the specified atmospheric supply conditions using the classical nucleation theory, with surface tension fitted to the experiments of Peters and Paikert [46], and the Hertz-Knudsen droplet growth law. Good agreement between the predictions of the asymptotic theory and the static pressure measurements of Schnerr [47] is achieved, despite the fact the predicted pressure distribution is a few percent higher than the measured ones downstream of the onset where two-dimensional effects seem to be important.

The discussed asymptotic solution of transonic nozzle flows with nonequilibrium condensation remains valid as long as the amount of heat released by condensation $q = g\,L$ does not exceed the critical value defined by (6.36) of the previous section. If, with the same working fluid and nozzle geometry and for fixed initial specific humidity ω_0 and fixed reservoir temperature T_0, the initial relative humidity is increased beyond a certain limit, the amount of heat released by condensation can exceed the critical amount moving the flow Mach number toward unity. The flow is then thermally choked and becomes supercritical where the inclusion of a normal shock wave becomes necessary. If the initial relative humidity is further increased with the rest of the conditions kept fixed, the normal shock becomes unstable and an unsteady periodic flow sets in near the throat. This unsteady flow pattern will be discussed later in this section. Supercritical nozzle flows with stationary normal shock waves induced by homogeneous condensation were discussed by Pouring [48] and Barschdorff [34]. Pouring considers the supercritical flow of moist air expansion in a convergent–divergent nozzle, where he assumes that the pressure distribution is known empirically. Barschdorff uses an iterative numerical scheme for the solution of the differential equations describing nozzle flows with nonequilibrium condensation of wet steam, (6.21)–(6.24), where he replaces the singularities by normal shock waves with continuous downstream acceleration to supersonic flow. He concludes that the condensation of wet steam for supercritical flow occurs closer to the throat than the shock-free (subcritical flow) condensation. A complete description of supercritical flows with a stationary normal shock wave together with classification of possible regimes has been obtained by Delale et al. [38] using the asymptotic method described earlier. For this case the reconsideration of the condensation rate equation is necessary because the structure of the physically distinct condensation zones exhibited for subcritical flows alters significantly downstream of

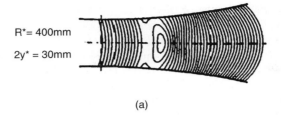

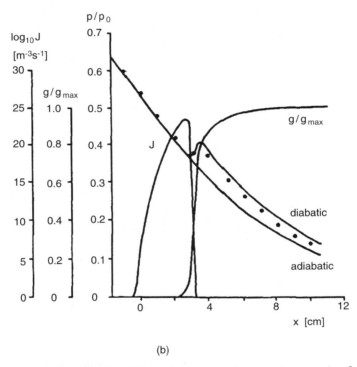

(a)

(b)

Fig. 6.11. (a) The Mach number ($M \geq 1$) contours of moist air expansion from the numerical 2-D simulation of Schnerr and Dohrmann [50] in the circular arc nozzle with throat radius of curvature $R^* = 400\,\text{mm}$ and throat height $2\,y^* = 30\,\text{mm}$ under the subcritical flow atmospheric supply conditions with relative humidity $\phi_0 = 36.4\%$, initial specific humidity $\omega_0 = 6.64\,\text{g\,kg}^{-1}$ and stagnation temperature $T_{01} = 296.6\,\text{K}$ (the increment between any two successive contours is $\Delta M = 0.02$). (b) The asymptotic predictions of the distributions of the pressure, the nucleation rate and the condensation mass fraction along the nozzle axis under the specified subcritical flow supply conditions [37]. *Filled circle* represents static pressure measurements of Schnerr [47]

the normal shock. The location of the normal shock, denoted hereafter by the axial coordinate z, lies downstream of the onset point x_k ($z > x_k$). In general both nucleation and droplet growth can prevail downstream of the shock. In reformulating the condensation rate equation, it should be noticed that all thermodynamic functions such as r^*, B, Ω, etc. will be discontinuous across the normal shock due to the increase in the pressure and temperature, as will be discussed later, but the condensation mass fraction g remains continuous (the continuity of g, not $\mathrm{d}g/\mathrm{d}x$, across the normal shock is a consequence of the assumption that droplets pass through the shock unaltered). The condensation rate equation (6.42) will now be valid only upstream of and at the shock location ($x \leq z$). Downstream of the shock location ($x > z$) the condensation rate equation can be constructed in a similar fashion, but now taking into account the contribution from downstream nuclei production as well, to yield

$$
\begin{aligned}
g(x) = \int_{x_s}^{z^-} &\left(r^*(\xi) \;+\; \lambda \int_{\xi}^{z^-} \Omega(\eta)\,\mathrm{d}\eta \;+\; \lambda \int_{z^+}^{x} \Omega(\eta)\,\mathrm{d}\eta \right)^3 \\
&\times\;\; \Sigma(\xi)\,A(\xi)\,\exp\left[-K^{-1}\,B(\xi) \right]\mathrm{d}\xi \\
+ \int_{z^+}^{x} &\left(r^*(\xi) \;+\; \lambda \int_{\xi}^{x} \Omega(\eta)\,\mathrm{d}\eta \right)^3 \Sigma(\xi)\,A(\xi)\,\exp\left[-K^{-1}\,B(\xi) \right]\mathrm{d}\xi
\end{aligned}
$$

$$(6.43)$$

for $x \geq z$ where (z^-) denotes the location just upstream of the shock front and (z^+) denotes the location just downstream of the shock front. The first integral in (6.43) characterizes the contribution from condensation nuclei produced upstream of the normal shock location, whereas the second integral characterizes the contribution from condensation nuclei produced downstream of the normal shock. In supercritical flows with stationary normal shock, four distinct flow regimes, as shown in Fig. 6.10b–e, can be distinguished, depending on the location of the normal shock in relation with the behavior of the normalized activation function B. Figure 6.10b shows the case, called Regime I, in which the onset zone (OZ) is embedded in the further growth zone (FGZ), where $\mathrm{d}B/\mathrm{d}x = O(1)$. In this case the normal shock location ($z < x_1$) proceeds OZ and the nucleation rate, before reaching the maximum, decreases significantly by the jump at the shock location. Downstream of the normal shock a relatively thin subsonic nucleation zone followed by a droplet growth zone, where the flow is accelerated back to supersonic speeds, takes place. In Fig. 6.10c we have a supercritical flow, called Regime II, in which the onset zone (OZ) is embedded in the rapid growth zone (RGZ) where $\mathrm{d}B/\mathrm{d}x = O(K^{1/2})$, but ends at the shock front z prior to the relative onset point $x_1(z < x_1)$. Downstream is the local subsonic zone in NZ and DGZ followed by an ultimate supersonic flow in DGZ. Figure 6.10d shows the supercritical flow Regime III, where the normal shock proceeds the relative onset point ($z < x_1$). In this case the normal shock is embedded in NZ followed by DGZ, where the flow is accelerated back to supersonic speeds. Finally, Fig. 6.10e shows the case of a supercritical flow, called Regime IV, where the normal shock is embedded in the droplet growth zone ($z > x_1$). Once again the usual structure of a local subsonic zone followed by a supersonic zone occurs downstream of the shock. In a thermally choked flow any of the earlier supercritical regimes, may occur, the particular regime depending on the location z of the normal shock fixed by a nucleation rate, sufficient for realizing the onset of condensation and by the nozzle geometry downstream of this onset zone. It should be mentioned that, as the relative humidity is increased, the location z of the normal shock moves upstream toward the throat and the supercritical Regime I is realized. The asymptotic expressions for the condensation rate equation in each supercritical flow regime and the complete solution together with the normal shock relations and a shock fitting technique are given in Delale et al. [38]. Figure 6.12 shows the predictions of the asymptotic theory for supercritical expansion (Regime I) of moist air with atmospheric supply conditions through the slender nozzle with specified geometric configuration using the classical nucleation theory and the Hertz-Knudsen droplet growth law. Excellent agreement is achieved between the asymptotic predictions and measured static pressure measurements of Schnerr [47].

6.4.2 Two-Dimensional Transonic Nozzle Flows with Nonequilibrium Condensation

The quasi-one-dimensional theory of steady-state condensing nozzle flows discussed earlier will remain valid as long as two-dimensional wall effects can be neglected (i.e., for slender nozzles). In nozzles with significant wall curvature, 2-D condensation structures without shock waves as well as with "X-shocks" or with normal shocks extending from wall to wall are possible. Moreover, depending on the angle between the onset Mach iso-line and the characteristic line at the wall [47], different condensation structures with subsonic or supersonic heating fronts can be observed in the diverging section of nozzles. Two-dimensional nozzle flows with nonequilibrium condensation were experimentally investigated by Bratos and Meier [49] and by Schnerr [47]. Figures 6.13 and 6.14 from Schnerr [47] show, respectively, schlieren photographs of an "X-shock" formed by homogeneous condensation from excessive heat release (supersonic heating front) and of an "X-shock" formed by the intersection of two weak oblique adiabatic shocks and located in front of the condensation zone (subsonic heating front). Two-dimensional structures arising from nonequilibrium condensation were also numerically simulated by Schnerr and Dohrmann [50]. They used the Euler equations in conservation form

$$\frac{\partial \mathbf{U}}{\partial t} + \frac{\partial \mathbf{F}}{\partial x} + \frac{\partial \mathbf{G}}{\partial y} = \mathbf{Q}, \tag{6.44}$$

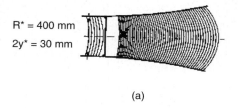

(a)

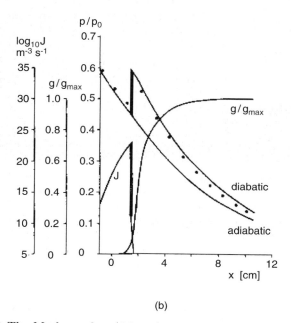

(b)

Fig. 6.12. (a) The Mach number ($M \geq 1$) contours of moist air expansion from the numerical 2-D simulation of Schnerr and Dohrmann [50] in the circular arc nozzle with throat radius of curvature $R^* = 400$ mm and throat height $2\,y^* = 30$ mm under the supercritical flow atmospheric supply conditions with relative humidity $\phi_0 = 73.4\%$, initial specific humidity $\omega_0 = 7.3\,\mathrm{g\,kg^{-1}}$ and stagnation temperature $T_{01} = 287.2\,\mathrm{K}$ (the increment between any two successive contours is $\Delta M = 0.02$). **(b)** The asymptotic predictions of the distributions of the pressure, the nucleation rate and the condensation mass fraction along the nozzle axis under the specified supercritical flow (Regime I) supply conditions [38]. *Filled circle* represents static pressure measurements of Schnerr [47]. Both the predicted and observed shock locations are at $z = 12$ mm

where

$$\mathbf{U} = \begin{pmatrix} \rho \\ \rho\,u \\ \rho\,v \\ \rho\,e \end{pmatrix}; \qquad \mathbf{F} = \begin{pmatrix} \rho\,u \\ \rho\,u^2 + p \\ \rho\,u\,v \\ (\rho\,e + p)\,u \end{pmatrix};$$

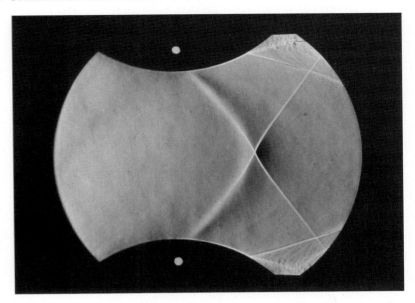

Fig. 6.13. Schlieren photograph showing the wall curvature effects of a 2-D nozzle flow with an "X-shock" formed by homogeneous condensation [47] (circular arc nozzle with throat radius of curvature $R^* = 100\,\text{mm}$ and throat height $2\,y^* = 60\,\text{mm}$ under the subcritical flow atmospheric supply conditions with relative humidity $\phi_0 = 36.6\%$, initial specific humidity $\omega_0 = 8.3\,\text{g\,kg}^{-1}$ and stagnation temperature $T_{01} = 300.3\text{K}$)

$$\mathbf{G} = \begin{pmatrix} \rho\,v \\ \rho\,u\,v \\ \rho\,v^2 + p \\ (\rho\,e + p)\,v \end{pmatrix} \quad \text{and} \quad \mathbf{Q} = \begin{pmatrix} 0 \\ 0 \\ 0 \\ \rho\,\dfrac{\mathrm{d}(L\,g)}{\mathrm{d}t} \end{pmatrix} \tag{6.45}$$

with $\mathbf{Q}$ denoting the heat supply due to homogeneous condensation. In (6.44), (6.45) ρ is the mixture density, u and v are, respectively, the cartesian x- and y-components of the velocity, p is the pressure, e is the specific internal energy of the mixture, L is the latent heat of condensation, g is the mass fraction, and t is the time. The rate of change of the condensation mass fraction is given by the integro-differential equation

$$\frac{\mathrm{d}g}{\mathrm{d}t} = \frac{4}{3}\,\pi\,\rho_1\,r^{*\,3}\,\frac{J}{\rho} + \int_{-\infty}^{t} \frac{4}{3}\,\pi\,\rho_1\,\frac{J(\tau)}{\rho(\tau)}\,\frac{\mathrm{d}r^3(\tau)}{\mathrm{d}\tau}\,\mathrm{d}\tau, \tag{6.46}$$

where ρ_1 is the density of the condensed phase, J is the nucleation rate, r^* is the critical radius of condensation nuclei and $\mathrm{d}/\mathrm{d}t$ is the total or material derivative. It should be mentioned that, aside from normalization, the earlier

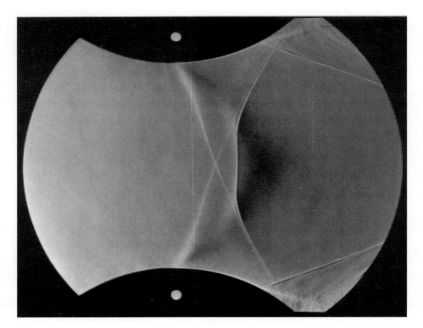

Fig. 6.14. Schlieren photograph showing the wall curvature effects of a 2-D nozzle flow with an "X-shock" formed by the intersection of two adiabatic weak oblique shocks prior to condensation [47] (circular arc nozzle with throat radius of curvature $R^* = 100\,\text{mm}$ and throat height $2\,y^* = 120\,\text{mm}$ under the subcritical flow atmospheric supply conditions with relative humidity $\phi_0 = 63.2\%$, initial specific humidity $\omega_0 = 9.2\,\text{g}\,\text{kg}^{-1}$ and stagnation temperature $T_{01} = 293.0\,\text{K}$)

equation for the condensation rate equation is equivalent to (6.42) employed in the previous subsection. Furthermore, by introducing the "surface averaged" droplet radius of Hill [33], the integro-differential equation (6.46) can be transformed into a system of four ordinary differential equations coupled to the two-dimensional flow equations (6.44),(6.45). Schnerr and Dohrmann employed the classical nucleation theory and the Hertz-Knudsen droplet growth law, with some poorly known coefficients like the surface tension and condensation coefficients obtained by fitting to the experiments of Peters and Paikert [46]. They solved (6.44)–(6.46) numerically by transforming them onto an equidistant mesh. For the numerical method, they used an explicit time finite volume method (FVM) to second-order accuracy (second-order upwind scheme). They simulated both subcritical and supercritical flows. They employed different nozzles (circular arc and hyperbolic) and varied the initial relative humidity over a wide range ($0.32 \leq \phi_0 \leq 0.65$) and compared their results with the experiments of Schnerr [47] where they reach very good

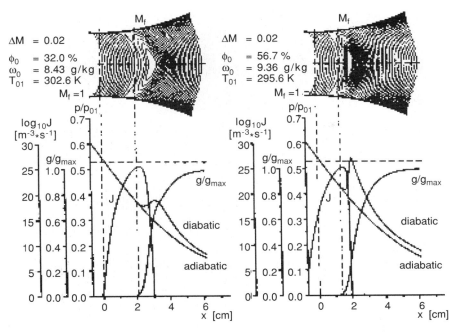

Fig. 6.15. Left: The Mach number contours and the distributions of the pressure, the nucleation rate and the condensation mass fraction for moist air expansion from the numerical 2-D simulation of Schnerr and Dohrmann [50] in the circular arc nozzle with throat radius of curvature $R^* = 127\,\mathrm{mm}$ and throat height $2\,y^* = 40\,\mathrm{mm}$ under the subcritical flow atmospheric supply conditions with relative humidity $\phi_0 = 32.0\%$, initial specific humidity $\omega_0 = 8.43\,\mathrm{g\,kg^{-1}}$ and stagnation temperature $T_{01} = 302.6\,\mathrm{K}$ (the increment between any two successive contours is $\Delta M = 0.02$); Right: The Mach number contours and the distributions of the pressure, the nucleation rate and the condensation mass fraction for moist air expansion exhibiting an "X-shock" from the numerical 2-D simulation of Schnerr and Dohrmann [50] in the circular arc nozzle with throat radius of curvature $R^* = 127\,\mathrm{mm}$ and throat height $2\,y^* = 40\,\mathrm{mm}$ under the atmospheric supply conditions with relative humidity $\phi_0 = 54.7\%$, initial specific humidity $\omega_0 = 9.36\,\mathrm{g\,kg^{-1}}$ and stagnation temperature $T_{01} = 295.6\,\mathrm{K}$ (the increment between any two successive contours $\Delta M = 0.02$)

agreement. Their results are shown in Figs. 6.15 and 6.16 For low initial relative humidities (around 32%), they obtain a subcritical flow where the flow is everywhere supersonic (Fig. 6.15). As the relative initial humidity was increased to moderate values (around 55%), they observed a weak "X-shock" with the onset shifted toward the throat, as is also shown in Fig. 6.15. Finally, as the relative initial humidity was increased further (64.8%), a stationary normal shock stretching nearly from wall to wall were observed (Fig. 6.16). For higher initial relative humidities, the flow was seen to be unsteady, as will be discussed later.

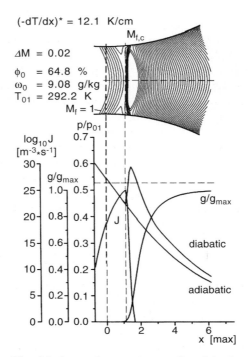

Fig. 6.16. Above: The Mach number contours of moist air expansion from the numerical 2-D simulation of Schnerr and Dohrmann [50] in the circular arc nozzle with throat radius of curvature $R^* = 127\,\text{mm}$ and throat height $2\,y^* = 40\,\text{mm}$ under the atmospheric supply conditions with relative humidity $\phi_0 = 64.8\%$, initial specific humidity $\omega_0 = 9.08\,\text{g}\,\text{kg}^{-1}$ and stagnation temperature $T_{01} = 292.2\,\text{K}$ (the increment between any two successive contours is $\Delta M = 0.02$); Below: The distributions of the pressure, the nucleation rate and the condensation mass fraction along the nozzle axis under the specified supply conditions exhibiting an "X-shock" from the 2-D numerical simulations of Schnerr and Dohrmann [50]

6.4.3 Unsteady Self-Excited Periodic Flows

The discussion of condensing nozzle flows have so far been for steady states. If the latent heat release to the flow becomes even stronger (e.g., by increasing the initial relative humidity further), the flow becomes unsteady and different modes of self-excited periodic oscillations alter the flow patterns completely. The phenomenon of self-excited oscillations in slender nozzles of moist air was discovered by Schmidt [51] and further investigated by Barschdorff [52, 53]. A similarity law for the dimensionless frequency of self-excited oscillations was derived by Zierep and Lin [54]. The first numerical computations of un-steady periodic flows in one-dimensional nozzles were carried out by Saltanov and Tkalenko [55] for moist air and by Guha and Young [56] for wet steam. They identified two oscillation modes. Later, two-dimensional calculation of unsteady wet-steam flow with periodic oscillations was carried out by White and Young [57]. Condensation induced oscillations for moist air expansions in nozzles were recently investigated in depth by Adam and Schnerr [58] both experimentally and numerically. Depending on the nozzle geometry, they found different types of self-excited oscillations caused by condensation. Their numerical results, which are in good agreement with the results of experiments, are summarized in Fig. 6.17. In relatively slender nozzles, as shown in Fig. 6.17a, they distinguish three different modes of oscillations. Near the stability limit of steady-state solutions, the frequency of self-excited periodic flow first decreases with increasing relative humidity (with the nozzle geometry and all other inlet conditions kept fixed) from a finite value until a well-distinguished minimum. Between the stability limit and the minimum frequency, they identify two modes of oscillations which they call Mode 3 and Mode 2. Past the minimum, the frequency of the periodic flow increases monotonically with increasing relative humidity (Mode 1). Figure 6.18 shows the schlieren streak photographs of all three modes observed in their experiments. The flow is from left to right and time increases from top to bottom. At the left, a shock formed by excessive heat release moves upstream, accelerates through the throat, and weakens in the convergent section. Finally it disappears at the left end of the photograph and a new shock builds up downstream. In this mode the whole flow field is affected by the oscillation. The streak photograph in the middle for a lower value of the initial humidity shows a situation in which the shock stops before reaching the nozzle throat, weakens as it returns in the main flow direction and dies out while a new shock forms downstream (Mode 2). In this case only the divergent nozzle section is affected by the flow oscillation and the mass flow rate through the throat remains constant. The frequency minimum separates the flow oscillations of Mode 1 and Mode 2. At the right of Fig. 6.18 is a flow oscillation just above the stability limit for steady-state (Mode 3). Here no periodic shock formation is observed. A shock with a nearly constant strength oscillates. The frequency in Mode 3 increases more strongly for decreasing initial relative humidity in comparison with Mode 2. The transition from steady-state to the periodic oscillations of the shock in this case occurs with finite frequency. The stability limit of stationary shock waves in supercritical flows through slender nozzles has been analyzed in detail by Delale et al. [40]. They show how the quasione-dimensional steady-state condensation shock structures (discussed earlier by the asymptotic theory) are altered by a catastrophic change in the phase portraits of the system of differential equations where the spiral singularity turns into a nodal point singularity as the initial relative humidity is increased beyond a certain limit. By using a variational analysis they find the break-up of structural stability of the system of differential equations (6.21)–(6.24) in the limit where nucleation has been quenched by orders of magnitude so that no sufficient heat is available to accelerate the flow in the subsonic heating zones to supersonic speeds (corresponding to a global solution of (6.21)–(6.24) with a nodal point solution behind the shock).

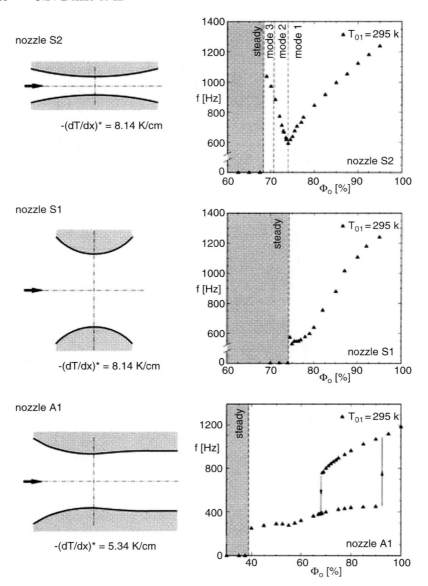

Fig. 6.17. Frequency dependency of self-excited flow oscillations on the initial relative humidity for moist air expansions in different nozzles (**a**) slender circular arc nozzle S2; (**b**)strongly curved circular arc nozzle S1; (**c**) nozzle A1 with parallel outflow [58]

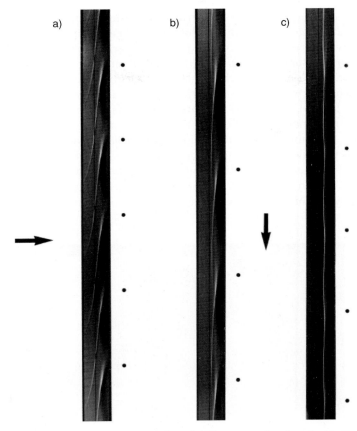

Fig. 6.18. Schlieren streak photographs of different modes (modes 1, 2, and 3) of condensation induced flow oscillations in the slender circular arc nozzle S2 [58]

They identify this limit as the stability limit of supercritical steady-state solutions and establish a criterion for the critical Mach number beyond which self-excited flow oscillations (presumably Mode 3) set in. They reach excellent agreement for the stability limit in comparison with the experiments of Adam and Schnerr [58] and Lamanna [27] in slender nozzles. Detailed numerical simulations for moist air expansions in the slender nozzle of Fig. 6.17a for the system of (6.44)–(6.46) were carried out using a MUSCL-type finite volume method with van Leer flux splitting by Adam and Schnerr [58]. Their results exactly showed the frequency minimum separating Mode 1 and Mode 2 flow oscillations. Wall curvature effects on the modes of flow oscillations are demonstrated in Fig. 6.17b. In this case the stability limit is shifted to higher values of the initial relative humidity and the strong minimum disappears. Two-dimensional numerical simulations by Adam and Schnerr in this case show that flow oscillations and unsteady shock formation are concentrated in regions near the centerline of the nozzle whereas at the nozzle walls the flow is stabilized by the locally stronger temperature gradients. A completely different, yet unresolved, flow pattern is observed, both experimentally and numerically, in a slender nozzle with low geometric divergence with a constant outflow Mach number as shown in Fig. 6.17c. A bifurcation of the frequency with respect to initial relative humidity showing hysteresis is observed. For initial humidities greater than 68%, two frequency branches are observed by a sudden jump of the frequency from one branch to the other. Increasing the initial relative humidity above 95% along the lower branch seems not possible and the frequency in this case jumps to the upper branch. Moreover, decreasing the initial relative humidity below 95% along this high frequency branch yields high frequencies until the initial relative humidity reaches a value of 68.5% where the frequency now jumps back to the lower

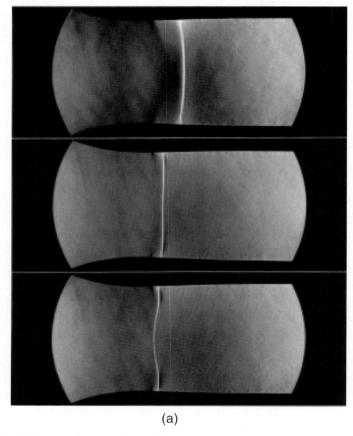

(a)

Fig. 6.19. Schlieren photographs of (**a**) the symmetric mode and (**b**) the nonsymmetric mode of flow oscillations resulting from bifurcation in nozzle A1 [58]

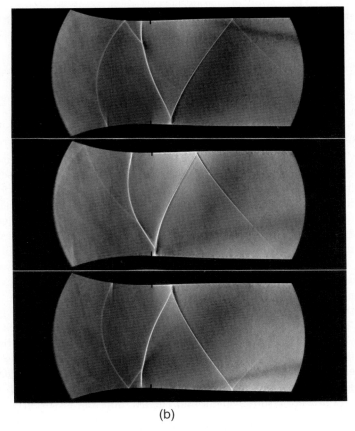

(b)

Fig. 6.19. continued

frequency branch completing a hysteresis loop. To date, this phenomenon has not physically been resolved yet. The schlieren photographs of one flow oscillations of each frequency branch are shown in Fig. 6.19. The sequence in Fig. 6.19a shows a symmetric oscillation of the lower branch, similar to Mode 1 discussed before. The sequence in Fig. 6.19b of a flow oscillation in the upper frequency shows a completely different phenomenon yet unresolved. A complex system of oblique shocks form and moves through the nozzle opposite to the main flow. The flow is no longer symmetric with respect to the nozzle axis. An unsteady Mach reflection near the throat is observed prior to the increased curvature of the upstream shock which then separates from the remaining shock system. This shock weakens and dies out as it accelerates into the oncoming flow. The flow repeats itself alternately at the upper and lower nozzle walls. This phenomenon awaits further investigation.

6.4.4 Condensation Induced Shock Waves in Prandtl–Meyer Flows

Condensation phenomena can also be investigated in Prandtl–Meyer flows around a sharp corner. They have direct applications at the trailing edge of the blades in steam turbines and can affect the downstream wake structure. The subject was investigated in the past by Smith [59] and Frank [60,61] experimentally and by Kurshakov et al. [62] both experimentally and numerically. The experiments mainly concentrated on measurements of the onset conditions, using a Mach–Zehnder interferometer, whereas numerical solutions used the method of characteristics. Neither seems to be sufficient in satisfactorily describing the phenomenon. A theoretical investigation of the subject using asymptotic solution of the rate equation along streamlines combined with the numerical method of characteristics has recently been made for both subcritical and supercritical flows (flows with embedded oblique shock waves) by Delale and Crighton [63,64]. A typical configuration of Prandtl–Meyer flows with homogeneous condensation around a sharp corner O between two uniform states, designated as 1 and 2, separated by an expansion fan is shown in Fig. 6.20. As soon as the flow crosses the saturation Mach line, clusters of vapor molecules of critical size (condensation nuclei) are produced by homogeneous nucleation. These nuclei then grow to form droplets of various sizes. Consequently, a nonequilibrium two-phase dispersed droplet flow with latent heat release from condensation is created. Assuming no velocity slip and the same temperature for the droplets and the surrounding gas, the nonequilibrium two-phase mixture can be written in the form of a single-phase flow of a homogeneous mixture with pressure p, density ρ, temperature T, flow speed u, flow direction θ, and enthalpy h. The equations of motion for the steady two-dimensional supersonic flow of the homogeneous mixture can then be written in natural coordinates $(s,\,n)$ as

$$\frac{1}{\rho}\frac{\partial \rho}{\partial s} + \frac{1}{u}\frac{\partial u}{\partial s} - \frac{\partial \theta}{\partial n} = 0, \tag{6.47}$$

$$\rho\,u\,\frac{\partial u}{\partial s} = -\frac{\partial p}{\partial s}, \tag{6.48}$$

$$\rho\,u^2\,\frac{\partial \theta}{\partial s} = \frac{\partial p}{\partial n} \tag{6.49}$$

and

$$\rho\,\frac{\partial h}{\partial s} = \frac{\partial p}{\partial s}, \tag{6.50}$$

where $h = c_{\mathrm{pm}}\,T - g\,L$. The system of equations is completed by the equation of state of the mixture given by (6.20) and the nonequilibrium condensation rate equation

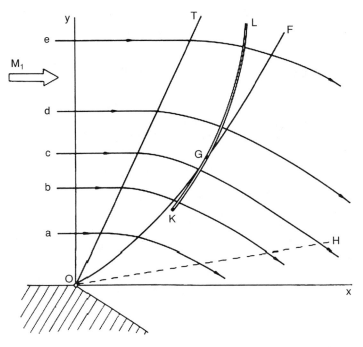

Fig. 6.20. Different classes of streamlines in Prandtl–Meyer flows with an embedded oblique shock wave KGL [64](OT and OH are, respectively, the isentropic tail and head of the expansion fan and OGF is the subcritical nucleation wave front). (**a**) A typical streamline along which the flow field remains subcritical, (**b**) a typical streamline which intersects the shock front in the droplet growth zone (DGZ), (**c**) a typical streamline which intersects the shock front in the nucleation zone with growth (NZ), (**d**) a typical streamline which intersects the shock front in the rapid growth zone (RGZ), and (**e**) a typical streamline which intersects the shock front in the further growth zone (FGZ)

$$g(s) = \int_{s_c}^{s} \left[r^*(\xi) + \lambda \int_{\xi}^{s} \Omega(\eta)\,\mathrm{d}\eta \right]^3 \frac{\Sigma(\xi)}{\rho(\xi)\,u(\xi)} \exp\left[-K^{-1} B(\xi) \right] \mathrm{d}\xi$$

$$(6.51)$$

along streamlines constructed in a manner similar to that of quasi-one-dimensional nozzle flows where the nozzle area is now replaced by the local normal separation $\Delta n \sim 1/(\rho\,u)$ of two arbitrarily close streamlines (the symbols K, B, λ, Σ, Ω, r^* are precisely the same as in (6.42)). The asymptotic solution for subcritical Prandtl–Meyer flows with homogeneous condensation follows by combining the flow equations (6.47)–(6.50) together with the thermal equation of state, (6.20), and the asymptotic solution of the condensation rate equation (6.51) along streamlines in the double limit as $K \to 0$ and $\lambda \to \infty$. Similar to the case of quasi-one-dimensional nozzle flows, the behavior of the activation function B shown in Fig. 6.10 (now x replaced by the streamwise coordinate s) identifies the distinct condensation zones along streamlines. For $s \leq s_c$, where s_c denotes the streamwise coordinate where the streamline meets the saturation line, the activation function B is large enough corresponding to practically vanishing nucleation rates. It decreases (corresponding to increasing nucleation rates) until it reaches a turning at $s = s_1$, the relative onset point, where $dB/ds = 0$ and the nucleation rate is practically maximum. In the interval $s_c \leq s \leq s_1$, where the flow is nearly frozen, one identifies the same physically distinct condensation zones of subcritical quasi-one-dimensional nozzle flows along streamlines. These zones, whose definitions and structures are discussed earlier, are the initial growth zone (IGZ) where the flow is isentropic, the further growth zone (FGZ), where $dB/ds = O(1)$, the rapid growth zone (RGZ), where $dB/dx = O(K^{1/2})$ and the onset zone (OZ), embedded in RGZ, where the influence of condensation begins to be felt. The corresponding asymptotic solutions for each nearly frozen zone along streamlines can be found in Delale and Crighton [63]. Along any streamline downstream of the onset zone ($s > s_1$), the effect of heat addition to the flow becomes dominant showing an increase in the pressure, density, and temperature of the mixture. Similar to the case of quasi-one-dimensional nozzle flows, two asymptotically and physically distinct condensation zones are distinguished: The nucleation zone with droplet growth (NZ) and the droplet growth zone (DGZ). In these zones the nearly frozen approximation fails and the asymptotic solution of the condensation rate equation along streamlines has to be coupled to the equations of motion with heat addition in characteristic form. The construction of the global solution of subcritical Prandtl–Meyer flows in the heat addition zones can be found in [63]. Using the condensation model given by Schnerr and Dohrmann [50], which has been verified in nozzle experiments, Delale and Crighton found that their results for the onset of condensation agreed only qualitatively with the experiments of Smith [59], i.e., the onset front of condensation was concave with respect to the oncoming flow and moved toward the oncoming flow as the free stream relative humidity was increased. Furthermore, their computations with the subcritical asymptotic algorithm showed that characteristics intersecting emanating from the corner intersected in the heat addition zones, demonstrating clearly the need for the inclusion of an embedded oblique shock wave. Prandtl–Meyer flows with an oblique shock wave embedded in the expansion fan will be called supercritical. Figure 6.20 shows a typical supercritical flow with an embedded shock front KL due to excessive heat release by condensation. It has been recently demonstrated by Delale and van Dongen [42] that such a shock wave shows a supersonic to supersonic transition and can, therefore, be assumed to be weak (strong shock waves with supersonic to subsonic transition occur in supercritical flows in ducts and nozzles, e.g., see Schnerr [47] and Delale et al. [38]). The existence of an embedded shock KL, with portion KG lying in the heat addition zones and with portion GL lying in the nearly frozen zones, distinguishes the different classes of streamlines, designated by a–e in Fig. 6.20, with distinct condensation zones structured by the variation of the activation function B along them. In Fig. 6.20, (a) denotes a streamline where subcritical flow prevails. In this case the condensation zones along the streamline are precisely the same as those discussed earlier. On the other hand, (b)–(e) in Fig. 6.20 show streamlines intersecting the embedded shock front KL. In particular, streamlines (b) and (c) show the condensation zones along those streamlines that intersect the portion KG of the embedded shock lying in the heat addition zones whereas streamlines (d) and (e) show the condensation zones along those streamlines which intersect the portion GL of the embedded shock lying in the nearly frozen zones. The embedded shock location s_z along each streamline, designated by b–e in Fig. 6.20, lies, respectively, in the droplet growth zone (DGZ), in the nucleation zone with growth (NZ), in the rapid growth zone (RGZ) and in the further growth zone (FGZ). The onset zone (OZ), where the observed onset of condensation falls, precedes the embedded shock location s_z in each case. The complete solution of supercritical Prandtl–Meyer flows with homogeneous condensation requires a shock formation theory due to excessive latent heat release. Such a theory together with the evaluation of the embedded shock origin and a shock fitting technique is discussed in detail in Delale and Crighton [64]. A complete solution of the problem is then determined using the equations of motion together with the asymptotic solution of the rate equation in different regimes along streamlines supplemented by the oblique shock relation [65]. The results of the computations using the condensation model of Schnerr and Dohrmann [50] for the experiments of Smith [59] are shown in Fig. 6.21. Good agreement between the onset conditions of the experiments and those of computations is achieved by adjusting the curvature of the embedded shock at the shock origin K. Before concluding this section, it should be mentioned that the experiments of Frank [60, 61] showed oscillatory instabilities for higher values of the initial relative humidities of the oncoming flow for moist air expansions around a sharp corner. However, in these experiments the flow was bounded by solid walls from both sides, the lower wall formed the sharp corner and the upper was a plane wall. In principle, this configuration forms a half-nozzle with a centered expansion at the throat and it is quite understandable that the observed oscillatory flows resemble those encountered in Laval nozzles.

6.4.5 Condensation Induced Shock Waves in Unsteady Rarefaction Waves

The shock tube offers an alternative technique for investigating flows with condensation. The experimental conditions in this case can be better controlled. Wegener and Lundquist [5] were the first to use the shock tube for condensation studies and the first streak photographs displaying condensation fronts in the rarefaction fan were obtained by Glass and Patterson [8]. Since then, shock tube experiments for the condensation of various vapors have been carried out [8–11, 66, 67, 69]. For flows with homogeneous condensation, a mixture of a condensable vapor and a carrier gas is initially at rest in the

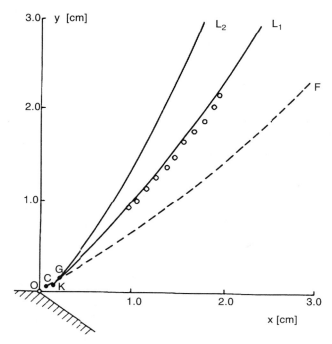

Fig. 6.21. The embedded shock fronts KGL_1 and KGL_2 evaluated, respectively, for the values of the embedded shock curvature $(d^2\beta/d\mu^2)^+_K = 0.01$ and 1.0 at the shock origin K, for the experiments of Smith [58] under nozzle supply temperature $T'_{01} = 284\,\mathrm{K}$, supply specific humidity $w_0 = 6.8\,\mathrm{g\,kg^{-1}}$ and supply relative humidity $\phi_0 = 0.41$ [64](o circles denote the experimental onset conditions and CGF is the subcritical nucleation wave front)

driver section (region 4 in Fig. 6.22) of a shock tube. Just after the diaphragm at $x = 0$ is instantly removed, the mixture expands into the channel and the condensable vapor is cooled after reaching saturation (line OS in Fig. 6.22) to high supersaturation ratios by homogeneous nucleation until condensation becomes visible (onset of condensation) in the rarefaction fan. A typical onset of condensation front EF is shown in Fig. 6.22. The characteristics or waves emanating from point O begin to curve due to latent heat release by condensation as they cross the onset front EF. Characteristics of the same family then intersect in regions of intense heat addition forming shock waves that propagate in the driver section. Using the homogeneous flow model, the equations of motion of unsteady compressible flow of a mixture of condensable vapor and a carrier gas can be written as

$$\frac{\partial \rho}{\partial t} + \frac{\partial}{\partial x}(\rho u) = 0, \qquad (6.52)$$

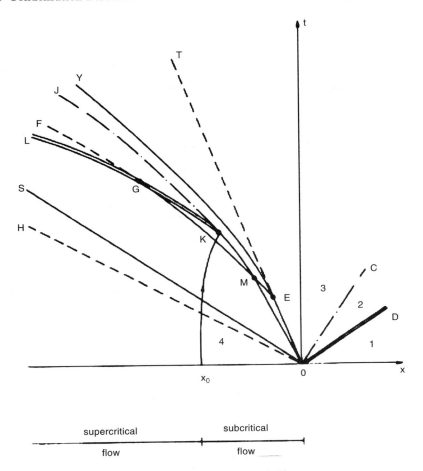

Fig. 6.22. Wave diagram for shock tube flows with nonequilibrium condensation. Regions 1 and 4, respectively, correspond to the initial states in the driver section and in the channel, OH and OT are, respectively, the head and the tail of the frozen rarefaction wave, EF is the condensation wave front [68] (onset of condensation), OMJ is a typical characteristic emanating from the origin O and penetrating into the heat addition zones KP that portion of the pathline initially at x_0 in the driver section

$$\frac{\partial u}{\partial t} + u\,\frac{\partial u}{\partial x} = -\frac{1}{\rho}\,\frac{\partial p}{\partial x}, \tag{6.53}$$

and

$$\frac{\mathrm{d}h}{\mathrm{d}t} = \frac{1}{\rho}\,\frac{\mathrm{d}p}{\mathrm{d}t}, \tag{6.54}$$

where ρ, p, and u are, respectively, the mixture density, the mixture pressure, and the flow velocity, $h = c_{\mathrm{pm}} T - g L$, t is the time and x is the axial coordinate along the tube with origin located at the diaphragm separating the driver section and the channel. The thermal equation of state is the same as (6.20). Similar to the case of nozzle and corner expansion flows, the system of equations is further supplemented by the nonequilibrium condensation rate equation, constructed in a similar manner,

$$g(t) = \int_{t_s}^{t} \left[r^*(\tau) + \lambda \int_{\tau}^{t} \Omega(\eta) \, d\eta \right]^3 \frac{\Sigma(\tau)}{\rho(\tau)} \exp\left[-K^{-1} B(\tau) \right] d\tau \quad (6.55)$$

along pathlines. The above set (6.52)–(6.55) were solved numerically by Sislian and Glass [9] using a second-order scheme characteristics method. They calculated the case of homogeneous condensation for water vapor–nitrogen mixture. They showed that characteristics emanating from the origin remained straight as long as the effect of homogeneous condensation were negligible. After penetrating the condensation zone the characteristics began to curve significantly toward the head of the rarefaction waves due to latent heat release from condensation. Their computations showed the formation of shock waves in the condensation zones. They employed Lax's method of implicit artificial viscosity to account for the presence of shock waves. Their results did not agree with those of experiments. Kotake and Glass [11] incorporated heterogeneous condensation by varying the contact, but their results were not satisfactory either. Recently, Delale et al. [68] used the asymptotic solution of the rate equation along pathlines and the method of characteristics in the heat addition zones to determine the solution for subcritical flows (flows without a shock wave in the rarefaction wave) using the condensation model of Schnerr and Dohrmann [50]. Their results for the onset of condensation showed agreement with those of Peters' experiments [69] for low vapor pressures, but did not agree with the results of Barschdorff's experiments [67]. They concluded that shock waves formed in the rarefaction waves due to intersecting characteristics were responsible for the discrepancy. They also qualitatively discussed supercritical flows with shocks, but they did not present any quantitative results for this case. Condensation phenomena in a rarefaction wave has also recently been investigated by van Dongen et al. [44] for the case of homogeneous nucleation in humid air. The Euler equations for an ideal two-phase two-component mixture are supplemented by the condensation model developed given by Prast [70] and Lamanna [27] and are solved numerically by a finite volume method developed by Sun et al. [71]. Figure 6.23 shows the numerical results for the pressure and nucleation rate against the reduced coordinate $x/(c\,t)$, where c_0 is the frozen speed of sound of the mixture. A condensation induced left-running compression wave that steepens into a shock wave is observed. The shock wave tends to take over the expansion wave. As time proceeds, the maximum supersaturation ratio is reduced. Consequently, the upstream moving shock locally quenches the nucleation rate, which is replaced further downstream. Finally, we should mention that a

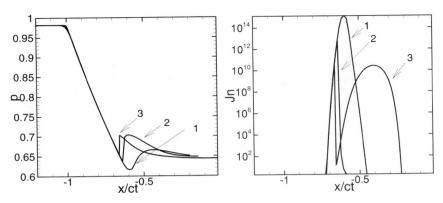

Fig. 6.23. (a) Snapshots of the pressure (bar) versus the reduced coordinate at (1) t = 10 ms, (2) t = 37 ms, and (3) t = 88 ms. (b) The nucleation rate $(m^{-3} s^{-1})$versus the reduced coordinate at (1) t = 10 ms, (2) t = 37 ms, and (3) t = 88 ms (humid nitrogen in the driver section of Fig. 6.24 with initial relative humidity ϕ_4 = 0.82, initial temperature T_4 = 298 K, initial pressure p_4 = 1.0 bar and with initial pressure p_1 = 0.4 bar in the channel) [44]

numerical, analytical, and experimental investigation of rarefaction waves for the heterogenous condensation of humid nitrogen was carried out by Smolders et al. [72, 73]. They have shown, by asymptotic analysis, that the self-similar character of the expansion wave in this case is restored in the long time (i.e., as $t \to \infty$) with the equilibrium speed of sound being the relevant scaling parameter.

6.4.6 Future Perspectives

In this study we tried to summarize the essential features of condensation induced shock waves in steady and unsteady nozzle flows, in Prandtl–Meyer expansions and in the rarefaction waves of shock tubes. Despite the fact that nozzle flows with stationary shock waves and the limit of stability of such shock waves are well understood at all levels (theoretical, numerical, and experimental) nowadays, unsteady flows in nozzles with three modes of oscillations in slender nozzles need further investigations concerning the existence of the frequency minimum. In addition the bifurcation to nonsymmetric flow oscillations with oblique shock waves and the hysteresis nature of the frequency curves need further investigations. The built-up of model equations and analogies from different fields showing similar flow patterns may be helpful in understanding these phenomena.

Most of the applications of nonequilibrium condensing flows have so far been in steam turbine technology, in particular, for the flow of steam through nozzles and between turbine blades. Nonequilibrium condensation phenomena can also be employed in separating a vapor phase from a two-phase mixture by

228   C.F. Delale et al.

means of phase separation. Condensation in cryogenic equipment has not been
studied sufficiently and needs further investigation. Condensation studies with
or without shock waves can also be used to validate different condensation
models (e.g., see [74–76]) providing information on nucleation and droplet
growth together with droplet size spectrum.

References

1. Prandtl, L.: Allgemeine überlegungen die strömung zusammendrückbarer
   flüssigkeiten. In: Atti del V Convegno Volta, 1st edition vol XIV. p. 167 Roma
   Reale Accademia D'Italia (1936)
2. Hermann, R.: Luftfahrtforschung **19**, 201 (1942)
3. Oswatitsch, K.: ZAMM **22**, 1 (1942)
4. Stever, H.G.: Condensation in high speed flows. In: Fundamentals of Gas Dynamics, High Speed Aerodynamics and Jet Propulsion. pp. 526–573 Princeton
   University Press, Princeton, NJ (1958)
5. Wegener, P.P., Lundquist, G.: J. Appl. Phys. **22**, 233 (1951)
6. Wegener, P.P., Mack, L.M.: Adv. Appl. Mech. **5**, 307 (1958)
7. Wegener, P.P., Pouring, A.A.: Phys. Fluids **7**, 352 (1964)
8. Glass, I.I., Patterson, G.N.: J. Aeronaut Sci. **22**, 75 (1955)
9. Sislian, J.P., Glass, I.I.: AIAA J. **14**, 1731 (1976)
10. Glass, I.I., Kalra, S.P., Sislian, J.P.: AIAA J. **15**, 686 (1977)
11. Kotake, S., Glass, I.I.: Prog. Aerospace Sci. **19**, 129 (1981)
12. Volmer, M., Weber, A.: Z. Phys. Chem. A **119**, 277 (1926)
13. Farkas, L.: Z. Phys. Chem. A **225**, 236 (1927)
14. Becker, R., Döring, W.: Ann. Phys. **24**, 719 (1935)
15. Zel'dovich, Y.B.: Sov. Phys. JETP **12**, 525 (1942)
16. Frenkel, J.: Kinetic Theory of Liquids. Oxford University Press, New York
    (1946)
17. Lothe, J., Pound, G.M.: J. Chem. Phys. **36**, 2080–2085 (1962)
18. Feder, J., Russel, K.C., Lothe, J., Pound, G.M.: Adv. Phys. **15**, 111 (1966)
19. Kashchiev, D.: Nucleation: Basic Theory with Applications. Butterworth-
    Heinemann, Oxford (2000)
20. Dillmann, A., Meier, G.E.A.: J. Chem. Phys. **94**, 3872 (1991)
21. Delale, C.F., Meier, G.E.A.: J. Chem. Phys. **98**, 9850 (1993)
22. Kalikmanov, V.I., van Dongen, M.E.H.: J. Chem. Phys. **103**, 4250 (1995)
23. Courtney, W.G.: J. Chem. Phys. **35**, 2249 (1961)
24. Girshick, S.L., Chiu, C.P.: J. Chem. Phys. **93**, 1273 (1990)
25. Gyarmathy, G.: The spherical droplet in gaseous carrier streams : Review and
    synthesis. In: Hewitt G.F., Delhaye J.M., Zuber N.(eds.)Multiphase Science and
    Technology vol.1. pp. 99–279 Springer, Berlin Heidelberg New York (1982)
26. Young, J.B.: Intl.J. Heat Mass Transfer **36**, 2941 (1993)
27. Lamanna, G.: On nucleation and droplet growth in condensing nozzle flows.
    PhD Thesis, Eindhoven University of Technology (2000)
28. Shapiro, A.H.: The Dynamics and Thermodynamics of Compressible Fluid
    Flow, vols. 1 and 2. Ronald, New York (1953)
29. Bartlmä, F.: Berechnung des Strömungsvorgangs bei Überschreiten der kritis-
    chen Wärmezufuhr. DVL Bericht Nr. 168 (1961)

30. Zierep, J.: Strömungen mit Energiezufuhr. Braun, Karlsruhe (1990)
31. Anderson, J.: Modern Compressible Flow. pp. 103–104 McGraw-Hill, London New York (2004)
32. Landau, L.D., Lifshitz, E.M.: Fluid Mechanics, Pergamon, Oxford New York Toronto (1959)
33. Hill, P.G.: J. Fluid Mech. **25**, 593 (1966)
34. Barschdorff, D.: Forsch. Ing. **37**, 146 (1971)
35. Blythe, P.A., Shih, C.J.: J. Fluid Mech. **76**, 593 (1976)
36. Clarke, J.H., Delale, C.F.: Phys. Fluids **29**, 1398 (1986)
37. Delale, C.F., Schnerr, G.H., Zierep, J.: Phys. Fluids A **5**, 2969 (1993)
38. Delale, C.F., Schnerr, G.H., Zierep, J.: Phys. Fluids A **5**, 2982 (1993)
39. Delale, C.F., Schnerr, G.H., Zierep, J.: Z. Angew. Math. Phys. (ZAMP) **44**, 943 (1993)
40. Delale, C.F., Lamanna, G., van Dongen, M.E.H.: Phys. Fluids **13**, 2706 (2001)
41. Möhring, W.: On flows with heat addition in Laval nozzles. In: Müller, U., Roesner, K.G., Schmidt, B. (eds.) Recent Developments in Theoretical and Experimental Fluid Mechanics. pp.179–185 Springer, Berlin Heidelberg New York (1979)
42. Delale, C.F., van Dongen, M.E.H.: Z. Angew. Math. Phys. (ZAMP) **49**, 515 (1998)
43. Younis, S.: Stationäre Strömungen durch Lavaldüsen mit Wärmezufuhr. Diplomarbeit, Universität (TH) Karlsruhe (1987)
44. van Dongen, M.E.H., Luo, X., Lamanna, G., van Kaathoven, D.J.: On condensation induced shock waves. In: Proceedings of 10th Chinese Symposium on Shock Waves. pp. 1–11 Chinese Academy of Sciences, Yellow Mountain China (2002)
45. Erdelyi, A.: Asymptotic Expansions. Dover, New York (1956)
46. Peters, F., Paikert, B.: Exps. Fluids, **7**, 521 (1989)
47. Schnerr, G.: Exps. Fluids **7**, 145 (1989)
48. Pouring, A.A.: Phys. Fluids **8**, 1802 (1965)
49. Bratos, M., Meier, G.E.A.: Arch. Mech. **28**, 1025 (1976)
50. Schnerr, G.H., Dohrmann, U.: Theoretical and experimental investigation of 2-D diabatic transonic and supersonic flow fields. In: Zierep, J., Oertel, H. (eds.) Proceedings of the IUTAM Symposium Transsonicum III. pp. 125–135 Springer, Berlin Heidelberg New York (1989)
51. Schmidt, B.: Theorie und Konstruktion eines Interferometers nach Mach-Zehnder. Dissertation, Technische Hochschule Karlsruhe (1962)
52. Barschdorff, D.: Kurzzeitfeuchtemessung und ihre Anwendung bei Kondensationserscheinungen in Lavaldüsen. Heft 6, Universität Karlsruhe (1967)
53. Barschdorff, D., Fillipov, G.A.: Heat Transfer – Sov. Res. **2**, 76 (1970)
54. Zierep, J., Lin, S.: Forsch. Ing. **34**, 97 (1968)
55. Saltanov, G.A., Tkalenko, R.A.: Zh. Prikl. Mek. i Tekn. Fiz. (USSR) **6**, 42 (1975)
56. Guha, A., Young, J.B.: Time-marching prediction of unsteady condensation phenomena due to supercritical heat addition, In: Turbomachinery: Latest Developments in a Changing Scene, Inst. Mech. Engrs., pp. 167–177, London (1991)
57. White, A.J., Young, J.B.,: J. Propulsion Power **9**, 579 (1993)
58. Adam, S., Schnerr, G.H.: J. Fluid Mech. **348**, 1 (1997)

59. Smith, L.T.: AIAA J. **9**, 2035 (1971)
60. Frank, W.: ZAMM **59**, T223 (1979)
61. Frank, W.: Acta Mech. **54**, 135 (1985)
62. Kurshakov, A.B., Saltanov, G.A., Tkalenko, R.A.: Zh. Prikl. Mek. i Tekn. Fiz. (USSR) **5**, 117 (1971)
63. Delale, C.F., Crighton, D.G.: J. Fluid Mech. **359**, 23 (1998)
64. Delale, C.F., Crighton, D.G.: J. Fluid Mech. **430**, 231 (2001)
65. Clarke, J.H., Delale, C.F.: Q. Appl. Math. **XLVI**, 121 (1988)
66. Kawada, H., Mori, Y.: Bull. JSME **16**, 1053 (1973)
67. Barschdorff, D.: Phys. Fluids **18**, 529 (1975)
68. Delale, C.F., Schnerr, G.H., Zierep J.: J. Fluid Mech. **287**, 93 (1995)
69. Peters, F.: J. Phys. Chem. **91**, 2487–2489 (1987)
70. Prast, B.: Condensation in supersonic expansion flows : Theory and numerical evaluation. SAI, PhD Thesis, Eindhoven University of Technology (1997)
71. Sun, M., Takayama K.: J. Comp. Phys. **150**, 143 (1998)
72. Smolders, H.J., Niessen, E.M.J., van Dongen, M.E.H.: Comput. Fluids **21**, 63 (1992)
73. Smolders, H.J., van Dongen, M.E.H.: Shock Waves **2**, 255 (1992)
74. Peeters, P., Pieterse, G., Hruby, J., van Dongen, M.E.H.: Phys. Fluids **16**, 2567 (2004)
75. Holten, V., Labetski, D.G., van Dongen, M.E.H.: J. Chem. Phys. **123**, 104505-1-9 (2005)
76. Gharibeh, M., Kim, Y., Dieregsweiler, U., Wyslouzil, B.E., Ghosh, D., Strey, R.: J. Chem. Phys. **122**, 094512-1-9 (2005)

Liquefaction Shock Waves

Gerd E.A. Meier

Summary. Abruptly changing the gaseous state of a substance to the liquid one with a compression in a shock wave is a very special phenomenon. In a usual shock wave compressing an arbitrary pure gas, it is expected the gas to be heated adiabatically and not liquefied. But for gases with a high specific molar heat capacity in certain state regimes a strong shock wave can change the gas into a liquid. Weaker shocks in those gaseous fluids also can lead to nucleation and particle condensation depending on the magnitude of achieved supersaturation. The reason for this different behavior of a certain class of fluids is their molecular structure. If the molecules become large, the number of degrees of freedom for vibrations grows. Consequently the specific heat capacity measured on a molar basis is increasing with the number of atoms in the molecule. Adiabatic compression can liquefy such a gas because the large internal heat capacity can store all the heat provided by the compression work. So it is evident, high specific heat capacity of a substance allows for a lot of interesting physical phenomena like liquefaction in shock waves or as well complete evaporation in an expansion wave under special conditions. Accompanying phenomena in these adiabatic phase transitions are shock splitting, shock instabilities, vortical instabilities and nucleation processes of different kind. An important aspect is the widespread practical use of these substances in many applications, where unexpected state changes can be dangerous or even helpful.

7.1 Introduction

In the past thirty-years some progress has been made in understanding the properties and related phenomena of real gas flow [93, 167]. The richness of peculiarities has forced us to introduce a special name for this field: Real Gas Dynamics. In this field the unusual properties of the so called "Retrograde Substances" have also led to the finding of liquefaction shock waves which is the main topic of this article.

It was an idea of the late Philip Thompson [1] in 1975, a strong shock wave being able to change a gas into a liquid for substances with a high specific molar heat capacity. His idea was verified later in 1978 experimentally by a common group in a first shock tube experiment in Gottinge [3]. For the so-called "Retrograde Fluids," it is possible in certain state regimes to achieve complete phase transition from the gas state to the liquid state and vice versa by adiabatic pressure changes. Adiabatic change characterizes a process without heat transfer to the environment or external sources. A retrograde fluid in this definition is a fluid for which the specific heat capacity multiplied by temperature is greater than the latent heat of evaporation.

Although this definition can easily be extended to other phase changes, such as the graphite–diamond transition, in this article we refer only to the vapor/liquid transition of organic substances with high molecular weight and complex molecular structure. Their structure allows for many vibrational degrees of freedom and high specific heat on a molar basis. A typical test substance of this kind is PP3 (Perfluoro-dimethyl-cyclohexane) which has a specific heat of about 50 gas constants. Since this substance is not poisoness and also not burnable it is a preferred substance for tests. The properties of PP3 are similar to those of octane which is a more common type of a retrograde fluid.

The spectacular flow phenomena described in this article include liquefaction shock waves, shock splitting, and evaporation waves. Instabilities of those waves are mainly observed in supersonic flow and in strong pressure waves. These gasdynamic processes with velocities in the order of the sound speed can produce the large pressure changes resulting in the mentioned phase changes in a flow or wave.

7.2 Retrograde Fluids and their Role in Gasdynamics

7.2.1 Introductary Remarks

As mentioned earlier, a retrograde fluid in our definition is a fluid for which the specific heat capacity multiplied by temperature is greater than the latent heat of evaporation. Retrograde behavior is shown by fluids with a heat capacity higher than 11 gas constants [1]. For these substances, it is easily possible to achieve complete or partial phase transition from the gas to the liquid state and vice versa by means of adiabatic pressure changes in flows or waves, without heat transfer to external sources. Figure 7.1 shows the enormous difference of the shape of phase boundaries for normal and retrograde substances in a temperature/entropy diagram.

The gasdynamic flows encountering large pressure changes producing the phase changes described earlier, are realized in shock tubes and expansion ducts. Especially in cases with flow velocities in the order of the speed of sound all mentioned phenomena are easily observed. The flow phenomena described in this article include liquefaction and evaporation waves, wave instabilities and supersonic two-phase flow. Spectacular additional observations like nucleation processes, particle condensation, vortical instabilities, strange

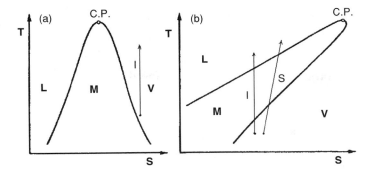

Fig. 7.1. Phase boundaries in a T/s - Diagram for (**a**) normal substances like water and (**b**) for retrograde substances like octane. Isentropic processes I can pass in (**b**) from gas states up into liquid states through the mixture states and vice versa. The arrows show isentropic (I) or shock (S) compression

two phase wave patterns and boundary layer effects or the whole field of mixtures of different fluids can only be mentioned. It is not possible to give a comprehensive description of the whole field of Real Gas Dynamics in this article. The reader may therefore see it as a sectional overlook and introduction to the field with additional references to the numerous papers cited here.

7.2.2 The Liquefaction Shock Wave History

The liquefaction shock is thus a new phenomenon, quite distinct from the well-known condensation of vapor in an expanding flow, for example, in a supersonic nozzle, which was studied by many authors in the past [see 101–159]. Virtually all former spontaneous condensation experiments of which we are aware have involved states of supersaturation reached in an expansion process, using regular fluids (mainly water) or, occasionally, slightly retrograde fluids (such as benzene). These experiments extend back at least to the time of Wilson in 1897 [101]. A useful summary for water was given by Wegener in 1969 [67, 68]. This form of rapid condensation has invariably been associated with supersaturated states achieved by rapid adiabatic expansion like for instance in supersonic Laval Nozzle flow. Mainly by the related heat addition to the flow by vapor condensation, it is often accompanied by a recompression shock wave, sometimes called a "condensation shock." In order to clearly distinguish the subject of this article from such shock waves, we have chosen the distinct name "liquefaction shock." Unfortunately, the designation "complete condensation shock" was used in the title of the work of Thompson and Sullivan from 1975 [1], which describes the elementary theory for the liquefaction shock. It is proposed to avoid this designation in the future.

The physical basis for the liquefaction shock can be easily understood, thinking of the shock process as an adiabatic compression. Such a compression will, if carried far enough, inevitably lead to liquefaction of the gas, provided that the gas temperature does not become too large. This will indeed be the case if the gas molecules have a large number of vibrational degrees of freedom (i.e., the gas has a large molar heat capacity) so that the work of compression can be stored as internal energy in the gas with only a modest temperature rise. It is instructive in this connection to consider an isentropic compression in a perfect gas, for which $T \sim P^{(\gamma-1)/\gamma}$; then as $\gamma \to 1$ then $T \to$ const. After the onset of liquefaction, these arguments apply to the storage of the latent heat released by condensation as well, i.e., the temperature rise will again be small if the molar heat capacity is large.

7.2.3 Fluid Properties of Retrograde Substances

The detailed calculations of Thompson and Sullivan yield minimum values for the heat capacity of the fluid which allow liquefaction [1]. These are conveniently expressed in terms of a standard heat capacity $\widetilde{c}_v = c_v^0(T)/R$, the normalized ideal-gas heat capacity evaluated at the critical temperature for the substance of interest. Partial shock liquefaction is possible for $\widetilde{c}_v > 11.2$ and complete shock liquefaction is possible for $\widetilde{c}_v > 24.1$ the former value happens to correspond to sulfur hexafluoride (SF_6) and the latter value corresponds roughly to hexane (C_6H_{14}). A typical value for currently-used, multiatomic experimental fluids, $\widetilde{c}_v = 50$. The fluids for which $\widetilde{c}_v > 11.2$ are always mentioned as "retrograde fluids," $\widetilde{c}_v < 11.2$ as "regular fluids." Retrograde behavior can occur over a restricted range of temperature, depending on the heat capacity c_v. Substances with $\widetilde{c}_v < 11.2$ for which adiabatic condensation of the vapor is only possible in an expansion process, referred to as regular, are for example water H_2O with $\widetilde{c}_v = 3.5$ or other substances with low molecular weight.

The physical distinction between retrograde and regular fluids lays in the number of internal degrees of freedom, in particular, vibrational modes of the molecule. This, in turn, represents the ability of the molecule to store internal energy. For a fluid near the lower limit of retrograde behavior $c_pT \approx L$, where L is the latent heat. Substances with $\widetilde{c}_v > 11.2$, for which adiabatic condensation of the vapor is possible in a compression process, referred to as retrograde, have large molecules like octane C_8H_{18} ($\widetilde{c}_v = 36.8$). Regular and retrograde processes are shown in the pressure/volume diagram of Fig. 7.1. It should be remarked that the establishment of a limiting value $\widetilde{c}_v = 11.2$ for retrograde behavior, and of similar limits, is based on the assumption of a particular kind of corresponding-states thermodynamic behavior, and therefore represents an approximation. The corresponding states model used was explicitly described by Thompson and Becker in 1979 [51].

The property calculations used for comparison with and conversion of experimental data are based on the "BWR 44" equation of state (denoting a Benedict-Webb-Rubin equation with 44 constants) of Yamada from 1973 [45] and were checked in some cases using the virial-type equation of Hobbs from 1976 [54]. The calculation of property jumps across a shock flow follows conventional shock theory with assumed equilibrium states upstream and downstream of the shock (in certain cases, with assumed metastable states, i.e., supersaturated vapor) as described by Thompson and Sullivan [1].

The exclusively plausible possibility of compression condensation is sometimes regarded with skepticism because Landau and Lifshitz make 1959 the following comment about the possibility of a liquefaction shock wave [162]:

"It should be emphasized that condensation discontinuities are a distinct physical phenomenon, and do not result from the compression of gas in an ordinary shock wave; the latter effect cannot lead to condensation, since the increase of pressure in the shock wave has less effect on the degree of super-saturation than the increase of temperature..."

This argument corresponds to the physical basis discussed in the preceding, but since it is incomplete, leads to an incorrect conclusion.

7.2.4 Gasdynamic Aspects of Fluid Properties

Ordinary gas dynamics treats the rapid motions of compressible fluids under pressure gradients with the usual approximations, including isentropic flow (except at shock discontinuities) and local thermodynamic equilibrium. Increasing the molar specific heat capacity serves to increase the adiabatic compressibility of the fluid. Surprisingly, this leads to drastic changes in the gasdynamic behavior of flows with phase changes, especially when coupled with departures from equilibrium. The enhancement of compressibility can be represented by the simple identity

$$(\partial p/\partial v)_s = \gamma(\partial p/\partial v)_T, \qquad (7.1)$$

where the ratio of specific heats $\gamma = c_p/c_v$ approaches the minimum value unity as $c_v \to \infty$. Thus, as c_v becomes large, isentropes approach isotherms in slope and form. Details become more clear in the idealized pressure/volume diagram of Fig. 7.2. Consider a saturated vapor, as at point g in Fig. 7.2: if the heat capacity is large enough, an isentropic compression, like an isothermal compression, will result in condensation of the vapor (Fig. 7.2a) – we call this behavior "retrograde." If the heat capacity is small, isentropic expansion will result in condensation (Fig. 7.2b) – we call this behavior "regular." Examples of fluids displaying retrograde and regular behavior are – as already mentioned – n-Octane and water, respectively. At some intermediate heat capacity, isentropic pressure change will just hold the vapor on the saturation boundary. These distinctions were known to van der Waals already [160] and are extensively discussed by Thompson et al. [1, 47, 49, 50].

The reversal of the adiabatic condensation behavior – that is, the change from regular to retrograde behavior with increasing heat capacity, is the basis for several new forms of wave behavior in fluids [77, 107]. Retrograde behavior corresponds to a positive slope of the saturated-vapor boundary in a conventional temperature–entropy diagram (see Fig. 7.1): That is, $(ds/dT)_\sigma$ and

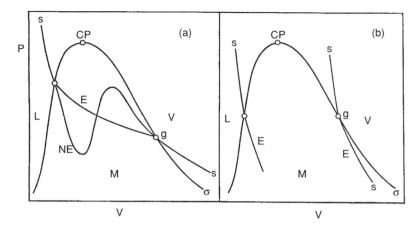

Fig. 7.2. Isentropes in the pressure–volume plane for retrograde and regular substances. L, M, and V designate liquid, mixture, and vapor, respectively. CP = critical point, E = equilibrium isentrope, NE = nonequilibrium isentrope. (**a**) Retrograde case. (**b**) Regular case (only equilibrium isentropes are shown)

therefore $(\mathrm{d}s/\mathrm{d}p)_\sigma$ are positive. It is convenient to derive the nondimensional "retrogradicity"

$$r(T) = (\partial T/\partial V)_p(\mathrm{d}s/\mathrm{d}p)_\sigma, \tag{7.2}$$

where $r > 0$ corresponds to retrograde behavior and $r < 0$ to regular behavior and the subscript σ indicates a path along the saturation boundary.

The existence and stability of shock waves is dependent on the curvature of the isentropes after Bethe [161] since the entropy change in a shock

$$s_2 - s_1 = \frac{1}{12T_1} \left(\frac{\partial^2 V}{\partial p_1^2} \right)_s (p_2 - p_1)^3 \tag{7.3}$$

is positive for a positive curvature only. Therefore, after Landau [8, 162] a so-called Fundamental Derivative Γ is a useful quantity measuring the curvature

$$\Gamma = -\frac{V}{2} \left(\frac{\partial^2 p}{\partial V^2} \right)_s \bigg/ \left(\frac{\partial p}{\partial V} \right)_s = \frac{c^4}{2V^3} \left(\frac{\partial^2 V}{\partial p^2} \right)_s = c \left(\frac{\partial(\rho c)}{\partial p} \right)_s . \tag{7.4}$$

It is easy to show from (7.4) that $\Gamma < 0$ near the critical point and at high temperatures on the vapor boundary, and for all states at the liquid boundary. For other phase transitions, e.g., solid–liquid or solid–solid, Γ can have either sign, $\Gamma > 0$ being typical in solids and ordinary fluids. Supposing that a fluid has a large heat capacity, the following properties are related to its gasdynamic behavior:

– Negative values of the fundamental nonlinearity parameter $\Gamma = c(\partial\rho c/\partial p)_s$ will be found at near critical temperatures, implying the possibility of single-phase rarefaction shocks [4]. Positive values will be found elsewhere.

– Singular values of Γ ($\Gamma = -\infty$) are found at the saturated-vapor boundary, as at the kink point g in Fig. 7.2a, implying the possibility of shock splitting [33–36] (see the following paragraphs) and two-phase rarefaction shocks.

– Internal energy storage which is large compared to the latent heat, implying the possibility of liquefaction shock waves, in which superheated vapor is compressed to a mixture or liquid downstream state.

Each of the phenomena mentioned earlier can be described, at least qualitatively, in terms of equilibrium states. In shock splitting, however, condensation is delayed and nonequilibrium, supersaturated states play an important role.

7.2.5 One-Dimensional Treatment of Real Gas Flows

For the calculations of the isentropic real gas flows the following set of equations can be used for a one-dimensional treatment in simple flow geometries:

$$\begin{array}{ccc} \text{Continuity} & \text{Momentum} & \text{Energy} \\ \mathrm{d}(\rho w F) = 0 & w\mathrm{d}w + V\mathrm{d}p = 0 & \mathrm{d}s = 0 \end{array} \tag{7.5}$$

with ρ for density, w for velocity, F for area function, V for volume, p for pressure, and s for entropy.

A comparatively simple thermal equation of state after Abbott [44] is

$$p_\mathrm{r} = \frac{T_\mathrm{r}}{Z_\mathrm{c}(V_\mathrm{r} - 1) - (O_c - 1)} - \frac{O_c^3 + (a_c O_c^2 (O_c - 1) + O_c^2)(T_\mathrm{r} - 1)}{Z_\mathrm{c}^2 (V_\mathrm{r} - 1)^2 + O_c Z_\mathrm{c}(V_\mathrm{r} - 1) - O_c^2 (O_c - 1)} \tag{7.6}$$

with p_r reduced pressure, T_r reduced temperature, and V_r reduced volume in the flow. O_c called the "Nondimensional Temperature Function," a_c called "Riedel Parameter," and Z_c called "Critical Compressibility" are typical constants of the fluid. For PP1 (C_6F_{14}) these constants are for instance

$$O_c = 0.838; \quad a_c = (\mathrm{d}p_s/\mathrm{d}T_s)_k = 8.18; \quad Z_\mathrm{c} = p_c V_c / R T_c = 0.267.$$

Besides the Riedel Parameter a_c these values are similar for most organic substances.

A suitable caloric equation of state is a simple formula for the reduced specific heat $\widetilde{c}_v$, depending on a temperature function $Y = B(T_\mathrm{r} - t)$, in the following form [46, 51]:

$$\begin{aligned} \widetilde{c}_v &= c_v/R = (Y^2/(1 + Y^2))c_e + c_a \quad \text{for } Y > 0, \\ \widetilde{c}_v &= c_v/R = c_a \quad \text{for } Y < 0. \end{aligned} \tag{7.7}$$

For PP1 the constants are for instance: B = 1.08; t = – 0.398; the so called limiting heat capacities have the values $c_a = 3$ and $c_e = 54$.

The above given set of equations is the most simple approach for calculating a one dimensional isentropic real gas flow. Equations (7.6) and (7.7) allow for the calculation of the state changes shown in some diagrams. If there are shock waves included, then the Rankine Hugoniot shock relations have to be included

$$\text{Momentum}\quad\text{Continuity}\quad\text{Total Enthalpy}$$
$$[p + \rho w^2] = 0, \quad [\rho w] = 0, \quad [h + w^2/2] = 0 \tag{7.8}$$

with the conditions for entropy $[s] > 0$ and for the shock Mach Numbers $M_1 \geqslant 1 \geqslant M_2$. The Rayleigh Line resulting from conservation of momentum and continuity in (7.8) is $(\rho w)^2 = [p]/[V]$ and the shock adiabate is $[h] = (V_1 + V_2)[p]/2$, using the convention that the brackets $[\,]$ indicate the difference of the included quantity across a shock wave.

The entropy increase, which is important to know for the isentropic flow downstream of the shock wave, can be calculated from

$$[s] = -\frac{1}{12 T_1}\left(\frac{\partial^2 p}{\partial V^2}\right)_s [V^3] + O\left([V]^4\right), \tag{7.9}$$

indicating once again the important role of curvature of the isentropes for the sign of $[s]$.

7.2.6 Initial Flow Experiments with Retrograde Fluids with Shock Waves

The following previous investigations seem to be the first experimental observations and proofs in the field of liquefaction shock waves. In an experiment, carried out at the California Institute of Technology in 1970, Sturtevant found scattering of laser light on the downstream side of a shock in Freon 113 ($\widetilde{c}_v$=17.4), indicating the presence of condensate droplets [2]. Dettleff, Thompson, and Meier gave in 1976 an account of experiments seeking to show the existence of liquefaction shock waves [3, 6]. In view of the later extensive experimental observations and theoretical investigations, there is no doubt about the existence of liquefaction shocks in retrograde substances left [7].

Consider a pressure wave propagating into a single-phase or two-phase fluid, i.e., liquid, vapor, or mixture, and suppose (in appropriate cases) that the wave has sufficient amplitude to bring the initial state across the saturation boundary. The pressure wave may be of the compression or rarefaction type and the fluid may be regular or retrograde. In addition, the saturation boundary in question may be the liquid or vapor boundary. This leaves eight possible cases for the resulting changes (in other systems, there may be several phase transitions) – these are shown in Table 7.1. By applying the method of characteristics to the simple wave, taking into account the discontinuity in sound speed (kink) at the phase boundary and the (positive or negative) values of Γ, one arrives at the various wave types shown in the table.

Table 7.1. Waves crossing the saturation boundary

	Regular	Retrograde
Compression		
liquid sat. boundary:	Liquid Compr. Shock	Liquid Compr. Shock
vapor sat. boundary:	Vapor Shock	Liquefaction Shock
Rarefaction		
liquid sat. boundary:	Wave splitting	Wave splitting
vapor sat. boundary:	Moisture Shock	Rarefaction Shock

First Liquefaction Shock Wave Experiments

Sudden compression of a superheated retrograde vapor can lead to shock liquefaction, provided that the initial state of the vapor is sufficiently close to saturation (a necessary condition on entropy is that $s < s_{cr}$). With increasing amplitude of the compression wave, the following hierarchy of compression discontinuities is observed (see also Fig. 7.1) [3, 7, 11]:

(a) A single vapor-phase shock (vapor $\rightarrow$ vapor).
(b) Shock splitting, a vapor-phase shock followed by a subsonic condensation discontinuity (vapor $\rightarrow$ supersaturated vapor $\rightarrow$ mixture).
(c) A partial liquefaction shock (vapor $\rightarrow$ mixture).
(d) A complete liquefaction shock (vapor $\rightarrow$ liquid).

A liquefaction shockfront includes a vapor phase compression region, possibly extending to the vapor spinodal, followed by a region of nucleation and droplet growth. For shocks with high moisture fraction downstream states, the thickness of droplets is estimated to be of the order of $50\,\mu\text{m}$ or less for typical conditions. The shockfront can be considered the compact form of the extended two-shock split system discussed in the following.

The existence of liquefaction shocks was first demonstrated in the experiments of Dettleff et al. [3, 6, 7]. Partial and complete liquefaction shocks were generated, mainly by reflection at the closed end of a shock tube (Figs. 7.3 and 7.4). A remarkable feature of the complete liquefaction shocks was the appearance of numerous tiny vortex rings (more than 3×10^7 per m^2 of shockfront in some experiments). The rings are generated within the shockfront and follow it ("chase the shock front") in the frame of the downstream liquid [3, 7].

Later liquefaction-shock experiments [11] allow the shockfront to be photographed as it emerges from the open end of the shock-tube into an observation chamber filled with the same gas as the shock tube. In Fig. 7.5, a shock has just emerged from the tube: the condensation front is clearly visible, boundary-layer condensate flowing from the tube marks the boundary of the outflow, including the large-scale ring vortex characteristic of such flows (see e.g., Skews [151]). The flat central liquefaction shock wave, generating a dense mixture state (case c), shows splitting at a certain radius. The outer vapor phase, nearly spherical forerunner shock, visible by refraction of background grid lines, is followed by a condensation discontinuity (case b).

a   b

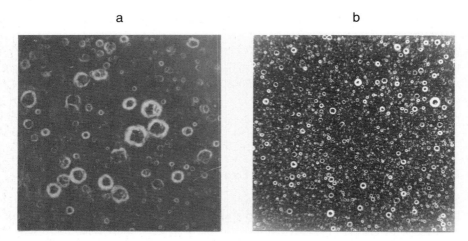

Fig. 7.3. The first views on a liquid layer at the glass end plate of a shock tube

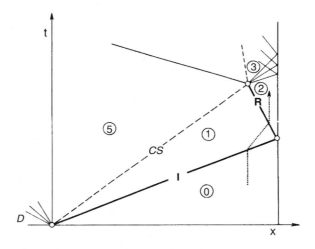

Fig. 7.4. Hypothetical $x - t$ wave diagram for the shock-tube flow. The closed end of the section is at the right. Arabian numerals designate test-fluid states. Other designations: I – incident shock, R – reflected (liquefaction) shocks, CS – contact surface, D – driver gas initial state. The initial shock I is precompressing the gas, the much stronger reflected shock wave R passes over all mixture states 1 right into the liquid state 2

Another unusual opportunity for a phase transition from a gas to a mixture state occurs if an obstacle moves with a sufficiently high velocity into a gas of high specific heat. Figure 7.6 shows four short exposure photographs of small projectiles of different Mach number penetrating into a resting gas (PP1). The projectiles are generating a bow shock in front, liquefying the gas. In case of

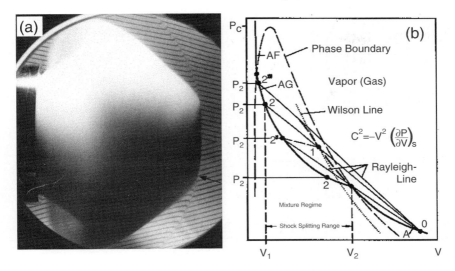

Fig. 7.5. (a) A typical liquefaction shock system emerging from a tube in FC-75 test fluid. Shock Mach number 1.56, initial conditions $p_0 = 1.09$ bar, $T_0 = 112°C$. The main liquefaction shockfront (normal to the shock tube axis) is still planar where expansion waves from the tube corner have not yet penetrated. Shock splitting occurs near the edge of this planar region, where the main shock is weakened by the expansion. (b) The different shocks illustrated in a p/V diagram

the smaller Mach numbers the transition is fairly smooth. For the higher Mach numbers the surface of the mixture region becomes turbulent. This seems to be the same effect of destabilization of the liquefying shock wave which is observed in the case of a shock wave emerging from a tube to be discussed later.

Many puzzles about liquefaction shock waves remain, for example, the relation of the tiny vortex rings to nucleation events, the physical mechanism and conditions for instability, the structure of the shockfront, and the relation of observed shockfront asperities to the vortex rings. Detailed knowledge about nucleation and droplet growth in conjunction with the wave dynamics is lacking.

Evaporation Waves in Retrograde Fluids

The rapid depressurization of superheated liquids has often been investigated for normal liquids like water [37, 143]. In this case only a small percentage of the liquid can be evaporated because of the limited reservoir of evaporation heat. In case of fluids with sufficiently high specific heat the whole liquid can be flash-evaporated in case of sudden depressurization [38–40]. The evaporation is after the experience of the recent experiments in expansion tubes of constant

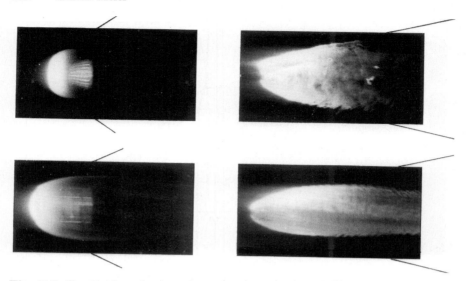

Fig. 7.6. Penetration of projectiles into a resting gas of high specific heat (PP1). Left upper frame shows a smooth a smooth liquefaction front at a Mach Number M = 1.76; the left lower frame a slightly distorted shock at a Mach Number M = 2.3; right upper frame turbulent liquefaction front at a Mach Number M = 4.4; righ lower frame a less turbulent front at a Mach Number M = 4.9

and increasing cross-section clearly divided in the two steps of nucleation and bubble growth.

Figure 7.7 shows the pressure time history of an evaporation experiment which is started by the rupture of a membrane at the top of the duct. The breaking of the membrane starts a nucleation wave which is running with the high sound velocity of the liquid (forerunner wave). The evaporation wave with the bubble growth can follow only with the smaller sound velocity of the two phase system. The final complete evaporation is fulfilled by reflected evaporation waves visible at the right end of the pressure traces. From these observations it becomes very clear that all very fast processes of phase transition are governed by wave propagation phenomena.

Two Phase Supersonic Expansion Flow

In case of ordinary supersonic gas flow phase transition is familiar and often observed for the case of water condensation in the supersonic part of wet air and steam Laval-nozzle flow [102, 108]. For fluids with high specific heat one can predict also in this case extraordinary behavior if the initial state of the Laval-nozzle flow is that of a superheated liquid (Fig. 7.8). One would expect a partial evaporation of the liquid in the convergent part of the nozzle because of the substantial pressure drop, a fairly low mixture velocity in the sonic throat and the final evaporation in the supersonic part of the nozzle flow. All

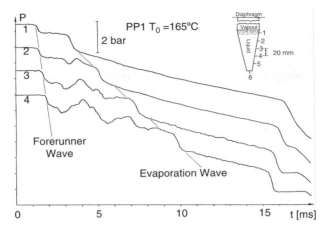

Fig. 7.7. Pressure traces of an expansion with phase transition in a divergent duct

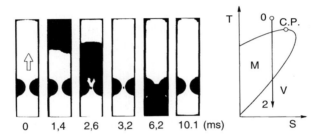

Fig. 7.8. Time sequence of photographs of a Laval nozzle flow with evaporating liquid of high specific heat (PP1)

this has been proven by experiments [38, 39]. Figure 7.8 shows a sequence of high speed frames photographed during an evaporation of a saturated liquid in the lower part of the duct, through a Laval nozzle visible in the center of the duct. The white parts are pure liquid or pure gas, while the black parts mark the nontransparent mixture states. Once the flow is established by generating a mixture state above the nozzle in 2.6 ms after start, it is nearly constant in rate and pressure over a longer time like at $t = 3.2$ ms, slowly transforming the liquid in the lower part into a mixture like at $t = 6.2$ ms. The flow rate of the critical two phase flow is constant then, because it depends only on the initial conditions like ordinary gas flow does and is controlled by the sound velocity of a corresponding mixture state with the maximum stream function. The end of the evaporation process is clearly visible as the disappearance of the dark flow regime at $t = 10.1$ ms, when all mixture in the lower vessel part is evaporated into gas. Evaluations of these experiments show a linear dependence of the flow rate of the initial temperature.

With the assumption of small changes of the sound velocity in the subsonic part of the nozzle flow one can calculate the values of the critical velocity from the experimental data and compare with the equilibrium sound velocity for a mixture state close too the liquid saturation [38]. Despite the strong assumption, the comparison shows a good agreement of both results. This proves to a certain extent that the equilibrium sound velocity is determining the critical flow of a two phase mixture through the throat of a Laval nozzle.

7.3 Experimental Details of Liquefaction Shock Wave Studies

7.3.1 Objectives of the Experiments

A liquefaction shock wave is a compression discontinuity for which the upstream state is vapor and the downstream state is liquid. This definition for a liquefaction shock given in the first and second paragraphs implies the complete liquefaction of a superheated vapor, i.e., a complete transition from gas to liquid across the shock. Depending on experimental conditions, it is of course possible that either the upstream state or the downstream state, or both, consist of a mixture of both phases. In the experiments reported here, all combinations of states have been found: the emphasis in this paragraph is, however, on the complete liquefaction shock, and to a lesser extent on the partial liquefaction shock for which the downstream fluid is a mixture of liquid droplets and gas, i.e., a fog. In the great majority of the experiments, the upstream fluid was a dry vapor.

The experiments reported here have four main objectives:

 - The proof of the existence of the liquefaction shock.
 - The determination of the extent to which the shock transition agrees with the predictions of classical shock-discontinuity theory (i.e., the Rankine-Hugoniot equations).
 - To observe whatever associated phenomena might arise in the shock process.
 - The developments of specific experimental methods.

To a considerable extent, these objectives may be regarded as having been met, the first one in particular. It will be seen, however, that several questions of interest remain.

7.3.2 Experimental Arrangement of the First Tests

The first experiments were carried out in a cylindrical shock tube. Following the suggestion of Thompson and Sullivan [1], the experiment was arranged to produce the liquefaction shock as the reflected shock from the closed end of the shock tube test section, as shown in the hypothetical $x - t$ wave diagram in Fig. 7.4. The initial shock is precompressing the gas, the much stronger reflected shock wave passes over all mixture states right into the liquid state (see also Fig. 7.1). Because the density of the liquid 2 is large, the reflected liquefaction shock, which then resembles the surface of a liquid, travels slowly away from the end wall, reaching a maximum distance $d \simeq 2$ cm from the end wall in a typical experiment. Under these conditions, experimental observation of the shock is relatively convenient (Fig. 7.3). Many experimental details not given here can be found in Dettleff et al. [3, 7].

The driver section of the shock tube was 2.50 m long, the test section 2.426 m long. The shock tube is made of brass, with a wall thickness of 5 mm and an internal diameter of 80 mm. In order to allow stable initial temperatures up to $T_0 = 160°C$, the shock tube was surrounded with electrical resistance heating and covered with a flexible foam insulation with a total thickness of $h \simeq 60$ mm.

The diaphragm material was Dupont "Capton" polyamide, in various thicknesses and combinations in the range 25–300 µm. Diaphragm burst could be initiated by an arrowhead plunger mounted on the driver-section side of the diaphragm. This plunger was released by an electrical signal. This method of starting the experiment has the advantage of a precise choice of initial conditions in pressures and time.

Most of the measuring sensors are mounted in the observation section, which forms the final 300 mm length of the shock tube, as shown in Fig. 7.9. Almost all of the shock measurements, on both partial and complete liquefaction shocks, were carried out here. Two distinct window arrangements are available for optical measurements and photography: narrow side windows 12 mm × 180 mm permit a transverse view through the observation section, and a large circular end window, with inner frame diameter of 70 mm, can be mounted as shown in Fig. 7.9 to replace the optional metal end plate (a).

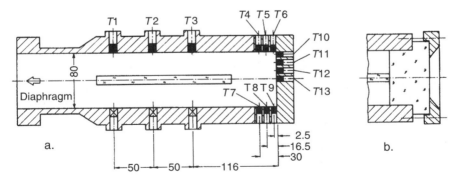

Fig. 7.9. Observation section of the shock tube. The designations T1 to T12 refer to the (pressure) transducer locations

Table 7.2. Test Fluids ($\tilde{M}$ molecular weight, T_b boiling temperature, T_c critical temperature, p_c critical pressure, and ϱ_l is the saturated-liquid density at 20°C)

Formula	Name	$\tilde{M}$	$T_b(°C)$	$T_c(°C)$	$p_c(\text{bar})$	$\tilde{c}_v$	$\varrho_l(\text{g cm}^{-3})$
C_8F_{16}	PP3	400	102	241.5	19.2	53.9	1.84
$C_{11}F_{20}$	PP9	512	160	313.4	16.9	73.4	1.98
$(C_4F_9)_3N$	FC40	650	150	277	14.8	90.0	1.87

The transducer mounting locations T1 trough T12 can be used for pressure transducers, temperature probes, or other devices.

Three different test fluids, all retrograde substances, all fluorocarbons, were used in the experiments: C_8F_{16} (ISC Ltd. Designation PP3), $C_{11}F_{20}$ (ISC Ltd. designation PP9), and a substance of unknown exact chemical composition, but similar to $(C_4F_9)_3N$ (MMM Co designation FC 40). The two ISC substances PP3 and PP9 are cyclic compounds, perfluoro-dimethylcyclohexane and perfluoro-methyldecalin, respectively. Properties as given by the manufacturers are summarized in Table 7.2.

The advantage of using fluorinated substances is that they do not present any health hazards and are not flammable: the disadvantages are their environmental problems and their relatively high cost. With reasonable precautions, there would be no difficulty in using normal hydrocarbons such as decane $C_{10}H_{22}$, nor is there any reason to expect that the experimental results would be significantly different.

The majority of the experiments reported here were performed with PP3. All of the substances are of low purity, and no attempt was made to improve the purity, by freezing or otherwise. Each sample was degassed before use, by heating under intermittent exposure to vacuum.

Here some remarks about the experimental procedure: The shock tube was during test weeks continuously heated in order to keep the steady initial temperature T_0 constant. After mounting the diaphragm, the test section was evacuated to a pressure less than 1 Pa. The desired quantity of degassed test fluid was drawn into an evacuated sample bottle and weighed on a precision balance: the test fluid was then drawn into evacuated test section. Because the last portion of liquid drawn into the test section could not be visually observed, it had to be checked by an additional proof that all of the measured mass m_0 went into the tube. The initial state (0) was therefore based on the pressure P_0, measured with a Kistler 7261 quartz piezoelectric transducer (decay time $\approx$ 30 min), and on the temperature T_0, measured with the temperature-control thermocouples in the tube wall as well as with an independent miniature thermocouple inside the tube. The measured pressure P_0 was found in every case to be consistent with that calculated from the equation of state for $P_0 = P_0(T_0, m_0/V)$, where V is the volume of the test section.

After the filling of the shock tube was completed and the triggering, measuring, and recording devices had been prepared, the driver section was

Table 7.3. Measured properties

m_0, P_0, T_0	initial-state properties
$P(t), T(t)$	static pressure and temperature
$I(t)$	transmitted-light intensity in state (1)
v_I, v_R	velocity of incident and reflected shocks
n	index of refraction in state (2)

pressurized with nitrogen driver gas from a storage bottle. Diaphragm burst was effected in the first experiments simply by continuously increasing the pressure P_D until the diaphragm bursts spontaneously. In the later experiments by piercing the diaphragm with the spring-loaded plunger a more precise operation was possible. Following the diaphragm burst, the experimental events of interest were recorded on high speed digital data recorders within about 10 ms. About three experimental runs (shots) per hour could be achieved this way. The main experimentally measured properties and the variables are shown in Table 7.3.

Not all of the earlier properties were measured in any given run. The transient pressures were measured by Kistler quartz piezoelectric transducers type 603B, membrane diameter 3.5 mm. Vibro-Meter type TA-3C charge amplifiers converted the piezoelectric charge to an output voltage signal. Transient temperatures were measured at transducer positions T12 and T13 with very thin iron-constant thermo-couple elements designed to produce the fastest possible temperature response, consistent with a lifetime of at least several shots. Both designs used wires tapered by etching until the small-end diameter was about 10 μm. The two ends were then welded at once by means of a current pulse to form the thermocouple. The response time of the elements was about 2 ms in air ($P \simeq 5$ bar) and about 0.1 ms in the test fluids.

The transmitted-light intensity $I(t)$ behind the incident shock wave was measured at the transducer position T1 in Figure 7.9. A beam from a helium–neon laser ($\lambda = 533$ nm, beam diameter 0.6 − 0.7 mm) was directed through the narrow side windows, perpendicular to the axis of the shock tube, onto a 2 mm diameter photodiode.

The velocity v_1 of the incident shock wave was determined from the $P(t)$ signals from positions T1 and T3, and checked from time to time with the signals from T1, T2, and T3 and T4. The velocity v_R of the reflected shock wave could not be determined by this method because of the limited distance of travel, and especially because of strong boundary-layer interactions ahead of the shock. Consequently, a special probe built for the measurement of v_R was mounted at position T4. Passage of the incident and reflected shocks over the thermocouple film at the end of this probe produced successive jumps in its output signal. Comparing the time of these jumps with the time at reflection, as indicated by the jump in the output signal of the pressure transducer mounted at position T12, allowed the determination of the velocity v_R and with less accuracy, the value of v_1.

The index of refraction n of the fluid in state (7.2) was measured photographically according to the altered apparent length L of an immersed object. The value of n follows from the elementary formula

$$n^2 - 1 = (\alpha^2 - 1)\cos^2 \Phi, \tag{7.5}$$

where $\alpha = L_0/L_{\mathrm{f}}$ and Φ is the angle of incidence of the observation beam. This measurement was made in relatively few experiments.

Most of the time-dependent variable output signals (P, T, I) were stored in 4 transient recorders Datalab type DL 905. The recorded $y(t)$ data were then plotted via a digital to analog converter with an $y - t$ plotter. Certain data were also recorded on a storage oscilloscope during the test immediately.

By photography various kinds of the phenomena in the observation section have been recorded. These were mainly conventional flow photographs where the visibility of the object details depends on the presence and light scatter of density variations and phase boundaries, as for example, in the case of liquid droplets in a gas. The cameras used for individual pictures were Rolleiflex SL 66, $6\,\mathrm{cm} \times 6\,\mathrm{cm}$ film format. In a few experiments, two cameras were arranged to make stereoscopic pictures. In every case, two General Radio Strobotac flash lamps with about $1\,\mu\mathrm{s}$ flash duration were used for illumination, the flash being triggered through a variable time delay from a pressure transducer at T1, T2, and T3. The time instant of the flash exposure was registered in the transient recorder from a photodiode signal.

Motion pictures were made with a Fastax 16 mm camera (maximum framing rate 8,500 frames s^{-1}) and two Variolux light sources for illumination. Synchronization between this camera and experiment was achieved by initiating the plunger-actuated diaphragm burst from a camera output signal which was automatically triggered after the camera start.

In all of the photographic and optical observations, thermally insulating covers remained over the windows until about 30 s before the diaphragm burst to keep the wall temperature constant at initial conditions.

7.3.3 Experimental Results for Partial Liquefaction Shocks

For partial liquefaction shocks, nucleation and droplet condensation behind the incident shock was studied. In these experiments, the upstream state 0 consisted of superheated gas and the equilibrium downstream state 1 varied from superheated gas to a gas–liquid mixture, with a maximum of 10% moisture. The extent of liquefaction was thus minimal, with an expected downstream state consisting of liquid droplets in gas. The test substance was PP3. The three different initial Mach numbers used are shown in the first row of Table 7.4.

In order to get some idea of the uniformity of the incident shock behavior, its position was recorded as a function of time at four different pressure-transducer locations. The average shock Mach number was determined

Table 7.4. Estimated condensate properties for different shock mach numbers

M_I	$\rho_k\,(\mathrm{kg\,m^{-3}})$	$M_k(\%)$	$r \times 10^8\,(\mathrm{m})$	$N \times 10^{-17}\,(\mathrm{m^{-3}})$	$\Delta\,(\mathrm{mm})$
2.5	1.63	2.7	8.1	6.3	20
3.6	3.26	5.0	8.7	10.5	10
4.7	5.57	7.9	9.2	15.1	5

between each pair of transducer locations, based on the difference in arrival time. At the highest Mach number, spontaneous condensation was observed in state 1 behind the incident shock.

The idea behind increasing the Mach number M_1 of the incident shock gradually is, first bringing the shock state 1 closer to the saturation boundary and then having state 2 inside the mixture region, where spontaneous condensation is to be expected.

Before a saturated state 1 is reached, however, condensation is observed in the boundary layer in the form of a thin layer of fog near the wall. Considering that the wall temperature remains close to the initial temperature T_0 throughout the experiment, the layer of fluid immediately adjacent to the wall undergoes nearly isothermal compression, and thus, with increasing M_I, reaches saturation well before the bulk fluid in state 1. The thickness of the fog layer was measured by directing a collimated flashed light beam through the side windows. A bright column was formed by scattering where the beam passed through the fog layer: the height h of the column was then measured photographically through the rear window. By flashing the light source with a measured delay from the time of arrival of the incident shock, and making use of the measured shock velocity, the distance d behind the shock corresponding to the measured height h could be determined.

The onset of spontaneous condensation in state 1 was measured by light transmission measurements also. Measurements of the transmitted light intensity $I(t)$ behind the incident shock were made for a light beam traversing the shock tube. A sample $I(t)$ trace, with the corresponding pressure records, is shown in Fig. 7.10.

The sudden dip in intensity at the time of arrival of the shock can be interpreted as a deflection of the light beam by the shockfront; the following overshoot in intensity can be interpreted as an increase in the index of refraction in the fluid across the shock. All of the measured intensities, when plotted semilogarithmically against time, yielded approximately straight lines, with the corresponding form for $I(t)$:

$$I(t)/I_0 = \mathrm{e}^{-\beta t}. \tag{7.6}$$

The value of the decay constant β is interpreted as light attenuation (scattering) by the boundary fog layer and the condensation. A discontinuity in the slope of $\beta(M_1)$ is interpreted as the onset of spontaneous condensation in the core flow.

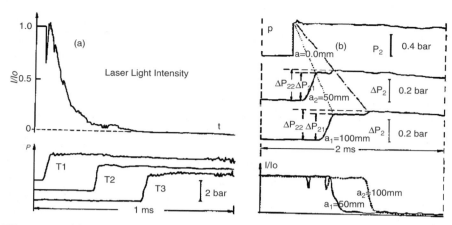

Fig. 7.10. (a)Transmitted light intensity $I(t)$ behind the incident shock together with the corresponding pressure records $P(t)$ at T1, T2, and T3 for strong condensation shocks; (b) pressure records $P(t)$ at a fixed location for a weaker shock split into a forerunner wave and a condensation discontinuity

Based on the idea that the decay constant β in (7.6) can be expressed as the sum $\beta_{\rm bl} + \beta_{\rm cf}$ of a boundary-layer and a core-flow contribution, estimates of the condensation history behind the incident shock can be made [7]. The model implied by (7.6) is that of an attenuation which increases from zero linearly with time, without ever reaching an upper limit. In conjunction with an assumed Rayleigh light scattering model (see e.g., Born and Kerker [168, 169]), valid for the wavelength $\lambda \gg r$, this yields for the droplet number density N and droplet radius r

$$Nr^6 = kt, \tag{7.7}$$

where the constant of proportionality k depends on the value $\beta_{\rm cf}$ determined from Fig. 7.10. Equation (7.7) is clearly incorrect for large times, since it predicts that the mass density of condensate $\varrho_k \sim Nr^3$ grows without limit. This corresponds to the absence of a finite lower bound for $I(t)$ in (7.6), reflecting the inability to measure small light intensities with the measuring device employed. If, however, one assumes that (7.7) holds up until some large time t_1 at which condensation is complete, with the density of condensate ϱ_k predicted by the thermodynamic model, then (7.7) together with mass conservation can be used to predict numerical values for N and r. Arbitrarily setting t_L to $0.5\,\text{ms}$ yields the estimates shown in Table 7.4, for an initial state $t_0 = 130°\text{C}$ and $P_0 = 0.67$ bar.

In this table, the third column indicates the mass percent of condensate M_k in state 1. The quantity Δ, intended as a crude measure of the thickness of the condensation zone, shows the distance downstream of the shockfront at which the transmitted light intensity has fallen to 10% of 1st maximum value.

These results are in agreement with calculations of nucleation and droplet growth by Bratos and Meier using classical nucleation theory [9]. A better approach to the nucleation problem was developed later by Dillmann and Meier [141, 142].

Quite aside from the limitations of the measurements and models described, it should be borne in mind that these estimates apply to the feeble end of the liquefaction shock spectrum. It may be expected that the structure of a complete liquefaction shock is quantitatively different.

Especially the shock splitting under the influence of curvature changes of the isentropes is a remarkable finding. The wave splitting, associated with the discontinuity in slope of the isentrope at the saturated-vapor boundary, as first suggested by Bethe in 1942 [161], was observed for the shock waves as for the evaporation waves crossing the phase boundary [33–36].

7.3.4 Experimental Results for Complete Liquefaction Shocks

Complete liquefaction behind the reflected shock R was studied in the shock-tube described earlier at the end wall. In order to show the transitions, with increasing Mach number M_I, from a single-phase gas $\rightarrow$ gas shock to a gas $\rightarrow$ mixture shock to a gas $\rightarrow$ liquid shock, measurements over a corresponding range of shock strengths M_I will be reported in this paragraph. The emphasis, however, is on the experimental confirmation of the existence of a complete liquefaction shock, that is, a shock transition from superheated gas 1 to compressed liquid 2.

The various test fluids and the corresponding initial states used are shown in Table 7.5. More than 80% of the runs were made with PP3.

A typical sequence of thermodynamic states for the case of complete liquefaction is shown in Table 7.6. For the purpose of the liquefaction shock

Table 7.5. Test fluids and initial states

Substance	$T_0(^\circ C)$	$P_0(\text{bar})$	$m_0(\text{g})$
PP3	130	0.51	73
PP3	130	0.67	99
PP3	130	0.99	149
PP3	80	0.24	38
PP9	130	0.24	43
FC-40	150	0.33	73

Table 7.6. Typical sequence of states in a shock wave in PP3

State	$P(\text{bar})$	$T(^\circ C)$	$T_{\text{sat}}(^\circ C)$	$\rho(\text{kg m}^{-3})$	$c(\text{m s}^{-1})$	M_s
0	0.67	130	90	8.33	89.3	$\downarrow$
1	4.02	158	155	55.6	88.2	2.43
2	24.0	229	–	1,190	–	2.19

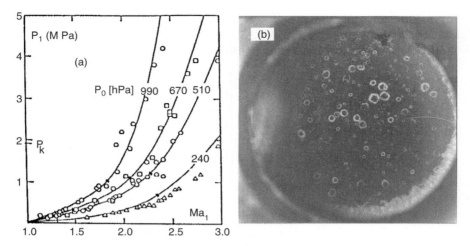

Fig. 7.11. (a) Pressure p_1 downstream of the incident shock as a function of incident shock Mach number M_1. *Points* represent the measurements and the *solid curves* the calculations. Transverse marks on the curve indicate the onset of spontaneous condensation in state 1. (**b**) View on the liquid layer through the (glass) end wall of the shock tube

experiment, the incident shock should bring the test gas to a slightly super-heated (dry) state 1 while in motion toward the shock-tube end wall. Such a state is indicated in Table 7.6, which shows a typical sequence of states for the case of complete liquefaction. The properties shown for states 1 and 2 are calculated.

A comparison of the measured pressures p_1 behind the incident shock with the calculated values is shown in Fig. 7.11a. For sufficiently large shock Mach numbers M_1, the state (1) becomes wet, as discussed in Sect. 7.4 and shown in the Fig. 7.11b. The onset of condensation in state (1) is not manifested in the pressure measurements of Fig. 7.10a themselves.

The pressure distribution on the wall near the end of the observation section is not uniform, corresponding to the three-dimensional interaction flow. The approaching incident shock is first registered by a transducer trace at T4 (see Fig. 7.9), which also shows the arrival of the upstream branch of the reflected shock. The pressure distribution near the end wall is almost axially symmetric, as indicated by comparative traces from symmetrically placed transducers. The traces from transducers T11 and T12 show an approximately uniform pressure distribution in the neighborhood of the shock-tube axis, consistent with a one-dimensional flow model for this region. On the basis of continuity, however, a low-speed inward displacement of liquid would be expected during the liquefaction phase, leading to a reflected-shock velocity w somewhat greater than that predicted by a one-dimensional model.

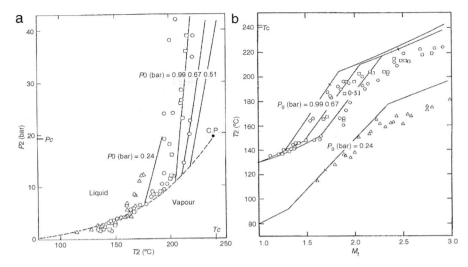

Fig. 7.12. (a) Pressure at the end wall. (b) Temperature at the end wall

So, the measured pressures p_1 near the end wall for all of the PP3 experiments, plotted as a function of the incident-shock Mach number in Fig. 7.11a, were determined as the mean pressure of a plateau. The agreement between the measured and predicted pressures in Fig. 7.11a is reasonably good, but there is no vivid indication of the onset of condensation and, at a larger Mach number M_1, of complete liquefaction.

The measured temperatures T_2 at the end wall are more satisfactory in this respect, as shown in Fig. 7.12b. For any given set of initial conditions, each calculated curve has a distorted, angular S-form: the lower and upper bars of the "S" correspond to states 2 in the dry vapor and compressed liquid states, respectively, and the steep diagonal corresponds to liquid–vapour mixture states. The vertical separation ΔT between the upper and lower bars is a rough measure of the latent heat of evaporation.

Similar to the calculated "S" curve, the existence of complete liquefaction states along the upper part of the "S" thus appears to be verified experimentally by the measurement values, but with a deviation ($\sim 10°C$) from the predicted end temperature T_2. We cannot offer a definitive explanation for this deviation. It seems likely that it is due mainly to inaccuracies in the (estimated) material properties of the test substance: the essential parameter latent heat depends on the slope of the vapour–pressure curve and the volume change on evaporation (via the Clapeyron equation), and on the heat capacity, all of which are inaccurately known. A further inaccuracy lies in the equation of state, which is not accurate at high densities. It is, in any case, considered unlikely that the temperature deviation is attributable to measurement error or to nonequilibrium effects.

The achievement of measured complete-liquefaction end states is indicated clearly in Fig. 7.12a. As by definition, the compressed-liquid states lie above the vapour–pressure curve and to the left of a vertical line for the liquid temperature. Also the velocity of the reflected liquefaction shock was measured with the probe already described. The resulting velocities confirm the theoretical prediction fairly well.

7.3.5 Photographic Observations of the Liquefaction Shock Wave

Using the equipment described earlier, conventional photographs and motion-picture films were made through the circular end window. The various lighting and camera arrangements are shown in Fig. 7.13.

The most important photographically observed results are visualized in the frames of Fig. 7.14 and can be summarized as follows:

1. A clear phase, interpreted as a liquid, is observed in region 2 in Fig. 7.4.
2. Torus-form two-phase objects like rings appear in region 2 and appear to be formed within or immediately behind the liquefaction shock.
3. The shock-tube boundary-layer vorticity in the test gas becomes concentrated in a large ring, interpreted as a vortex, surrounding region 2 at the end of the shock tube. This corner vortex is the dominant visible structure in the shock-boundary-layer interaction.
4. In the terminal phase of the liquefaction, a small ring is naturally selected and grows rapidly on the shock-tube axis until it meets the growing corner vortex.
5. The region which includes the shockfront and the phase transition, roughly designated as the shockfront, varies in appearance from smooth to rough and feathery, depending on the substance and experimental conditions.

The sequence of motion picture frames in Fig. 7.14, is showing the entire liquefaction history. Frame a corresponds to a time shortly before the arrival of the incident shock and the absence of any backscattered light, i.e., of a visible fog, is characteristic of a dry or supersaturated state 1. The incident

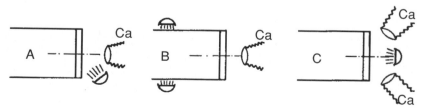

Fig. 7.13. Schematic diagrams of the various (A,B,C) lighting/camera arrangements used for photography. Ca designates the camera location

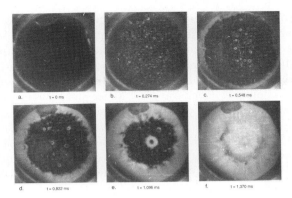

Fig. 7.14. Frame sequence of liquefaction at the end wall

shock is reflected between frames a and b. Small rings described earlier can already be seen in frame b. Some are observed to grow with time. The large corner vortex described in (3) can be seen in frame d and becomes larger thereafter. In frame e a ring near the shock-tube axis has been selected and thereafter grows rapidly until it meets the corner vortex ring (frame f). In frame f, liquid is no longer visible. The arrival of the contact-region between driver gas and test gas is placed between frames e and f, based on the pressure record in Fig. 7.10.

The appearance of the small torus-form rings behind the liquefaction shock was unexpected. There are several reasons for believing that these rings are, in fact, vortex rings. This hypothesis is supported by the following qualitative arguments:

- The visual impression is that of a torus, rather than, for example, that of a spheroidal "droplet" or "bubble." This impression is especially strong in the stereoscopic photographs.
- The ring-vortex form with a visible two-phase core is consistent with the dynamical pressure distribution in a ring vortex.
- The observed flow patterns and inferred growth and decay of the rings are similar to those found in ring-vortex experiments. In particular, a period of initial growth and ring fusion followed by wave-form instability and subsequent decay is observed. The larger rings are clearly seen to lie well above a background of small rings.

It was found that the ring objects could be artificially produced by cementing small obstacles onto the inside surface of the end window. For example, a thin wire bent into a V-form and fastened with the sharp vertex of the V pointing toward the diaphragm of the shock tube produced repeatable rings centered on the vertex.

The existence of the small shock-associated ring vortices, in particular, has been inferred rather than proven. Accepting the plausibility of this phenomenon, one naturally seeks a mechanism for the generation of the vortices in or behind the shockfront: is the liquefaction shock wave perhaps unstable. The experiments with the open end shock tube and the bullets show instability of the wave front clearly. Instabilities leading to nonuniform flows behind the discontinuity have been observed in detonations by White in 1961 [170] and in ionizing shock waves by Meier and Sandeman in 1975 [134]; and by Glass and Liu in 1978 [123] also. A theoretical basis for instability in the present case is not clear.

A simple physical argument given by Zel'dovich suggests a basis for nonuniform flow. The shock is not a true discontinuity and must have some small-scale internal structure. There is, however, no continuous sequence of homogeneous thermodynamic states connecting the equilibrium end states 1 and 2, as there would be for gaseous argon, for example. Thus, states 1 and 2 are connected by discontinuous states similar to dropwise condensation. The resulting situation of large density differences in a region of strong pressure gradient (deceleration) and condensation may well be unstable. Formation of a locally dense region – effectively, a droplet – would lead to the formation of vortices under the action of the shock pressure gradient, as predicted by the vorticity-growth equation.

7.3.6 Shock-Boundary-Layer Interaction Region

The interaction of the shock-tube boundary layer and the reflected liquefaction shock produces a complicated three-dimensional flow. Interaction flows of this kind are already familiar for reflected shocks in gases, where the thickness of the interaction region is found to increase with decreasing ratio of specific heats γ, as discussed by Davies and Wilson in 1968 [109] and by Honda et al. in 1975 [120]. This behavior can be associated with the large density jump across the shock and is thus consistent with the enhanced interaction effects found in our experiments. The main liquefaction shock divides into two branches near the wall, forming a "λ-shock" or "bifurcation." One oblique-shock branch extends far upstream. Vapor from region (1) is compressed by this shock into a wedge-shape region as an opaque two-phase mixture: the outflowing mixture is assumed to roll up into the corner vortex. An outward radial flow would be consistent with the presence of the corner vortex and its image in the end wall. The direction of any such radial flow could not be established on the basis of the experiments, however. A recirculation region or "separation bubble" lies upstream of the corner vortex, geometrically consistent with the flow deflexion of the leading oblique shock. Photographs of the ring-shaped region referred to here as the corner vortex are shown in Fig. 7.14c–e.

7.3.7 Experimental Results for Liquefaction Shocks from an Open Shock Tube

An especially designed shock tube with an open end allows the liquefaction shock to be photographed as it emerges from the opening into an observation

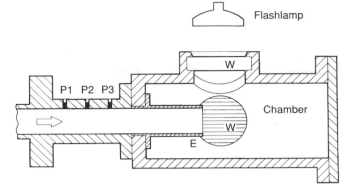

Fig. 7.15. The "Open Shock Tube" test section

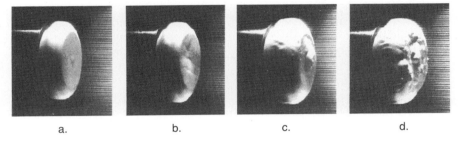

a.   b.   c.   d.

Fig. 7.16. Liquefaction shockfront sequence with progressive instability of a shock in FXC-75 with various forms of instability in FXC-75 test fluid

chamber. Figure 7.15 show the details of the end section. The test chamber is filled with the same test gas (FC 75) as the last part of the shock tube.

The smooth shockfront shown already in Fig. 7.5 can be contrasted with the irregular ones shown in Fig. 7.16b–d, with its cracks, lumps, and asperities. It is found that a transition from smooth to chaotic shock fronts occurs over a narrow range of increasing shock Mach numbers over the range 2.1–2.3. This progressive instability has been related to existing theories of shock stability by Thompson et al. [13, 55, 60]. The results are especially interesting when coupled with the preliminary experiments with projectiles (Fig. 7.6) where for higher Mach Numbers instabilities occur also.

7.4 Concluding Remarks

It appears that the liquefaction shock wave does exist in a form consistent with classical shock-wave theory. The variety of associated phenomena found in the experiments was not anticipated, nor has each one been thoroughly investigated yet.

Table 7.7. Some special phenomena of real gas dynamics

Waves	Flows
Liquefaction shocks	Inclined condensation shocks
Condensation shocks	Stagnation liquefaction
Nucleation waves	Splitting field shocks
Rarefaction shocks	Complete evaporation
Evaporation waves	Mach number discontinuities
Wave splitting	Phase change instabilities

Table 7.8. Recommended future real gas research in physics and fluid mechanics

Problems left	Research topics
Equations of state	Multi Phase Flows
Nucleation	3-D Flows
Cluster formation	Phase Transition in Mixtures
Instabilities	Nonequilibrium Flows
Relaxation effects	Inclusion of Solid Phase

The expected, observed, and described phenomena, quite unusual in ordinary gas dynamics and supporting the proposed title Real Gas Dynamics for this field, are summarized in Table 7.7.

For the future, also with respect to the practical importance of the field, some evident problems and research areas are recommended for activity in Table 7.8.

Acknowledgments

The author would like to express his gratitude to the late Philip A. Thompson for his friendship and devotion to our common research topic. Only by his initial ideas and proposals we could start the research in the field of Real Gas Dynamics. The work reported in this article was performed together with him and G.C. Carofano, H. Chaves, G. Dettleff, D.E. Hobbs, Y.-G. Kim, H. Lang, E. Puttendörfer, H.-D. Speckmann, D.A. Sullivan, and C. Yoon. Several scientists at the Max-Planck-Institut für Strömungsforschung in Göttingen and also the technicians of the Max-Planck-Institut and the Rensselaer Polytechnic Institute in Troy provided valuable support. Mrs. K. Hartwig and Mr. I. Meier assisted in typing the text. The used and cited literature reflects the important work of the past and also the present continuation of research in this interesting and practically important field.

References

Papers Related to Liquefaction Shock Waves

1. Thompson, P.A and Sullivan: On the possibility of complete condensation shock waves in retrograde fluids. J. Fluid Mech. 70, part 4, pp. 639–649, 1975
2. Sturtevant, B.: Lecture to the Division of Fluid Dynamics Meeting, American Physical Society 1970. See also Amer. Phys. Soc. Bull. 15, 1546
3. Dettleff, G., Thompson, P.A., Meier, G.E.A., Speckmann, H.D.: An experimental study of liquefaction shock waves. Journal of Fluid Mechanics 95, part 2, 279–304, 1979
4. Thompson, P.A. and Lambrakis, K.C.: Negative shock waves. J. Fluid Mech. 60, part 1, pp. 187–208, 1973
5. Thompson, P.A., Carofano, G.C., and Kim, Y.-G.: Shock waves and phase changes in a large-heat-capacity fluid emerging from a tube. J. Fluid Mech. 166, pp. 57–92, 1986
6. Dettleff, G. Experimente zum Nachweis der VerflssigungsstoßßSwelle in retrograden Gasen. Dissertation, Georg-August-Universitat Göttingen, 1978
7. Dettleff, G., Meier, G.E.A., Speckmann, H.-D., Thompson, P.A., Yoon, C.: Experiments in shock liquefaction. In: Proceedings of 13th International Symposium on Shock Tubes and Waves, Trainorand, C.E., Hall, J.G., eds., State University of New York Press, Albany, 1982
8. Thompson, P.A.: A fundamental derivative in gasdynamics, Phys. of Fluids 14 No. 9, pp. 1843–1849, 1971
9. Bratos, M., Meier, G.E.A.: Shock wave induced condensation in retrograde vapour. In: Archives of Mechanics 31, 5, pp. 679–693, Warschau 1979
10. Meier, G.E.A., Thompson, P.A.: Real Gas Dynamics of Fluids with High Specific Heat. In: Lecture Notes in Physics, Vol. 235, pp. 103–114, Springer-Verlag 1985
11. Thompson, P.A., Carofano, G.C., Kim, Y.G.: Shock-waves and phase-changes in a large-heat-capacity fluid emerging from a tube, Journal of Fluid Mechanics CS 166: 57–92 Mai 1986
12. Meier, G.E.A.: Zur Realgasdynamik der Fluide hoher spezifischer Wärme. Habilitationsschrift, Universität Göttingen, 1987
13. Thompson, P.A., Strock, T.W., and Lim, D.S.: Estimate of shock thickneses based on entropy production. Phys. Fluids 26, pp. 48–49, 1983
14. Thompson, P.A., Strock, T.W., and Lim, D.S.: Estimate of shock thickneses based on entropy production. Phys. Fluids 26, pp. 48–49, 1983
15. Meier, G.E.A., Thompson, P.A.: Method for converting a retrograde substance to the gaseous state. In: United States Patent No. 4,522,183, 1985
16. Hesselink, L., Sturtevant, B.: Propagation of weak shocks through a random medium, Journal of Fluid Mech. 196: 513–553 Nov. 1988
17. Gulen, S.C., Thompson, P.A., Cho, H.J.: An experimental study of reflected liquefaction shock-waves with near-critical downstream states in a test fluid of large molar heat-capacity, JFM 277: 163–196 Oct. 25 1994
18. Gulen, SC: On the possibility of Shock-induced condensation in the thermodynamically unstable region, Journal of Non-Equilibrium Thermodynamics 19 (4): 375–393 1994

19. Kobayashi, Y., Watanabe, T., Nagai, N.: Vapor condensation behind a shock wave propagating through a large molecular-mass medium, Shock Waves 5 (5): 287–292 Feb 1996
20. Maerefat, M., Fukijawa, S., Akamatsu, T., et al.: An experimental study of nonequilibrium vapor condensation in a shock tube, Exp in Fluids 7 (8): 513–520 1989
21. Brown, B.P., Argrow, B.M.: Two-dimensional shock tube flow for dense gases Journal of Fluid Mech. 349: 95–115 Oct 25 1997
22. Bates, J.W., Montgomery, D.C.: The D'yakov–Kontorovich instability of shock waves in real gases, Physical Review Letters 84 (6): 1180–1183 Feb. 7 2000
23. Fan, H.T.: Symmetry breaking, ring formation and other phase boundary structures in shock tube experiments on retrograde fluids J. of Fluid Mechanics 513: 47–75 Aug 25 2004

Papers Related to Negative Shock Waves

24. Lambrakis, K.C., and Thompson, P.A.: Existence of real fluids with a negative fundamental derivative. Phys. of Fluids 15, No. 5, pp. 933–935, 1972
25. Borisov, A.A., Borisov, Ai.A., Kutateladeze, S.S., and Nakoryakov, V.E.: Rarefaction shock wave near the critical liquid–vapour point. J. Fluid Mech. 126, pp. 59–73, 1983
26. Cramer, M.S., Kluwick, L.T. Watson, and Pelz, W.: Dissipative waves in fluids having both positive and negative nonlinearity. J. Fluid Mech. 169, pp. 323–336, 1986
27. Cramer, M.S.: Structure of weak shocks in fluids having embedded regions of negative nonlinearity, Physics of Fluids 30 (10): 3034–3044 Oct 1987
28. Kluwick, A., Koller, F.: Propagation of small amplitude periodic waves in gases with large specific heat, Zeitschrift fr angewandte Mathematik und Mechanik 68 (5): T306–T307 1988
29. Cramer, M.S., and Sen, R.: Shock formation in fluids having embedded regions of negative nonlinearity. J. Fluid Mech. 169, pp. 323–336, 1986
30. Cramer, M.S.: Negative nonlinearity in selected fluorocarbons, Physics of Fluids A-Fluid Dynamics 1 (11): 1894–1897 Nov 1989
31. Kluwick, A., Czemetschka, E.: Spherical and cylindrical waves in media with positive and negative nonlinearity, ZAMM 70 (4): T207–T208 1990
32. Bulgakova, M.: Possibility of rarefaction shock wave under short pulse laser ablation of solids, Phys. Rev. E 60 (4): R3498–R3500 Part A Oct 1999

Papers Related to Wave Splitting

33. Thompson, P.A., Kim, Y.G.: Direct observation of shock splitting in a vapor–Liquid system, Physics of Fluids 26 (11): 3211–3215 1983
34. Chaves, H., Lang, H., Meier, G.E.A., Speckmann, H.-D.: Adiabatic phase transitions and wave splitting in fluids of high specific heat. In: Lecture Notes in Physics, Vol. 235, pp. 115–124, Springer-Verlag, 1985
35. Thompson, P.A., Chaves, H., Meier, G.E.A., et al.: Wave splitting in a fluid of large heat-capacity, JFM 185: 385–414 Dec 1987
36. Cramer, M.S.: Shock splitting in single-phase gases, Journal of Fluid Mechanics 199: 281–296 Feb. 1989

Evaporation Waves

37. Kutateladze, S.S., Nakoryakov, V.E., Borisov, A.A.: Rarefaction waves in liquid and gas–liquid media, Annual Review of Fluid Mechanics 19: 577–600 1987
38. Chaves, H., Kowalewski, T.A., Kurschat, T. et al.: Similarity in the behavior of initially saturated or subcooled liquid jets discharging through a nozzle, Chem. Phys. 126 (1): 137–143 Oct 15 1988
39. Kurschat, T., Chaves, H., Meier, G.E.A.: Complete adiabatic evaporation of highly superheated liquid jets, JFM 236: 43–59 Mar 1992
40. Simoes-Moreira, J.R., Shepherd, J.E.: Evaporation waves in superheated dodecane, JFM 382: 63–86 Mar 10 1999
41. Cramer, M.S., Crickenberger, A.B.: Prandtl–Meyer function for dense gases, AIAA Journal 30 (2): 561–564 Feb.1992
42. Vidal, P., Khasainov, B.A.: Analysis of critical dynamics for shock-induced adiabatic explosions by means of the Cauchy problem for the shock transformation, Shock Waves 9 (4): 273–290 Aug. 1999
43. Hahne, E., Barthau, G.: Evaporation waves in flashing processes Int. J. of Multiphase Flow 26 (4): 531–547 Apr. 2000

Papers on Equations of State and Flow Theory

44. Abbott, M.M.: Cubic equations of state, AIChE Journal 19 No. 3, pp. 596–601, 1973
45. Yamada, T.: An improved generalized equation of state. A.I.Ch.E.J. 19, 286–291, 1973
46. Thompson, P.A. and Sullivan, D.A.: Similarity principle for the elementary Gasdynamics of real fluid isentropic flow, Phys. of Fluids 20, No. 7, pp. 1064–1071, 1977
47. Thompson, P.A.: A Simple formula for the heat capacity of polyatomic gases, with constants for 143 substances, Journal of Fluids Engineering, Vol. 100, 413–418, 1978
48. Städtke, H.: Gasdynamische aspekte der nichtgleichgewichtszweiphasenströmung durch Düsen, Ladewig, B., Berlin, pp. 7–133, 1975
49. Thompson, P.A., Sullivan, D.A.: Simple predictions for sonic conditions in a real-gas, Journal of Fluids Engineering-Transactions of the ASME 99 (1): 217–225 1977
50. Thompson, P.A. and Sullivan, D.A.: Exact and approximate equations for real-fluid isentropic flow. Journal of Fluids Engineering 100, pp. 413–418, 1978
51. Thompson, P.A. and Becker, F.: A one-parameter thermal-caloric corresponding-states model. Chem. Eng. Science 34, 93–99, 1979
52. Sullivan, D.A.: Fundamentals of real-fluid dynamics: Parts 1 & 2. Technical Information Series, General Electric, New York, 1979
53. Sullivan, D.A.: Historical review of real-fluid isentropic flow models. ASME, pp. 1–11, 1979
54. Hobbs, D.E.: A virial equation of state utilizing the principle of corresponding states. Dissertation, Rensselaer Polytechnic Institute, 1983

55. Thompson, P.A., Strock, T.W., Lim, D.S.: Estimate of shock thickness based on entropy production, Physics of Fluids 26 (1): 48–49 1983
56. Hagan, R., Slemrod, M.: The viscosity-capillarity criterion for shocks and phase-transitions, Archive for Ratio al Mechanics and Analysis 83 (4): 333–361, 1983
57. Dobran, F.: A nonequilibrium model for the analysis of two-phase critical flows in tubes. Stevens Inst. of Technology, Hoboken, New Jersey, pp. 141–149, 1985
58. Kouremenos, D.A. and Kakatsios, X.K.: Ideal gas relations for the description of the real gas isentropic changes. Forschung im Ingenieurwesen 51, Nr.6, pp. 169–174, 1985
59. Slemrod, M.: Dynamic phase-transitions in a Vanderwaals fluid, Journal of Differential Equations 52 (1): 1–23 1984
60. Thompson, P.A.: Shock-wave series for real fluids. Phys. Fluids 26, pp. 3471–3474, 1983
61. Biesheuvel, A. and van Wijngaarden, L.: Two-phase flow equations for a dilute dispersion of gas bubbles in liquid. J. Fluid Mech. 148, pp. 301–318, 1984
62. Prosperetti, A.: A generalization of the Rayleigh–Plesset equation of bubble dynamics. Phys. Fluids 25 No.3, pp. 69–77, 1984
63. Prosperetti, A. and Lezzi, A.: Bubble dynamics in a compressible liquid. Part 1. First-order theory. J. Fluid Mech. 168, pp. 457–478, 1986
64. Cramer, M.S., Kluwick, A., Watson, L.T., et al.: Dissipative waves in fluids having both positive and negative nonlinearity, Journal of Fluid Mech. 169: 323–336 Aug 1986
65. Cramer, M.S., Sen, R.: Exact-solutions for sonic shocks in Vanderwaals gases, Physics of Fluids 30 (2): 377–385 Feb.1987
66. Cramer, M.S., Sen, R.: Shock formation in fluids having embedded regions of negative nonlinearity, Physics of Fluids 29 (7): 2181–2191 Jul. 1986
67. Wegener, P.P.: Gasdynamics of expansion flows with condensation, and homogeneous nucleation of water vapor. In Nonequilibrium Flows, (ed. P.P. Wegener). New York: Marcel Dekker, 1969
68. Wegener, P.P. and Wu, B.J.C. Homogeneous and binary nucleation: New experimental results and comparison with theory, Faraday Disc. Chem. Soc. No 61, 77–82, 1976
69. Kotake, S. and Glass, I.I.: Survey of flows with nucleation and condensation, Institute for Aerospac Studies, Toronto, UTIAS 42, 1978
70. Oldervik, O. and Selanger, K.A.: Numerical model for transient pipe flow with nonequilibrium gas/liquid mixture, NHL 24 EE, pp. 59–71
71. Saltanov, G.A., Tsiklauri, G.V. and Shanin, V.K.: Shock waves in a flow of wet vapor with a high liquid phase content, High Temperature 8, 533–539, 1970
72. Boguslavskii, Y.Y. and Grigor'ev: Propagation of waves of arbitrary amplitude in a gas–liquid mixture. Sov. Phys. Acoust. 23 No. 4, pp. 364–365, 1977
73. Anderson, W.K.: Numerical study on using sulfur–hexafluoride as a wind-tunnel test gas, AIAA Journal 29 (12): 2179–2180 DEC 1991
74. Schnerr, G.H., Leidner, P.: Diabatic supersonic flows of dense gases, Physics of Fluids A-Fluid Dynamics 3 (10): 2445–2458 Oct. 1991
75. Cramer, M.S., Crickenberger, A.B.: The dissipative structure of shock-waves in dense gases, Journal of Fluid Mech. 223: 325–355 Feb. 1991

76. Chandrasekar, D., Prasad, P.: Transonic flow of a fluid with positive and negative nonlinearity through a nozzle, Physics of Fluids A-Fluid Dynamics 3 (3): 427–438 Mar.1991
77. Kluwick, A., Cox, E.A.: Propagation of weakly nonlinear-waves in stratified media having mixed nonlinearity, Journal of Fluid Mech. 244: 171–185 Nov. 1992
78. Cramer, M.S., Sen, R.: A general scheme for the derivation of evolution-equations describing mixed nonlinearity, Wave Motion 15 (4): 333–355 May 1992
79. Cramer, M.S., Tarkenton, G.M.: Transonic-flows of Bethe–Zeldovich–Thompson fluids, Journal of Fluid Mech. 240: 197–228 Jul.1992
80. Cramer, M.S., Tarkenton, L.M., Tarkenton, G.M.: Critical Mach number estimates for dense gases, Physics of Fluids A-FluidDynamics 4 (8): 1840–1847 Aug.1992
81. Cramer, M.S., Monaco, J.F., Fabeny, B.M.: Fanno processes in dense gases, Physics of Fluids 6 (2): 674–683 Part 1 Feb.1994
82. Kluwick, A., Scheichl, S.: Unsteady transonic nozzle flow of dense gases, Journal of Fluid Mech. 310: 113–137 Mar. 10 1996
83. Argrow, B.M.: Computational analysis of dense gas shock tube flow, Shock Waves 6 (4): 241–248 Oct. 1996
84. Ishii, R., Mizuno, M., Yuhi, M.: Shock waves in a van der Waals gas. 3. Two-dimensional shock waves, Transactions of the Japan Society for Aeronautical and Space Sciences 39 (125): 350–357 Nov. 1996
85. Bates, J.W., Montgomery, D.C.: Some numerical studies of exotic shock wave behavior, Physics of Fluids 11 (2): 462–475 Feb. 1999
86. Wang, C.W., Rusak, Z.: Numerical studies of transonic BZT gas flows around thin airfoils, Journal of Fluid Mech. 396: 109–141 Oct.10 1999
87. Fan, H.T.: Convergence to traveling waves in two model systems related to the dynamics of liquid–vapor phase changes, Journal of Differential Equations 168 (1): 102–128 Part 1 Nov. 20 2000
88. Suliciu, I.: On modeling phase transions by means of rate-type constitutive-equations shock-wave-structure, Int J Eng Sci 28 (8): 829–841 1990
89. Quartapelle, L., Castelletti, L., Guardone, A.,: et al. Solution of the Riemann problem of classical gasdynamics, Journal of Computational Physics 190 (1): 118–140 Sep. 1 2003
90. Guardone, A., Vigevano, L., Argrow, B.M.: Assessment of thermodynamic models for dense gas dynamics, Physics of Fluids 16 (11): 3878–3887 Nov. 2004
91. Guardone, A., Argrow, B.M.: Nonclassical gasdynamic region of selected fluorocarbons, Physics of Fluids 17 (11): Art. No. 116102 Nov. 2005
92. Bulgakova, N.M., Burakov, I.M.: Nonlinear hydrodynamic waves: Effects of the equation of state, Physical Review E 70 (3): Art. No. 036303 Part 2 Sep. 2004
93. Kluwick, A.: Internal flows of dense gases, Acta Mechanica 169 (1-4): 123–143 May 2004
94. Cinnella, P., Congedo, P.M.: Aerodynamic performance of transonic Bethe–Zel'dovich–Thompson flows past an airfoil, AIAA Journal 43 (2): 370–378 Feb. 2005
95. Schlamp, S., Rosgen, T.: Flow in near-critical fluids induced by shock and expansion waves, Shock Waves 14 (1–2): 93–101 Jun. 2005

96. Pinhasi, G.A., Ullmann, A., Dayan, A.: Modeling of flashing two-phase flow, Reviews in Chemical Engineering 21 (3-4): 133–264 2005
97. Le Metayer, O., Massoni, J., Saurel, R.: Modelling evaporation fronts with reactive Riemann solvers, Journal of Computational Physics 205 (2): 567–610 May 20 2005
98. Corli, A., Fan, H.T.: The Riemann problem for reversible reactive flows with metastability, SIAM Journal on Applied Mathematics 65 (2): 426–457 2005
99. Scheichl, S.: Linear and nonlinear propagation of higher order modes in hardwalled circular ducts containing a real gas, Journal of the Acoustical Society of America 117 (4): 1806–1827 Part 1 Apr. 2005
100. Cramer, M.S., Rayleigh processes in single-phase fluids, Physics of Fluids 18 (1): Art. No. 016101 Jan. 2006

Papers in Related Fields

101. Wilson, C.T.R.: Kondensation des Wasserdampfes in Gegenwart von staubfreier Luft und anderen Gasen. Beiblätter zu den Wiedemann Annalen der Physik und Chemie 21, 720–722, 1897
102. Oswatitsch, K.: Kondensationserscheinungen in Überschalldüsen. Zeitschrift für Angewandte Mathematik und Mechanik 22, pp. 1–14, 1942
103. Glass, I.I. and Patterson, G.N.: A theoretical and experimental study of shock-tube flows. Journal of the Aeronautical Sciences 22, No. 2, pp. 73–100, 1955
104. Oswatitsch, K.: Filmkondensation in Strömungen binärer Mischungen. Proc. of the 7th Lecture Series on Two-Phase Flow, Trondheim, pp. 121–156, 1983
105. Chen, G., Thompson, P.A. and Bursik, J.W.: Soundtrack measurements in vapor–liquid mixtures behind shock waves. Experiments in Fluids 4, pp. 279–282, 1986
106. Hiller, W.J., Jaeschke, M., Meier, G.E.A.: The influence of air humidity on pressure and density fluctuations in transonic jets. Journal of Sound and Vibration 17, pp. 423–428, 1971
107. Jaeschke, M., Hiller, W.J., Meier, G.E.A.: Acoustic damping in a gas mixture with suspended submicroscopic droplets. Journal of Sound and Vibration 43 (3), pp. 467–481, 1975
108. Bratos, M., Meier, G.E.A.: Two-dimensional, two-phase flows in a laval nozzle with nonequilibrium phase transition, Archives of Mechanics 28, 5-6, pp. 1025–1037, Warschau 1976
109. Davies, L. and Wilson, J.L.: The influence of shock and boundary layer interaction on shock tube flows. On Proc. 6th Int. Shock Tube Symp., Ernst–Mach-Inst., Freiburg i. B, 1968
110. Taylor, J.R. and Hornung, H.G.: Real gas and wall roughness effects on the bifurcation of the shock reflected from the end wall of a tube, Report, Australian National University, Canberra, Australia
111. Clarke, J.F.: Lectures on plane-waves in reacting gases, Annales de Physique 9 (2): 211–306 1984
112. Griffith, W.C.: Shock waves. J. Fluid Mech. 106, pp. 81–101, 1981
113. Thompson, P.A., Kim, Y.-G., Meier, G.E.A.: Flow visualization of a shock wave by simple refraction of a background grid. In: Optical Methods in Dynamics of Fluids and Solids, pp. 225–231, Springer-Verlag, 1985

114. Hagan, R. and Serrin, J.: Dynamic changes of phase in a van der Waals fluid. In: New Perspectives in Thermodynamics, Springer, Berlin, pp. 241–260, 1986

115. Grinfeld, M.: Dynamic phase-transitions-existence of cavitation waves, Proc of the Royal Society of Edinburgh Section A Mathematics 107: 153–163 Part 1-2 1987

116. Adamson, T.C., Jr. and Nicholls, J.A.: On the structure of jets from highly underexpanded nozzles into still air, J. of the Aero/Space Sciences, pp. 16–24, 1959

117. Blythe, P.A. and Shih, C.J.: Condensation shocks in nozzle flows. J. Fluid Mech. 76, 593–621, 1976

118. Oshima, Y. and Osaka, S.: 1975 Interaction of two vortex rings moving side by side. Nat. Sci. Rep. Ocharzonizu Univ. 26, 31–37

119. Elder, F.K. and de Haas, N.: Experimental study of the formation of a vortex ring at the open end of a cylindrical shock tube. J. Appl. Phys. 23. 1065–1069, 1952

120. Honda, M., Takayama, K., Onodera, O. and Kohama, Y.: Motion of reflected shock waves in shock tube. In Modern Developments in Shock Tube Research (ed. G. Kamimoto), Japan Shock Tube Res. Soc 1975

121. Mirels, H.: Boundary layer growth effects in shock tubes. In Shock Tube Research (ed. J. L. Stollery, A. G. Gaydon, P. R. Owen). London: Chapman and Hall 1971

122. Grolmes, M.A. and Fauske, H.K.: Axial propagation of free surface boiling into superheated liquids in vertical tubes, J. of Heat Transfer, pp. 30–34, 1979

123. Glass, I.I. and Liu, W.S.: Effects of hydrogen impurities on shock structure and stability in ionizing monatomic gases: Part 1, argon. J. Fluid Mech. 84, 55–77, 1978

124. Temkin, S. and Kim, S.S.: Droplet motion induced by weak shock waves. J. Fluid Mech. 96, part 1, pp. 133–157, 1980

125. Litwiniszyn, J.A.: Model for the initiation of coal–gas outbursts, International Journal of Rock Mechanics and Mining Sciences 22 (1): 39–46 1985

126. Caflisch, R.E., Miksis, M.J., Papanicolaou, G.C. and Ting, L.: Wave propagation in bubbly liquids at finite volume fraction. J. Fluid Mech. 160, pp. 1–14, 1985

127. Homer, J.B.: Studies on the nucleation and growth of metallic particles from supersaturated vapours. In Shock Tube Research (eds. J. L. Stollery, A. G. Gaydon and P. R. Owen), Proc. 8th Int. Shock Tube Symp. London: Chapman and Hall, 1971

128. Henry, R.E. and Fauske, H.K.: The two-phase critical flow of one-component mixtures in nozzles, orifices and short tubes. J. of Heat Transfer, pp. 179–187, 1971

129. Wallis, G.B.: Critical two-phase flow. Int. J. Multiphase Flow 6, pp. 97–112, 1980

130. Zysin, V.A., Parfenova, T.N., Pervushin, L.K. and Papov, G.A.: Critical Phenomena during the discharge of a self-evaporating liquid. J. of Heat Transfer 6 No. 1, 1974

131. Noordzij, L. and Van Wijngaarden, L.: Relaxation effects, caused by relative motion, on shock waves in gas-bubble/liquid mixtures. J. Fluid Mech. 66, part 1, pp. 115–143, 1974

132. Becker, E., Boehme, G., Burger, W.: Simple waves in heat conducting and relaxing gases, Zeitschrift für Angewandte Mathematik und Mechanik 55 (1): 31–40 1975
133. Bains, J.A., Breazeale, M.A.: Nonlinear distortion of ultrasonic-waves in solids – approach of a stable backward sawtooth, Journal of the Acoustical Society of America 57 (3): 745–746 1975
134. Meier, P. and Sandeman, R.J.: Interferometric studies of shockwaves into argon.. In Modern Developments in Shock Tube Research (ed. G. Kamimoto). Japan Shock Tube Res. Soc. 1975
135. Olson, J.D. 1975 The refractive index and Lorenz–Lorentz function of fluid methane. J. Chem. Phys. 63, 474–484
136. Slemrod, M.: Admissibility criteria for propagating phase boundaries in a van der Waals fluid, Achive for Rational Mech. and Analysis 81, pp. 301–315, 1983
137. Peregrine, D.H.: The fascination of fluid-mechanics, Journal of Fluid Mechanics 106 (May): 59–80 1981
138. Fowles, G.R. and Houwing, A.F.P.: Instabilities of shock and detonation waves, Phys. of Fluids 27, pp. 1982–1990, 1984
139. Kiefer, S.W.: Blast dynamics at Mount St. Helens on May 1980. Nature, Vol. 291, No. 5816, pp. 568–570, 1981
140. Turcotte, D.L., Ockendon, H., Ockendon, J.R., et al.: A mathematical-model of vulcanian eruptions, Geophysical Journal International 103 (1): 211–217 Oct. 1990
141. Dillmann, A., Meier, G.E.A.: A refined droplet approach to the problem of homogeneous nucleation from the vapor-phase, J. of Chem. Phys. 94 (5): 3872–3884 Mar 1 1991
142. Delale, C.F., Meier, G.E.A.: A Semiphenomenological droplet model of homogeneous nucleation from the vapor-phase, J. of Chem. Phys. 98 (12): 9850–9858 JUN 15 1993
143. Collins, R.L.: Chokes expansion of subcooled water and the I.H.E. flow model. J. of Heat Transfer 100, pp. 275–280, 1978
144. Fuchs, H. and Legge, H.: Flow of a water jet into vacuum. Acta Astronautica 6, 1213–1226 S., 1979
145. Datta, N. and Mishra, S.K.: On plane laminar two-phase jet flow with full expansion, Z. Flugwiss. Weltraumforschung 8, Heft 2, pp. 113–117, 1984
146. Russo, G.: Stability properties of relativistic shock-waves-applications, Astrophysical Journal 334 (2): 707–721 Part 1 Nov. 15 1988
147. Cramer, M.S., Best, L.M.: Steady, Isentropic flows of dense gases, Physics of Fluids A – Fluid Dynamics 3 (1): 219–226 Jan. 1991
148. Cramer, M.S., Fry, R.N.: Nozzle flows of dense gases, Physics of Fluids A-Fluid Dynamics 5 (5): 1246–1259 May 1993
149. Kluwick, A.: Transonic nozzle-flow of dense gases, Journal of Fluid Mechanics 247: 661–688 Feb. 1993
150. Schnerr, G.H., Leidner, P.: Real-gas effects on the normal shock behavior near curved walls, Physics of Fluids A-Fluid Dynamics 5 (11): 2996–3003 Nov. 1993
151. Skews, B.W., Atkins, M.D., Seitz, M.W.: The impact of a shock-wave on porous compressible foams, Journal of Fluid Mechanics 253: 245–265 Aug. 1993
152. Rusak, Z., Wang C.W.: Transonic flow of dense gases around an airfoil with a parabolic nose, Journal of Fluid Mechanics 346: 1–21 Sep. 10 1997

153. Cramer, M.S., Whitlock, S.T., Tarkenton, G.M.: Transonic and boundary layer similarity laws in dense gases, Journal of Fluids Engeneering-Transactions of the ASME 118 (3): 481–485 Sep. 1996

154. Monaco, J.F., Cramer, M.S., Watson, L.T.: Supersonic flows of dense gases in cascade configurations Journal of Fluid Mechanics 330: 31–59 Jan. 10 1997

155. Brown, B.P., Argrow, B.M.: Nonclassical dense gas flows for simple geometries, AIAA Journal 36 (10): 1842–1847 Oct. 1998

156. Kluwick, A., Cox, E.A.: Nonlinear waves in materials with mixed nonlinearity Wave Motion 27 (1): 23–41 Jan.1998

157. Cramer, M.S., Park, S.: On the suppression of shock-induced separation in Bethe–Zel'dovich–Thompson fluids, Journal of Fluid Mechanics 393: 1–21 Aug. 25 1999

158. Krysac, L.C., Maynard, J.D.: The role of convection during the self-focusing of nonlinear second sound pulses near the lambda point, Journal of Low Temperature Physics 110 (5–6): 949–962 Mar. 1998

159. Muralidharan, S., Sujith, R.I.: Shock formation in the presence of entropy gradients in fluids exhibiting mixed nonlinearity, Physics of Fluids 16 (11): 4121–4128 Nov. 2004

Historical Papers and Books

160. Van der Waals, J.D.: Lehrbuch der Thermodynamik, bearbeitet von Ph. Kohnstamm, Maas und Van Suchtelen, Leipzig 1908

161. Bethe, H.A.: The theory of shock waves for an arbitrary equation of state. Office Sci. Res. and Dev., Washington, Rep. No. 545, p. 57, 1942

162. Landau, L.D. and Lifshitz, E.M. Fluid Mechanics, p. 96. Pergamon 1959

163. Skripov, V.P.: Metastable liquids. A Halsted Press Book, 1974

164. Thompson, P.A.: Compressible fluid dynamics. McGraw-Hill Book Company, 1972

165. Walton, A.J.: Three phases of matter. Clarendon Press, Oxford, 1983

166. Zel'dovich, Ya.B. and Raizer, Yu.P.: Physics of Shock Waves and High-Temperature Hydrodynamic Phenomena. ed. W.D. Hayes and Probstein, Vol.2, pp. 750–756, Academic Press, 1967

167. Meier, G.E.A. and Obermeier, F. (Eds.): Flow of real fluids. Lecture Notes in Physics 235, Springer-Verlag Berlin, Heidelberg, New York, Tokyo (348 S.), 1985

168. Born, M.: Optik. Springer 1972

169. Kerker, M.: The Scattering of Light. Academic Press 1969

170. White, D.R. Turbulent structure of gaseous detonation. Phys. Fluids 4, 465–480, 1961

171. Benett, F.D.: Vaporization waves as a general property of high temperature matter. Ballistic Research Laboratories, Report No. 1272, pp. 3–15, 1965

Part III

Shock Waves Interacting with Solid Foams, Textiles, Porous and Granular Media

8

Experimental Studies of Shock Wave Interactions with Porous Media

Beric Skews

8.1 Introduction

Many early studies of the interaction of a shock or blast wave with a porous barrier or screen have had as their main interest methods of attenuating the wave so as to reduce downstream loads. When a shock wave strikes a sheet or slab of material a wave is reflected back upstream from the surface and a wave passes through to the rear face resulting in a transmitted wave into the space behind the sheet, or onto a backing plate if there is no cavity. The strength and nature of these waves depends on the nature, composition, and topology of the sheet. In this chapter the treatment will concentrate on materials with voids and interstitial spaces such as sheets with perforations, rigid and deformable slabs, and textiles. In the latter case both open and closed weave materials will be discussed. Both rigidly held and freely suspended specimens will be considered. In order to gain some understanding of the complex nature of shock interaction with permeable surfaces a variety of experimental investigations are described, ranging from shock impact on a rigidly held permeable barrier, through the effects of allowing the barrier to move under the shock loading, to considering the effects of a layer of textile overlying a surface in a shock environment. The nature of the flow through slabs of rigid porous materials, both with head-on and oblique impact are also considered.

Studies of shock impact on a surface covered with a layer of flexible foam have shown the interesting phenomenon of pressure amplification, i.e., the wall experiences a much higher pressure load during shock reflection than it would have experienced were it uncovered. Not only is there movement of the slab itself but there are also waves propagating within the interstitial spaces. Understanding of the mechanisms has received considerable attention over the past decade or so and is treated in Sect. 8.3.2. Similarly it is found that a textile or porous sheet covering a surface can also cause pressure amplification, although in that case the mechanisms are different. Some consideration will also be given to shock induced flow over particle beds.

The material of this chapter will deal with the physical processes that occur during the shock-induced flow rather than the design of blast protection systems.

8.2 Thin Barriers

There have been a number of empirical studies on the effect of barriers for the attenuation of blast, eg., [1, 22]. Most studies have looked at attenuation along a duct system. However oblique incidence is equally important.

8.2.1 Perforated Rigid Plates

Early detailed work on the effect of a shock wave interacting with a permeable sheet was done on perforated plates. The first such study was that conducted by Friend [2]. He used a uniform distribution of 6 mm diameter circular holes in a 3 mm thick sheet to give a blockage ratio of 50%. Most of his tests were conducted at high pressure ratios so that the perforations became choked, thereby giving him an extra boundary condition for analysis. Furthermore it was assumed that tangential momentum was conserved across the plate.

A comprehensive study was undertaking by Onodera and Takayama [11] on shock propagation through a plate with slits, giving two-dimensional inter-actions, for plates with a blockage ratio of 60%. Their interest was primarily in the range from glancing incidence up to transition to regular reflection. In their analysis they assumed that the flow into the plate was normal to the surface in a frame of reference fixed in the plate. They showed that the angle for transition was decreased by about 10° compared to an impervious plate at the same angle. These tests were extended by Skews and Takayama [17] to mainly cover the region from regular reflection up to head-on incidence. These tests are particularly of value since the flow behind the reflected shock is su-personic relative to the reflection point and thus the corner signal from the front edge of the model does not reach the reflection point. The data can thus more confidently be treated as being pseudostationary. These results showed a number of interesting features. Of most interest was that the inflow into the slits, in grid-fixed coordinates was not normal to the plate but at about 17° to the surface over the range of wall angles between the shock and the plate from zero to 60°, this being very different from the 90° assumed in previous work. Furthermore the flow was shown to exit approximately normal from the plate, showing that for the geometry tested tangential momentum is not conserved.

A much more comprehensive study has more recently been undertaken by Skews [14] in order to cover a much wider range of plate blockages. These studies used contact shadowgraph imaging rather than the holography used in the previous work as it was found to show up much more detail regarding the weak wavelets that are developed in the flow. The test piece used in the shock tube is shown in Fig. 8.1. The grid is mounted on a rotatable frame and the

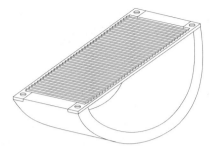

Fig. 8.1. Shock tube test piece mounted on a rotatable frame for changing incidence angle

(a)   (b)

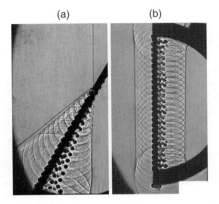

Fig. 8.2. Contact Shadowgraphs: (**a**) $\theta = 60°$, $M = 1.35$. (**b**) $\theta = 90°$, $M = 1.51$

transverse ribs are glued to narrow support strips on either side. Some typical experimental results are shown in Fig. 8.2. It is noted that both the reflected and transmitted waves develop from wavelets generated by the presence of the ribs, and soon become plane. Secondary waves, not noted in the previous studies, and parallel to the reflected and transmitted waves are also noted. A contact surface consisting of a series of vortices is visible behind the plate which separates the gas that has passed through the plate from that, originally below the plate, which has been put into motion by the transmitted wave. The pseudosteady assumption was found to describe the motion of these interfaces reasonably well. However, the source of the first row of vortices established in this work was found to arise at the trailing edges of the ribs, and not from the separation of the flow at the leading edge as had been assumed in the previous study.

There is a consistent pattern of reducing reflection angle for a given incidence angle and shock Mach number as the blockage decreases, as indicated in Fig. 8.3. The limiting conditions of 100% blockage representing a nonporous wall with regular reflection, and a wall with zero blockage, corresponding to the reflected wave being a sound wave, straddle these experimental values.

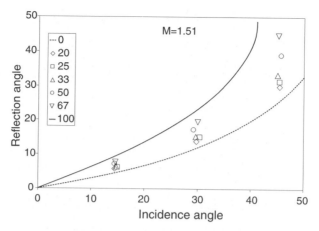

Fig. 8.3. Variation of reflected shock angle with incident shock angle for various values of blockage

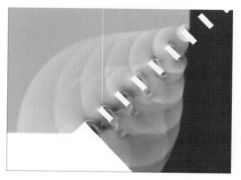

Fig. 8.4. Greyscale plot of sound speed for an incident shock Mach number of 1.36 and wall angle of 45°

The nature of the source of the multiple vortex pattern and the secondary reflected waves is strongly dependant on the grid geometry, and is conveniently studied through CFD simulation. A typical example is given in Fig. 8.4. On the upper surface a vortex is shed due to flow separation at the edge of each crossbar as the incident shock diffracts into the gap. A shear layer is sustained at this position due to the inclination of the shock induced flow into the gap. As the shock passes through the gaps it diffracts at the lower corners of the crossbar resulting in the leading lines of vortices visible in the images.

An important parameter to assess is the stagnation pressure loss across the plate. It is evident from Fig. 8.2a that the transmitted wave is nearly as strong as the incident wave. By taken the exit flow to be perpendicular to the plate and within the pseudostationary flow assumption, the downstream stagnation pressure may be calculated. It is found that the loss increases with

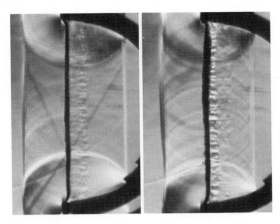

Fig. 8.5. Mach 1.39 shock impact on muslin and cotton specimens, showing reflected and transmitted waves, and the contact surface of flow through the material

increasing Mach number, increasing blockage, and decreasing wall angle. The effect of pore geometry has not yet been evaluated but will have an effect.

8.2.2 Textiles

Although the main interest in textile investigations is that of pressure amplification experienced on an adjacent underlying wall, as shown by Hattingh and Skews [6], it is instructive to do tests without the wall, as done for the grids described above, so as to characterize the flow and the transmitted wave system on the underside. Figure 8.5 shows typical head-on impact images for samples of muslin and cotton fabric. These materials vary significantly in permeability (154 and 33 $m^3 \, m^{-2} \, min^{-1}$ respectively), but have similar specific mass (95 and 104 $kg \, m^{-3}$).

The similarity with Fig. 8.2b is evident although the structure of the flow is less well defined because of the pore distribution. Thus the structure of the flow passing through the sheet does not appear as a set of discrete vortices, but rather as a turbulent front. The greater porosity and lower flow resistance for the muslin compared to the cotton is clear, by noting that this contact surface has penetrated much further into the downstream flow. The reflected wave is also weaker as evidenced by having travelled a smaller distance relative to that of the transmitted wave. Note also how the specimen is being pulled from the supporting frame where it was lightly held by double-sided tape.

When a surface is positioned a short distance behind the textile, as for a person wearing a layer of clothing, it has been found that the pressure experienced on the surface is higher than without clothing [6]. Figure 8.6 shows four sequential images of the details of the interaction and the movement of the cloth towards the back wall (the dark surface on the right of each image) at an incident shock Mach number of 1.37 and an initial gap between the textile

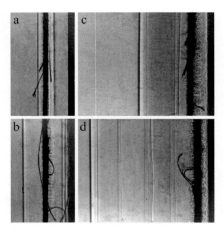

Fig. 8.6. Sequential images showing multiple shock development, cloth movement, and flow reversal

and the wall of 6 mm. The first image was taken just after initial impact and shows both the reflected and transmitted waves. In Fig. 8.6b, whilst there is no evidence of textile movement, the transmitted wave has reflected off the back wall. The contact surface showing the front of the gas that has passed through the specimen is also visible. At the time of Fig. 8.6c the cloth has started to move towards the wall, the turbulent region at the back of the specimen is larger, and the wave reflected off the back wall has passed back through the specimen and is moving upstream, following the primary reflected wave. A rereflected wave off the back face of the specimen is to be expected but is not visible in the gap between the wall and the textile. Nevertheless at later times, Fig. 8.6d, turbulent gas now starts to emerge from the front face of the specimen indicating that the pressure in the gap has risen to such an extent as to reverse the flow.

Furthermore, it is seen that a train of waves emerge back into the oncoming flow, indicating a series of wave reflections within the gap. This is confirmed by a wave diagram analysis based on both pressure traces and photographs, as presented in [6]. A typical pressure trace measured by a transducer in the back wall is given in Fig. 8.7, together with a trace of shock reflection without the covering, at the same Mach number. The trace with covering exhibits a series of steps joined by plateaus with slightly increasing pressure. The steps correspond to waves moving back and forward between the back of the textile and the back wall of the tube. The intermediate pressure increase is due to the slower piston action of the sheet approaching the wall. As noted in Fig. 8.6 the flow through the cloth reverses at some stage due to the higher pressure in the gap, and the textile will also start moving back away from the wall. When this happens depends on the porosity and mass of the specimen. The result will be that the air in the gap expands as noted in the pressure trace,

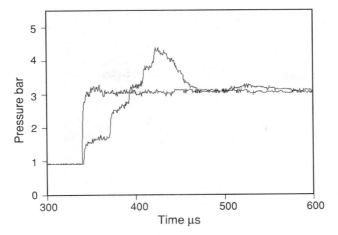

Fig. 8.7. Back wall pressure variation for a cotton specimen ($M = 1.37$, initial gap = 6 mm) compared with a trace with no textile present

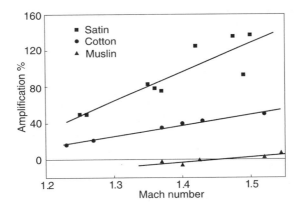

Fig. 8.8. Amplification for textile specimens with an initial 6 mm gap

even though there are still small steps due to a weak wave still oscillating in the gap. Eventually the pressure reaches the same value as that would result when the wall is uncovered. The peak pressure achieved is, however, significantly higher than for the uncovered wall.

Amplification percentages for three common textiles as a function of incident shock Mach number are given in Fig. 8.8. There is a steadily increasing trend of amplification with shock strength, with the least permeable, and heaviest, material giving the higher values. For low Mach numbers the very permeable material does not amplify the pressure. The satin permeability is $6\,\mathrm{m^3 m^{-2} min^{-1}}$ and the specific mass 134 kg m^{-3}. The values for the other two have been given previously. A much wider range of textiles have also

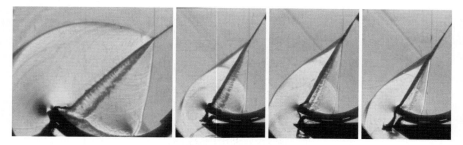

Fig. 8.9. Oblique shock wave impact on textile specimens, $M = 1.42$. (**a, b**) muslin, (**c**) cotton, (**d**) satin

been tested, with amplifications up to 400% being achieved for very heavy, essentially impervious, woven kevlar fabrics. Tests with varying gap indicate that the maximum amplification is with a gap in the range from 5 to 10 mm, depending on cloth type.

All the tests described above were for head-on incidence of the shock onto the sample. Clearly oblique incidence will introduce other effects. Figure 8.9 shows two results for muslin at 35 and 60° followed by 60° tests on cotton and satin. Care needs to be exercised in interpreting these images since the edges of the specimen adjacent to the shock tube windows tend to fold back allowing gas to leak through the gap. Thus in the cases of cotton and satin two transmitted shocks and two contact surfaces are visible. The leading one in each case is that due to the leakage and should be ignored. Some confirmation of which is the correct surface is obtained by calculating the contact surface angle that should result behind the transmitted wave. These results show strong similarity with the grid tests given in Sect. 8.2.1. One major difference is the significant movement of the textile as can be seen towards the bottom of the images as the material is pulled over the lower edge of the support frame. It is interesting that a sharp kink appears in the textile layer and that between the kink and the point of shock impingement the textile layer is nearly plane. This will make analysis easier, as the process at the higher wall angles would then appear to be pseudostationary. In the case of satin almost no air makes it way directly through the specimen. In the case of the muslin test at 35° the reflection is regular, although it would be Mach reflection for an impermeable surface, as in the case of the rigid grids, confirming that porosity delays transition. Wavelets emanating from the pores are also evident in this photograph.

When there is a backing surface behind the layer of textile pressure measurements still show the staircase pattern found for head-on incidence. Thus multiple reflections within the gap are indicated. An attempt to image these is given in Fig. 8.10. The transmitted wave and its reflection from the wall are clearly visible. However, the reflected wave then interacts with the contact

Fig. 8.10. Shock wave system within a gap for oblique impact

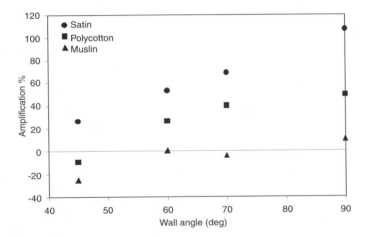

Fig. 8.11. The effects of incidence angle on amplification

surface and the subsequent interactions are unclear. Some photographs indicate the development of a turbulent flow in the gap as well as strong boundary layer effects.

Figure 8.11 gives the amplification results for three of the textiles tested. In all cases the maximum amplification occurs with head-on shock impact, with the heaviest, least permeable material, giving the highest value, of nearly 120%. These trends appear somewhat different from those obtained for the plastic sheets as discussed in Sect. 8.2.3 below, but it should be noted that the textile tests are for specimens less than 2 g in mass and very much lower permeability. Nevertheless, this will need further examination.

8.2.3 Movable Sheets

It is clear from the work on textiles that the fact that the sheet can move is an important factor relating to amplification. The two controlling parameters are clearly the specific mass and the porosity. Since it is not feasible to obtain textiles with identical porosity and different masses, and vice versa, special specimens were prepared for testing. Plastic sheet materials were used for the specimens. They were cut to size and then perforated by drilling holes of different sizes and spacing for different specimens to attain different porosity. To vary the mass of the samples a range of thicknesses were used from the same material. Samples were perforated using a high-speed drill to obtain smooth holes. Several groups of samples were manufactured each with the same porosity and different mass per unit area. Air permeability tests for the samples were done by the local Bureau of Standards.

The plastics were cut to fit the dimensions of the shock tube (180 mm high and 76 mm wide). The attachment of those samples, ahead of the base plate, that were thick enough and strong enough to support their own weight without buckling was done simply. A small piece of putty was placed at the bottom and the top walls of the test section and the sample was then positioned between the back wall of the test section and the putty. This arrangement allowed for the sample to be completely free of any obstructions during its motion towards the back wall after the impact of the shock wave. The air gap between the back wall and the sample was set to 3 mm for all tests. For samples that were too thin and which buckled when placed upright had to be supported on two 0.5 mm diameter pins positioned at the two upper corners of the shock tube back wall. The sample then hangs freely on these guide pins with very little resistance to backward motion.

Figure 8.12 gives the results for a selection of test plates of two different porosities and a series of sample masses. Tests were done at a Mach number of 1.42. A clear trend is evident, that the amplification reduces as the mass increases. This is to be expected since the acceleration towards the wall is reduced for the heavier samples and there is longer time for the gas initially in the air gap to escape back out through the perforations as the gap reduces in size. Similarly, comparing the two graphs, the higher porosity samples exhibit lower amplification. This is not only because the resistance to flow is less for the more porous sample, but also because the rereflected transmitted shock from the back wall results in a weaker returning wave when it strikes the back face of the specimen, and a stronger transmitted wave is sent back out into the external flow.

8.3 Thick Slabs

The material in the previous sections deals with test pieces where the thickness is very small in comparison with the frontal area and is of the same order as

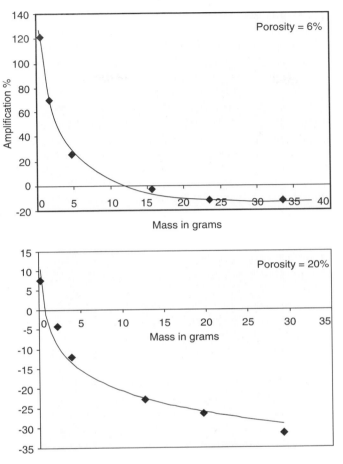

Fig. 8.12. Amplification by perforated plastic sheets mounted to enable free movement, and positioned in front of a wall

the pore size. However for thicker slabs there are two cases of interest: one where the skeleton can be considered as being rigid and one where it suffers considerable deformation during the loading. A particular case of interest is where the matrix collapse also contributes to the flow dynamics. Thus it is known that when a slab of open-cell polyurethane foam is placed adjacent to a wall, the wall pressure again exceeds that which would be experienced with no covering when impacted by a shock wave. Both these cases will be reviewed. There are two phenomena requiring consideration; one relating to the entry mechanism of the shock into the specimens, relating to reflection off the cell walls etc. and the other the transient flow in the bulk material dominated by viscous effects within the matrix.

8.3.1 Rigid Skeletons

One-dimensional studies by van der Grinten et al. [23] showed that the wave propagating through a slab consisting of bonded sand particles with the interstitial spaces filled with air, was fully dispersed with increasing rise time as it moved further into the slab. In this case the skeleton may be assumed to be rigid. When the slab was saturated with water, however, there is distinct detectable coupling between the two phases as both expand and compress and influence each other.

For much coarser materials Levy et al. [10] showed that a shock wave could penetrate into the slab, becoming weaker as it propagated until it also became dispersed. There are thus two time scales associated with the process: a short time scale associated with the wave interaction with the pore structure at the slab entrance and which feeds perturbations back into the external flow which then ultimately coalesce to form the reflected wave, and a longer time scale dominated by viscous effects as the compression wave propagates through the specimen. The tests were conducted on specimens of silicon carbide and of alumina, with pore sizes ranging from 10 to 40 pores in.$^{-1}$ and void fractions between 0.72 and 0.82.

Pressure traces from three transducers in front of, and four along the length of the specimen are shown in Fig. 8.13. (The horizontal lines are the calculated pressures for an equal strength incident shock and an uncovered rigid wall). It is noted that the reflected wave is followed by a compression and increases in strength as it moves away from the surface. This was interpreted as being due to the multiple interactions occurring within the pores at the early stage of the process sending compression waves back through the front face, which coalesce with the reflected shock causing it to strengthen. Once within the foam a weak shock wave is discernable for the most forward measuring point. It is followed by a continuous pressure increase. The leading shock wave becomes weaker as it propagates further into the specimen until it also becomes part of the dispersed compression resulting from the friction within the matrix. The final pressure is higher than that which would be

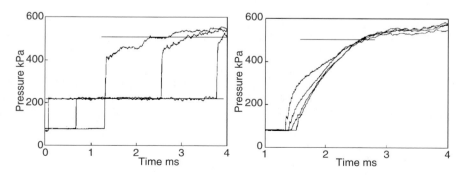

Fig. 8.13. Pressure variation at three positions ahead of, and four positions along the length of a 81 mm long specimen of 10 ppi SiC, $M = 1.54$

Fig. 8.14. Regular ($M = 1.22$, $\theta = 30°$), and Mach reflection ($M = 1.32$, $\theta = 20°$) using 10 ppi, SiC specimens

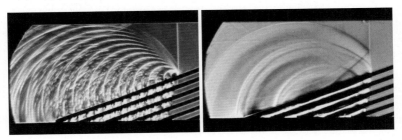

Fig. 8.15. Transverse bar layers ($M = 1.14$) and longitudinal bars ($M = 1.45$)

achieved by shock reflection off a rigid wall. No clearly demarcated evidence of reflection of the compression wave off the back wall of the shock tube is evident.

Using the same materials as in [10], two-dimensional studies have been reported [19]. The wave reflection pattern (Fig. 8.14) is significantly modified by the gas inflow, as is expected from tests on porous grids. The triple point trajectory angle is reduced for Mach reflection and likewise for the reflected wave angle in regular reflection. The Mach stem is not perpendicular to the surface. As expected, the reflection angle is suppressed compared to that for a rigid wall and the centers of the circular perturbations lie below the surface due to the gas inflow. However there a broad bands where these perturbations are more marked. This may be due to the transmitted wave reflecting off the base of the shock tube. Pressure measurements at the base of the wedge show a distinct pattern of oscillations [19], the source of which has not fully been determined.

In order to obtain some idea of the internal flow special models were manufactured containing either transverse or longitudinal bars giving a void fraction of 0.75 with the bar's position in successive layers being immediately below the gaps in preceding layers. Typical results are given Fig. 8.15. As expected the transverse bar case gives a reflection field similar to that for inclined grids. It is interesting to note that the transmitted wave, although segmented, has a planar envelope. There is no visual evidence of its reflection from the base of the test section. On the other hand the flow with longitudinal grids is much simpler and shows a limited number of reflected waves, evidently arising from each of the layers and thus are unlikely to correlate with the broad bands noted in Fig. 8.14 since in that case the pores are randomly distributed. For tests at the same Mach number and wall angle it is found that the transmitted wave is weaker for the transverse bar case indicating that boundaries of pores in a random structure that are transverse to the flow are likely to contribute more to shock attenuation.

8.3.2 Compressible Foams

Early studies of the interaction of shock waves with blocks of porous material were done to investigate attenuation and energy dissipation related to blast wave loading. Since it is well known that porous, open-cell foams have good properties in attenuating sound waves and are used for lining cavities and surfaces to reduce noise levels and for safety padding, they were a natural candidate to explore in the context of shock wave attenuation. Such attenuation does occur if a slab of such material is placed some distance ahead of a surface and a shock wave strikes the front surface [13]. However, if a small or no gap exists a significant pressure amplification can be produced at the rear interface.

Head-on Impact

The significant early one-dimensional studies of the amplification effects with open-cell foams were done in Russia [3,5]. A review of this and following work is given in [15]. The initial approach was to treat the foam and gas together as a single heavy gas phase (pseudogas), and to treat the situation as one of simple shock refraction between two different materials. This implied that the interface between the gas and foam was an impervious contact surface, and that a single wave propagated through the foam. Not withstanding the limitations of this assumption many of the main features of the interaction were captured. More detailed studies [20] showed that the interaction may be considered to consist of four main phases: the initial shock wave impact resulting in the formation of a reflected wave, a constant velocity compaction phase as the foam collapses towards the back wall, a stagnation phase as the foam and gas are brought to rest against the back wall, and a recovery phase as the pressure equilibrates throughout the system. An idealized wave diagram is given in Fig. 8.16. A more detailed study of the effect of foam properties for seven foams is given in [12] from which a number of the results given below are extracted. For these polyurethane foams the stiffness of the matrix structure is not a significant factor, stress levels for collapse of the cells being generally less than 4 kPa. Comparison of drag effects for shock loaded foams with steady flow results is treated in [7].

    The fact that gas moves into the foam immediately after the initial shock impact was demonstrated in a study of the reflected wave [21]. The initial reflected wave was shown to be a compression fan which comprised a succession of reflected waves arising as the gas was slowed down in the interstitial spaces

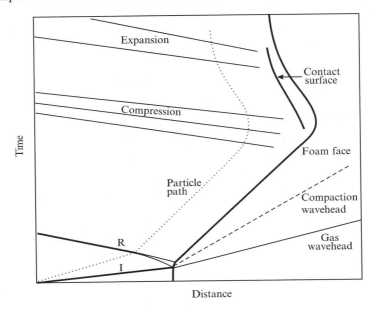

Fig. 8.16. Wave diagram of the main features for shock wave impact on a porous foam specimen

Fig. 8.17. Wavelets between the reflected shock moving to the left, and the specimen on the right following shock impact

within the front section of the foam. These waves rapidly coalesce into the single reflected shock wave. Figure 8.17 shows the wavelet pattern emerging from the face of the foam [12]. The foam is on the right and the reflected wave is the dark vertical line. The distribution and strength of these wavelets and the time they take to coalesce into a single front is determined by the structure and size of the interstitial spaces.

Figure 8.18 shows the pressure traces obtained from sidewall transducers in front of the foam block. At A the incident wave passes the first transducer and shows a typical step rise in pressure. The reflected wave, BCD, which passes back over the transducer has a longer rise time and shows significant rounding at CD as the wavelets forming a compression front coalesce into the reflected wave. The second transducer, positioned very close to the initial position of the foam face shows this rounding to be more extensive. The strength of the reflected wave after consolidation is indicated in Fig. 8.19 for two foams of similar permeability (10^{-4}m^2). It is noted that it is considerably

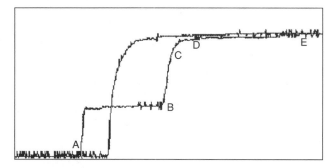

Fig. 8.18. Pressure traces from transducers 31 and 1 mm ahead of initial foam face position

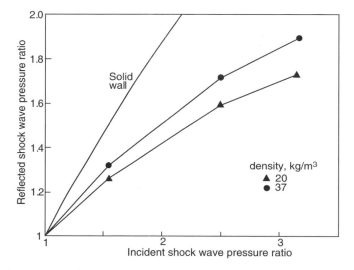

Fig. 8.19. Reflected wave strength for similar permeability foams

weaker than the reflection off a rigid wall. This is largely due to the receding foam face which experimentally is found to have a very similar velocity to that calculated to exist behind the reflected wave, as also indicated in the wave diagram above.

Shortly after impact the foam skeleton starts to collapse as the cell walls buckle, and an almost impervious compaction wave develops in the foam material. The foam face accelerates during the early phase but then after no further gas can enter, the face of the foam and the gas in front of it moves down the tube at a nearly constant velocity, whose value depends on the foam properties. The front few millimeters of foam at the face then recovers elastically as there is no longer a pressure gradient due to inflow.

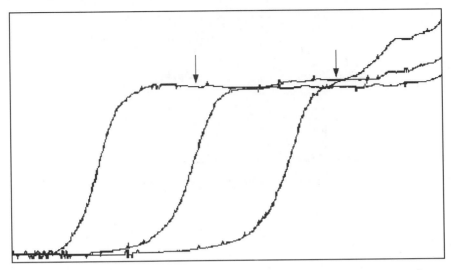

Fig. 8.20. Similarity of pressure traces at positions along the specimen. Arrows indicate when the foam face passes over the transducers

It has been shown [20] that, in contrast to the pseudogas model and as shown in the wave diagram, there are two waves which propagate through the material following the initial acceleration phase, the compaction wave propagating through the skeleton due to the collapsing cells, and a faster gas wave moving ahead of it. This part of the process, i.e., until the waves reach the end wall, is essentially pseudosteady as is noted from pressure traces taken at different positions down the tube, as shown in Fig. 8.20 where the shape of the trace is similar for different transducer positions. The reflection from the back wall becomes evident towards the end of these traces. However, the rise time of the gas compression front is slower the more permeable the foam. Measurement of the matrix compaction wave velocity is done by painting a grid of diagonal lines forming squares on the side surface of the foam so that the motion of individual points on the surface could be traced from successive photographs. An oblique image example is given in Fig. 8.21 where the compaction wave can be discerned about a third of the length along the specimen. This image also shows the dishing that occurs on the front face of the foam due to the friction of the sidewalls. An initial snug fit is required, however, since leaving a gap results in transverse collapse of the specimen [20]. The Poisson's ratio for such polyurethane foams is very small so it is not expected that the frictional force changes much in the process.

The back wall pressure amplification results when both the gas and foam movement are arrested at the back wall. The compression wave reflects and the momentum of the moving foam material is also transferred. Very high pressures may result. These can be sufficiently high that the heated com-

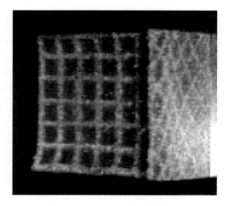

Fig. 8.21. Slab distortion due to shock loading, showing dishing of the front face and the compaction wave on the side face [20]

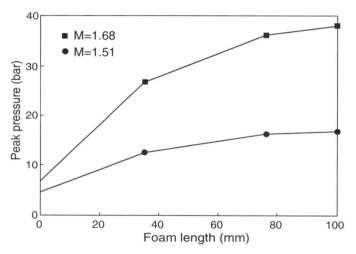

Fig. 8.22. Asymptotic behavior of peak pressure with increasing specimen length. Permeability 217×10^{-6} m^2, Density 36 kg m^{-3}

pressed gas is forced back out through the layer of highly compressed foam material and a contact surface emerges from the face of the foam as indicated on the wave diagram. This occurs shortly before the position of maximum foam compaction.

Besides dependance on foam properties the magnitude of the pressure peak is found to be dependent on foam length and incident shock Mach number. This increase in peak pressure with foam length is nonlinear (Fig. 8.22), asymptoting to a maximum at about 120 mm. The effect of Mach number is shown in Fig. 8.23 with the amplification increasing dramatically with increasing Mach number. The duration of the pressure pulse on the back wall,

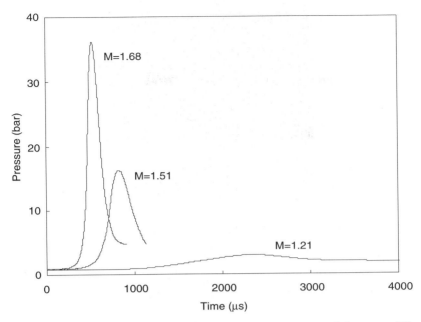

Fig. 8.23. Back wall pressure profiles for a 76 mm length of foam at different Mach numbers. Density $36 \, \mathrm{kg \, m^{-3}}$, Permeability $217 \times 10^{-6} \mathrm{m^2}$

however, decreases with increasing Mach number. At high Mach numbers significant amplification is achieved even for relatively short foam specimens, whereas for low Mach numbers the increase is more gradual. Amplifications of as high as six times have been recorded at a Mach number of 1.68, depending on the foam properties.

Oblique Impact

Usable results for oblique shock wave impact onto a foam wedge are not available. Attempts at such tests using foam wedges are limited because of the significant lateral collapse at the edges of the specimen during the loading, causing the foam distortion to become fully three-dimensional. Nevertheless schlieren images, such as that given in Fig. 8.24, show some interesting features. Firstly it is noted that the wavelets arising from the roughness of the surface have their centres well below the surface, thus allowing an estimate to be made of the gas penetration. Secondly, the reflected wave has what appears to be a multiwave structure, probably resulting from the emerging compression waves.

In an attempt to overcome the problem of unwanted three-dimensional collapse effects and also to avoid the frictional effects in testing a block of material in a shock tube, some experiments have been conducted using

Fig. 8.24. Shock wave impact on a foam wedge

Fig. 8.25. Foam cone collapse

conical specimens [18]. These are attached to the back wall of a shock tube using double-sided adhesive tape. This results in the base of the cone to be constrained to have a constant diameter but the rest of the cone can experience significant distortion as noted in the image in Fig. 8.25 taken near the time of maximum compaction. The specimen has an initial apex angle of 45° and is 180 mm long.

Some typical schlieren images are given in Fig. 8.26. The times are measured from when the shock strikes the rear wall. Considerable difficulty was encountered in trying to get a good image of the wave reflected off the foam surface. It is clearly very weak and this is made worse for visualization by it being axi-symmetric. The situation appears to be similar to that on a perforated wall in a transonic wind tunnel where wave cancellation occurs. A number of pictures using high schlieren sensitivity showed some shading behind the wave. The thin vertical line in all the frames of Fig. 8.26 is a reference marker. In the first frame the reflected shock wave propagating back up into the shock-induced flow moving from left to right, is clearly visible. The wave becomes highly curved as it diffracts back over the decreasing radius of the cone. The fact that the incidence angle to the surface becomes less than 90° is a very clear indication of inflow into the surface. At this early stage of the process there is already significant deformation of the cone. This is most noticeable near the rear wall where the material is constrained by its attachment to the wall. In the next frame the emergence of a contact surface over the rear three-quarters of the length of the cone is visible. Thereafter,

Fig. 8.26. Flow features of shock impact on a cone, showing the reflected wave (frame 1), contact surface emergence (frame 2), and reingestion (frame 3)

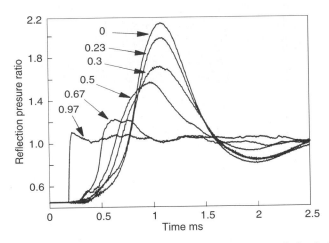

Fig. 8.27. Pressures on the back wall at various radii for axial shock loading of a foam cone

even though the deformation increases over parts of the surface this contact surface is reingested similar to that in the one-dimensional tests.

Both the gas wave and the material compaction wave within the specimen will reflect obliquely off the back wall. The effect of the reflection angle and interaction between the gas motion and that of the cell walls is not clear. Some idea of the behavior of the reflected wave can be obtained from the differences in the pressure traces taken at different radial positions behind the foam. Figure 8.27 shows the results for a specimen having a foam density of 28 kg m^{-3} impacted by a Mach 1.2 shock. The radial dimensions are nondimensionalised to the base radius. Multiple oscillations of the trace are evident which correlates with the distortion behavior. The cone surface also develops a decaying oscillatory behavior resulting from the internal wave motion.

Fully three-dimensional interactions will clearly be even more complex. Figure 8.28 is a single-frame multiexposure image showing six superimposed images of the deformation when the axis of a cone is perpendicular to the

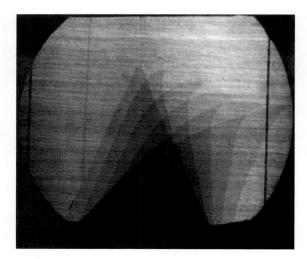

Fig. 8.28. Multiexposure image of transverse loading of a foam cone

shock propagation direction and the induced gas flow. Very large distortions
are evident and there is no evidence of emerging contact surfaces, thereby
indicating that the mechanism of collapse is somewhat different from other
cases with much lower internal pressures being generated. That this is so is
not too surprising as the skeleton of the material is not accelerated followed
by rapid stagnation as occurs for both the one-dimensional and axi-symmetric
cases.

Another method of removing the unwanted boundary constraints that has
been tried is to expose blocks of foam to blast waves generated from explosive
charges positioned within the block [8], giving, at early times, both spherical
symmetry and avoiding wave reflections from a foam/air interface, and thus
avoiding many of the extraneous effects. Wave time-of-arrival data within the
block is probed by contact gauges. Pressure measurement is more difficult as
the physical presence of the transducer embedded in the block could influence
the local collapse properties of the material.

8.4 Particle Beds

In all the test cases described above the porous material, whether rigid or
elastic, is made up of interconnected solid elements. There has been limited
work done on particle beds where the particles themselves are free to move.
There have been some studies on granular layers but these have been concerned more with compaction of the material and the particle contact rather
than the gas dynamic processes where the particles can separate, although
some attention has been given to attenuation effects. These studies have all
been one-dimensional. There has been work on the oblique reflection of shocks

Fig. 8.29. Particle lift-off phenomena at 0, 10, 30, 50, 70, and 90 ms after shock wave passage

from a thin dust layer lying on a surface, primarily relating to the effect on reflection angles [9], at times prior to particle motion, and on initial dispersion effects [4].

The material presented below identifies particle motion effects when a shock wave passes over the surface of a particle bed. The work was prompted by the observation that ground motions in high explosive field tests show motions in excess of that predicted by models of soil accounting for elasticity and roughness alone, and neglecting porosity. Upward velocities can develop in near surface soils subjected to air blast loading due to soil fluidisation caused by pore-air phenomena.

Figure 8.29 shows the results from some tests with a short bed length of 300 mm filled to a depth of 122 mm and exposed to a Mach 1.15 blast wave passing over the surface. The bed starts responding after about 10 ms, thus well after the wave has passed, and the surface particles lift simultaneously over the full bed length. The surface of the bed becomes fluidised and the blast wind starts to carry particle downstream. A wave then starts to develop, particularly at the leading edge. Pressure measurements at the base of the bed show a dispersed wave, similar to that found for rigid porous slabs. Similar tests for much longer beds are reported in [16], showing internal bed motion, lift-off mechanisms, and particle trajectories.

8.5 Conclusion

The interaction of shock waves with porous materials is generally very complex but can in some cases be dealt with in a global fashion. For example the reflected and transmitted waves through sheets, although they are formed from the merging of many discrete elements they can still be dealt with as an oblique wave in a pseudostationary frame of reference.

In general the passage of a wave through a porous material will result in wave attenuation. However, when positioned in the vicinity of a backing plate pressure amplification can occur, in some cases to a considerable degree. It is found that the mechanisms for this are different for a material in contact with the wall and when there is a small air gap.

Acknowledgment. Many of the results presented are the result of under-graduate and postgraduate student project work at the University of the Witwatersrand. I wish to express my appreciation to them. The financial support from the South African National Research Foundation for much of this work is also gratefully acknowledged.

References

1. Bowman, J.E., Niblett, G.F.: Proc Phys. Soc. B **68**, 1008 (1955)
2. Friend, W.H.: The interaction of a plane shock wave with a perforated plate. UTIA Technical Note. 25, University of Toronto, (1958)
3. Gel'fand, B.E., Gubin, S.A., Kogarko, S.M., Popov, O.E.: Sov. Phys., Appl. Maths Tech. Phys. **6**, 74 (1975)
4. Gerrard J.H.: Brit. J. Appl. Phys. **46**, 186 (1963)
5. Gvozdeva, L.G., Faresov, Y.M.: Zh. Tekh. Fiz. **55**, 773 (1985)
6. Hattingh, T., Skews, B.W.: Shock Waves **11**, 115 (2001)
7. Kitagawa, K., Jyounouchi, T., Yasuhara M.: Shock Waves **11**, 133 (2001)
8. Kleine, H., Diaconescu, G., Lee, J.H.S.: Blast wave propagation in foam. In: Sturtevant, B., Shepherd, J.E., Hornung, H.G. (eds.) Proceedings of the 20th International Syposium on Shock Waves, pp 1351–1356 World Scientific (1996)
9. Kobayashi, S., Adachi, T., Suzuki, T.: Regular reflection of a shock wave over a porous layer: theory and experiment. In: Brun, R., Dumitrescu, L.Z. (eds.) Shock Waves @ Marseille IV, Proceedings of the 19th International Symposium on Shock Waves, Springer, Berlin Heidelberg New York (1995) pp 175–180
10. Levy, A., Ben-Dor, G., Skews, B.W., Sorek, S.: Exp. Fluids, **15**, 183 (1993)
11. Onodera, H., Takayama, K.: Exp. Fluids **10**, 109 (1990)
12. Seitz, M.W., Skews, B.W.: Shock Waves **15**, 177 (2006)
13. Seitz, M.W., Skews, B.W.: Shock impact on porous plugs with a fixed gap between the plug and a wall. In: Sturtevant, B., Shepherd, J.E., Hornung, H.G. (eds.) Proceedings of the 20th International Symposium on Shock Waves, pp 1381–1386 World Scientific (1996)
14. Skews, B.W.: Exp. Fluids, **39**, 875 (2005)
15. Skews, B.W., Levy, A., Levi-Hevroni, D.: Shock Wave Propagation in Porous Media In: Ben-Dor, G., Igra, O., Elperin, T. (eds.) Handbook of Shock Waves, pp 545–596 Academic (2001)
16. Skews, B.W., Doyle, G.K.E.: Shock wave passage over porous particle beds. In: Proceedings of the 7th International Congress on Fluid Dynamics and Propulsion ASME and Cairo University, (2001) Paper 2001052
17. Skews, B.W., Takayama, K.: J. Fluid Mech. **314**, 27 (1996)
18. Skews, B.W., Sellschop, R.M., Legoete, T.: Axisymmetric tests of shock-wave impact on PUR foam. In: Sturtevant, B., Shepherd, J.E., Hornung, H.G. (eds.) Proceedings of the 20th International Symposium on Shock Waves, pp 1375–1380 World Scientific (1996)
19. Skews, B.W.: Shock Waves **4**, 145 (1994)
20. Skews, B.W., Atkins, M.D., Seitz, M.W.: J. Fluid Mech. **253**, 245 (1993)

21. Skews, B.W.: Shock Waves **1**, 205 (1991)
22. Tong, K.O., Knight, C.J., Srivastava, B.N.: AIAA J. **16**, 1298 (1960)
23. van der Grinten, J.G.M., van Dongen, M.E.H., van der Kogel, H.: J. Appl. Phys. **58**, 2937 (1985)

Linear Waves and Shock Waves in Flexible and Rigid Porous Media

David Smeulders and Marinus van Dongen

9.1 Introduction

Wave phenomena in permeable media are quite interesting from the point of view of gas dynamics. Such waves are strongly dispersive: a weak shock wave incident on a rigid gas-filled porous sample, for example, generates a compressive reflection wave with increasing pressure jump as it propagates backward. The incident shock that transmits into the pores has a sharp pressure jump that resembles a typical pressure jump across a shock wave, followed by a more gradual pressure increase which finally reaches asymptotic values. As the pressure wave propagates, the pressure jump decreases, but the final asymptotic value remains almost unchanged [1]. Because of the large compressibility contrast between the gas and the solid matrix, there is effectively no pressure transmission into the solid matrix. For flexible media, such as foams, additional phenomena become important. The deformations of the foam can be large, and when a shock wave reflects from a slab of foam adjacent to a solid wall, the end wall pressures observed are much higher than expected. A survey on experimental and theoretical investigations of wave propagation in foams is given by Skews [2]. Initially, the modeling of these wave phenomena was based on the assumption that the foam and the gas could effectively be considered as one phase. An example is the pseudogas model of Gel'fand et al. [3], that was later modified by Bazhenova and Gvozdeva [4]. Such models are rather successful, although not all phenomena observed can be explained. The differences between observation and theory are often attributed to the two-phase character of the wave phenomena. Baer [5] studied wave phenomena in a foam by applying a nonlinear two phase model, with viscous and viscoelastic interaction terms. By varying the parameters he was able to obtain a satisfactory numerical simulation of the experiments by Skews [2]. Although wave propagation in rigid and flexible porous media may be strongly nonlinear, we shall start with the application of the linear two-phase theory of Biot [6] to waves in foams. This will give insight into the effect of boundary conditions, on the importance of the interaction terms, on the coupling between the motion of the gas and the solid and on the physical and mathematical character of the waves. It is shown that in general two different wave modes exist each with their own frequency-dependent wave speeds and attenuation coefficients. Following van Dongen et al. [7], the theory will be applied to the interaction of a weak shock wave with a slab of foam adjacent to a solid wall. Experimental data will be given on the nonlinear permeability and stress–strain characteristics of flexible foams.

9.2 Acoustic Bulk Properties

The equations of motion for a fluid-saturated porous medium were formulated by Biot [6]. Zwikker and Kosten [8] and Frenkel [9] published earlier results, but some aspects were ignored in their papers. It is assumed that for long-wavelength disturbances ($\lambda \gg a$, where λ is the wavelength and a is a characteristic pore scale) we can define average values of the local displacements $\mathbf{u}_{\mathrm{s}}(\mathbf{r}, t)$ in the solid and $\mathbf{u}_{\mathrm{f}}(\mathbf{r}, t)$ in the fluid. If we consider a cube of unit size of bulk material (porosity $\phi_{\mathrm{f}} = 1 - \phi_{\mathrm{s}}$), we will denote τ_{ij} the tension forces per unit bulk area applied to that portion of the cube faces occupied by the solid. Conventionally, they are decomposed into intergranular stresses σ_{ij} and pore fluid pressure p:

$$\tau_{ij} = -\sigma_{ij} - \phi_{\mathrm{s}} p \delta_{ij}. \tag{9.1}$$

The total normal tension force per unit bulk area applied to the fluid faces of the unit cube is denoted τ:

$$\tau = -\phi_{\mathrm{f}} p. \tag{9.2}$$

In the case of isotropic materials, the stress–strain relations for the solid and the fluid may be written [6] as:

$$\tau_{ij} = 2\mu e_{ij} + A e_{kk} \delta_{ij} + Q \varepsilon_{kk} \delta_{ij}, \tag{9.3}$$

$$\tau = Q e_{kk} + R \varepsilon_{kk}, \tag{9.4}$$

where $e_{ij} = 1/2(\partial u_{\mathrm{s}i}/\partial x_j + \partial u_{\mathrm{s}j}/\partial x_i)$, $\varepsilon_{ij} = 1/2(\partial u_{\mathrm{f}i}/\partial x_j + \partial u_{\mathrm{f}j}/\partial x_i)$, and summation over repeated indices is assumed. A, Q, and R are generalized elastic parameters which can be related via Gedanken experiments [10] to porosity, bulk modulus of the solid K_{s}, bulk modulus of the fluid K_{f}, bulk modulus of the porous drained matrix K_{b}, and shear modulus μ of both the drained matrix and of the composite:

$$\begin{aligned}
A &= K_{\mathrm{b}} - 2\mu/3 + K_{\mathrm{f}}(\phi_{\mathrm{s}} - K_{\mathrm{b}}/K_{\mathrm{s}})^2/\phi_{\mathrm{eff}}, \\
Q &= \phi_{\mathrm{f}} K_{\mathrm{f}}(\phi_{\mathrm{s}} - K_{\mathrm{b}}/K_{\mathrm{s}})/\phi_{\mathrm{eff}}, \\
R &= \phi_{\mathrm{f}}^2 K_{\mathrm{f}}/\phi_{\mathrm{eff}}.
\end{aligned} \tag{9.5}$$

We have introduced an effective porosity $\phi_{\mathrm{eff}} = \phi_{\mathrm{f}} + K_{\mathrm{f}}/K_{\mathrm{s}}(\phi_{\mathrm{s}} - K_{\mathrm{b}}/K_{\mathrm{s}})$. Denoting the solid and fluid densities ρ_{s} and ρ_{f}, respectively, the linearized momentum equations are written as [6]:

$$\phi_{\rm s}\rho_{\rm s}\frac{\partial^2}{\partial t^2}u_{{\rm s}i} = -\frac{\partial\sigma_{ji}}{\partial x_j} - \phi_{\rm s}\frac{\partial p}{\partial x_i} + f_i, \tag{9.6}$$

$$\phi_{\rm f}\rho_{\rm f}\frac{\partial^2}{\partial t^2}u_{{\rm f}i} = -\phi_{\rm f}\frac{\partial p}{\partial x_i} - f_i. \tag{9.7}$$

The interaction term is specified in its linear form:

$$f_i = \left[b_0\frac{\partial}{\partial t} + \phi_{\rm f}\rho_{\rm f}(\alpha_\infty - 1)\frac{\partial^2}{\partial t^2}\right](u_{{\rm f}i} - u_{{\rm s}i}). \tag{9.8}$$

The viscous damping factor $b_0 = \eta\phi^2/k_0$, with η the dynamic fluid viscosity and k_0 the permeability. The tortuosity is denoted α_∞. The equations of motion resulting from momentum conservation and the stress–strain relations can now be written as:

$$\mu\nabla^2\mathbf{u}_{\rm s} + (A + \mu)\nabla\nabla\cdot\mathbf{u}_{\rm s} + Q\nabla\nabla\cdot\mathbf{u}_{\rm f} = \frac{\partial^2}{\partial t^2}(\rho_{11}\mathbf{u}_{\rm s} + \rho_{12}\mathbf{u}_{\rm f})$$
$$+ b_0\frac{\partial}{\partial t}(\mathbf{u}_{\rm s} - \mathbf{u}_{\rm f}), \tag{9.9}$$

$$Q\nabla\nabla\cdot\mathbf{u}_{\rm s} + R\nabla\nabla\cdot\mathbf{u}_{\rm f} = \frac{\partial^2}{\partial t^2}(\rho_{12}\mathbf{u}_{\rm s} + \rho_{22}\mathbf{u}_{\rm f})$$
$$- b_0\frac{\partial}{\partial t}(\mathbf{u}_{\rm s} - \mathbf{u}_{\rm f}), \tag{9.10}$$

where the density terms are given by

$$\begin{aligned}\rho_{12} &= -(\alpha_\infty - 1)\phi_{\rm f}\rho_{\rm f}, \\ \rho_{11} &= \phi_{\rm s}\rho_{\rm s} - \rho_{12}, \\ \rho_{22} &= \phi_{\rm f}\rho_{\rm f} - \rho_{12} = \alpha_\infty\phi_{\rm f}\rho_{\rm f}.\end{aligned} \tag{9.11}$$

9.2.1 Thermal Effects

It is clear that in the case of air-saturated porous samples thermal effects can no longer be neglected due to the relatively low heat capacity of air. Let us consider a harmonic change of pressure (angular frequency ω). As the air in the pores is compressed its temperature T rises; when the rise of temperature is appreciable, heat conduction becomes important and the air tends to cool off even before the expansion has started. When maximum compression is reached the temperature will already be decreasing as heat flows from the air to the surrounding solid. It is obvious, that is this case maximum temperature will be reached somewhat before maximum compression is established. Therefore, the temperature of the air at a given volume will be somewhat greater during the compression part of the cycle than during the expansion part. Since there is a direct relation between volume and pressure of the air, at a given volume the pressure exerted on the air during the compression will be greater than the corresponding pressure during the expansion. Hence more energy is required to compress the air than is regained during the subsequent expansion. The work done by the air during one cycle of its vibration is negative and represents a net flow of heat into the surrounding solid phase. Defining the thermal diffusivity a_g of the air phase, we may expect the oscillations of the air to occur isentropically when the thermal depth of penetration $\delta_T = \sqrt{2a_g/\omega}$ is much smaller than any characteristic thermal length scale Λ'. The relation between the air pressure p_g and the air density ρ_g is then given by the polytropic relation

$$\frac{dp_g}{p_g} = n\frac{d\rho_g}{\rho_g}, \tag{9.12}$$

with the polytropic constant n being equal to the specific heat ratio $\gamma = c_p/c_v$ in this case. When, on the other hand, the conductivity of heat is so complete that isothermal conditions prevail, n is equal to unity. This is the case when $\delta_T \gg \Lambda'$. In the intermediate case however, dp_g/p_g is no longer in phase with $d\rho_g/\rho_g$; it is this difference in phase that gives rise to a thermal damping mechanism described by n being a complex-valued quantity. The functional dependence was discussed by Champoux and Allard [11]:

$$n(\omega) = \frac{\gamma}{\chi(\omega)}, \qquad \chi(\omega) = \gamma - (\gamma - 1)\left[1 - \frac{i\omega_c'}{\omega}\sqrt{1 + \frac{1}{2}iM'\omega/\omega_c'}\right]^{-1}, \tag{9.13}$$

where $\omega_c' = \eta\phi_f/(k_0'\rho_f Pr) = a_g\phi_f/k_0'$ is the thermal characteristic frequency with Pr the Prandtl number and k_0' the so-called thermal permeability [12]. The thermal shape factor M' is defined as

$$M' = \frac{8k_0'}{\phi_f\Lambda'^2}. \tag{9.14}$$

Assuming $M' = 1$, which is a realistic value for most permeable media, we find that the characteristic thermal length scale is related to other material properties by $\Lambda' = \sqrt{8k_0'/\phi_f}$. The reduced frequency ω/ω_c' can now be related to the thermal penetration depth by $\omega/\omega_c' = \frac{1}{4}(\Lambda'/\delta_T)^2$. The characteristic thermal length is a purely geometrical quantity defined as twice the pore volume-to-surface ratio [11]. For cylindrical ducts with radius a, $\Lambda' = a$, and $k_0' = \phi_f a^2/8$. For spherical pores with radius R, the thermal length scale $\Lambda' = \frac{2}{3}R$.

Subsequently, a frequency-dependent complex-valued fluid (air) bulk modulus may be defined as

$$K_f(\omega) = \frac{K_a}{\chi(\omega)}, \tag{9.15}$$

where $K_a = \gamma p_0 = \gamma K_{f0}$ is the adiabatic fluid bulk modulus. In the limiting case for high frequencies, $\chi(\omega) \to 1$, so that adiabatic behavior is predicted. In the limiting case for low frequencies $\chi(\omega) \to \gamma$, and isothermal behavior is obtained.

9.2.2 One-Dimensional Field Equations

All essential features of wave propagation in porous media are brought out by considering the one-dimensional situation. Introducing $P = A + 2\mu$, (9.9) and (9.10) can be rewritten as:

$$\left[\rho_{11}\frac{\partial^2}{\partial t^2} + b_0\frac{\partial}{\partial t} - P\frac{\partial^2}{\partial x^2}\right] u_{\mathrm{s}} = \left[Q\frac{\partial^2}{\partial x^2} + b_0\frac{\partial}{\partial t} - \rho_{12}\frac{\partial^2}{\partial t^2}\right] u_{\mathrm{f}}, \quad (9.16)$$

$$\left[\rho_{22}\frac{\partial^2}{\partial t^2} + b_0\frac{\partial}{\partial t} - R\frac{\partial^2}{\partial x^2}\right] u_{\mathrm{f}} = \left[Q\frac{\partial^2}{\partial x^2} + b_0\frac{\partial}{\partial t} - \rho_{12}\frac{\partial^2}{\partial t^2}\right] u_{\mathrm{s}}. \quad (9.17)$$

Combination of both equations yields that

$$\left[\rho_{22}\frac{\partial^2}{\partial t^2} + b_0\frac{\partial}{\partial t} - R\frac{\partial^2}{\partial x^2}\right]\left[\rho_{11}\frac{\partial^2}{\partial t^2} + b_0\frac{\partial}{\partial t} - P\frac{\partial^2}{\partial x^2}\right] u_{\mathrm{s}}$$

$$= \left[Q\frac{\partial^2}{\partial x^2} + b_0\frac{\partial}{\partial t} - \rho_{12}\frac{\partial^2}{\partial t^2}\right]^2 u_{\mathrm{s}}. \quad (9.18)$$

Developing this expression, we find that

$$\left[(\rho_{11}\rho_{22} - \rho_{12}^2)\frac{\partial^4}{\partial t^4} - \Gamma\frac{\partial^4}{\partial t^2\partial x^2} + (PR - Q^2)\frac{\partial^4}{\partial x^4}\right.$$

$$\left. + b_0\rho\frac{\partial}{\partial t}\left(\frac{\partial^2}{\partial t^2} - \frac{H}{\rho}\frac{\partial^2}{\partial x^2}\right)\right] u_{\mathrm{s}} = 0, \quad (9.19)$$

where $\Gamma = P\rho_{22} + R\rho_{11} - 2Q\rho_{12}$, $\rho = \rho_{11} + \rho_{22} + 2\rho_{12}$, and $H = P + R + 2Q$. Dividing by $\rho_{11}\rho_{22} - \rho_{12}^2$, we find that

$$\left[\frac{\partial^4}{\partial t^4} - (V_+^2 + V_-^2)\frac{\partial^4}{\partial t^2\partial x^2} + V_+^2 V_-^2\frac{\partial^4}{\partial x^4} + \theta^{-1}\frac{\partial}{\partial t}\left(\frac{\partial^2}{\partial t^2} - V_0^2\frac{\partial^2}{\partial x^2}\right)\right] u_{\mathrm{s}} = 0, \quad (9.20)$$

with

$$V_{+,-}^2 = \frac{\Gamma \pm \sqrt{\Gamma^2 - 4(\rho_{11}\rho_{22} - \rho_{12}^2)(PR - Q^2)}}{2(\rho_{11}\rho_{22} - \rho_{12}^2)}, \quad (9.21)$$

$$V_0^2 = \frac{H}{\rho}. \quad (9.22)$$

We will show that $V_{+,-}$ are real-valued high-frequency phase velocities, and V_0 is a low-frequency phase velocity. Moreover,

$$\theta^{-1} = \frac{b_0\rho}{\rho_{11}\rho_{22} - \rho_{12}^2} \quad (9.23)$$

is the frequency (inverse time) which characterizes the transition from low-frequency viscosity dominated flow to high-frequency inertia dominated flow. Relation (9.20) can be rewritten as

$$\left[\left(\frac{\partial^2}{\partial t^2} - V_+^2 \frac{\partial^2}{\partial x^2} \right) \left(\frac{\partial^2}{\partial t^2} - V_-^2 \frac{\partial^2}{\partial x^2} \right) + \theta^{-1} \frac{\partial}{\partial t} \left(\frac{\partial^2}{\partial t^2} - V_0^2 \frac{\partial^2}{\partial x^2} \right) \right] u_\mathrm{s} = 0.$$
(9.24)

We now consider harmonic waves of the form $\exp \mathrm{i}(\omega t - kx)$, where k is the wavenumber. Substitution into (9.24) and introducing the complex wave speed $c = \omega / k$, yields that

$$c^4 \left(1 - \frac{\mathrm{i}\theta^{-1}}{\omega} \right) - c^2 \left(V_+^2 + V_-^2 - \frac{\mathrm{i}\theta^{-1}}{\omega} V_0^2 \right) + V_+^2 V_-^2 = 0.$$
(9.25)

From (9.25), the full solution is obtained as

$$c^2 = \frac{V_+^2 + V_-^2 - \mathrm{i}(\omega\theta)^{-1} V_0^2 \pm \sqrt{D}}{2[1 - \mathrm{i}(\omega\theta)^{-1}]},$$
(9.26)

with

$$D = (V_+^2 - V_-^2)^2 + 2\mathrm{i}(\omega\theta)^{-1}[2V_+^2 V_-^2 - V_0^2(V_+^2 + V_-^2)] - (\omega\theta)^{-2} V_0^4$$
$$= -(\omega\theta)^{-2} V_0^4 \left\{ -O(\omega^2) - 2\mathrm{i}\omega\theta V_0^{-4} \left[2V_+^2 V_-^2 - V_0^2(V_+^2 + V_-^2) \right] + 1 \right\}.$$
(9.27)

The last part of (9.27) will appear helpful for investigating low-frequency behavior. The result of the dispersion relation (9.26) is that there are two distinct longitudinal modes which we call mode 1 (first wave) and mode 2 (second wave). The solution (9.26) is somewhat modified by introducing a frequency-dependent friction factor. It corrects for the fact that above the characteristic frequency θ^{-1}, deviations from low-frequency Stokes' flow become important owing to inertia effects. In the limit of high frequencies, the viscous skin depth $\delta = \sqrt{2\nu/\omega}$ (ν denotes the kinematic viscosity η/ρ_f) eventually becomes much smaller than a characteristic viscous length scale Λ. The steady-state friction factor b_0 in (9.23) is then replaced by a more realistic frequency-dependent friction factor $b_0 F(\omega)$, where the viscous correction factor $F(\omega)$ is given by [13]:

$$F(\omega) = \sqrt{1 + \frac{1}{2} \mathrm{i} M \omega / \omega_\mathrm{c}}.$$
(9.28)

Here $\omega_\mathrm{c} = \eta \phi_\mathrm{f} / (\alpha_\infty k_0 \rho_\mathrm{f})$ is the frequency for which inertia and viscous forces are of equal importance in rigid porous solids, and M is the viscous shape factor defined as

$$M = \frac{8 \alpha_\infty k_0}{\phi_\mathrm{f} \Lambda^2}.$$
(9.29)

It was shown [13,14] that for many porous materials the shape factor $M = 1$ in good approximation, so that the characteristic viscous length scale is directly related to other material properties by $\Lambda = \sqrt{8 k_0 \alpha_\infty / \phi_\mathrm{f}} = \sqrt{8\nu/\omega_\mathrm{c}}$. We remark that there is direct relation between the characteristic frequencies ω_c and θ^{-1}:

$$\omega_\mathrm{c} = \theta^{-1} \frac{\phi_\mathrm{s}\rho_\mathrm{s} + (1 - \alpha^{-1})\phi_\mathrm{f}\rho_\mathrm{f}}{\phi_\mathrm{s}\rho_\mathrm{s} + \phi_\mathrm{f}\rho_\mathrm{f}}.$$
(9.30)

In the limiting case of large tortuosity, $\omega_\mathrm{c} = \theta^{-1}$.

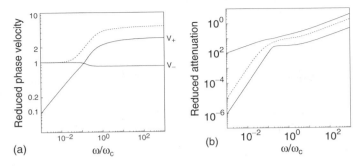

Fig. 9.1. Reduced phase velocities (**a**) and attenuation coefficients (**b**) versus reduced frequency of modes 1 and 2 for permeable foam. *Solid lines*: $K_\mathrm{p}/K_\mathrm{f0} = 2$; *dashed curves*: $K_\mathrm{p}/K_\mathrm{f0} = 0$. Also the limiting velocities V_+ and V_- for the high frequencies are indicated in the plots. The reducing velocity is V_0; the reducing length is V_0/ω_c

Table 9.1. Properties of a gas-saturated foam

porosity	ϕ_f	0.98	
tortuosity	α_∞	1	
permeability	k_0	$90.91 \cdot 10^{-11}$	m^2
isothermal gas bulk modulus	K_f0	0.1	MPa
solid density	ρ_s	$1,200$	$\mathrm{kg\,m}^{-3}$
solid compressibility	$1/K_\mathrm{s}$	0	1/Pa
gas viscosity	η	$1.8 \cdot 10^{-5}$	$\mathrm{kg\,m}^{-1}\,\mathrm{s}^{-1}$
gas density	ρ_f	1.2	$\mathrm{kg\,m}^{-3}$
gas specific heat ratio	γ	1.4	
gas thermal diffusivity	a_g	$1.87 \cdot 10^{-5}$	$\mathrm{m}^2\,\mathrm{s}^{-1}$

In Fig. 9.1a, b some results are given for the phase velocities $V = \omega/\mathrm{Re}(k)$ and damping coefficients $d = -\mathrm{Im}(k)$ of a flexible foam material. Two different cases are considered: $K_\mathrm{p}/K_\mathrm{f0} = 2$ and 0, respectively. The latter case corresponds to foams without any mechanical strength. Then only one wave mode remains that is comparable with the wave propagating in dusty gases, for example. Other parameters are given in Table 9.1. The density ratio $\rho_\mathrm{s}/\rho_\mathrm{f} = 1,000$, which is a realistic value for flexible foams according to Gibson and Ashby [15]. Limiting velocities $V_{+,-}$ are indicated along the high-frequency axis and will be discussed later. We note from Fig. 9.1a, b that mode 1 is propagatory over the entire frequency domain and significantly less damped than mode 2. Mode 2 is diffusive ($\sim\omega^{1/2}$, for both phase velocity and damping) at low frequencies and becomes propagatory for higher frequencies. The phase velocity of mode 2 is less than that of mode 1, at low frequencies. At high frequencies, $V_2 > V_1$. We remark that both modes are unambiguously defined. Although there is a specific frequency where $V_1 = V_2$, this is not the case for the attenuation, so that the complex-valued wave speeds ω/k

are never identical. Conventionally, the modes 1 and 2 are called "fast" and "slow" wave, respectively. We notice here that this terminology is not generally appropriate. The "fast" wave may become slower than the "slow" wave, as is the case here for high frequencies. Strictly speaking, the terms "fast" and "slow" are only applicable in the high-frequency limit, where they can be identified with in-phase and out-of-phase behavior, as we will see. We choose to define mode 1 as the wave that has finite phase velocity in the low-frequency limit. The phase velocity of mode 2 tends to zero, in the limiting case for low frequencies.

The Biot theory not only provides the phase velocities and attenuation values, but also yields the relation between the complex fluid and solid displacement amplitudes for the modes $j = 1, 2$ as can be seen from (9.17):

$$\beta_j = \frac{u_{fj}}{u_{sj}} = \frac{Q/c_j^2 - \rho_{12} - ib_0 F/\omega}{-R/c_j^2 + \rho_{22} - ib_0 F/\omega}. \tag{9.31}$$

In Fig. 9.2a, b these ratios are plotted versus frequency. From Fig. 9.2a it can be seen that the mode 1 fluid motion is locked-on to the solid's ($\beta_1 = 1$), at low frequencies. Increasing the frequency causes the oscillation to become out-of-phase. Mode 2 has out-of-phase behavior at low frequencies and in-phase behavior at high. The behavior of the foam without mechanical strength will be discussed in Sect. 9.2.5

9.2.3 Low-Frequency Limit

When the frequency is much lower than the characteristic frequency θ^{-1}, the viscous effects in the fluid dominate inertial effects and the viscous skin depth

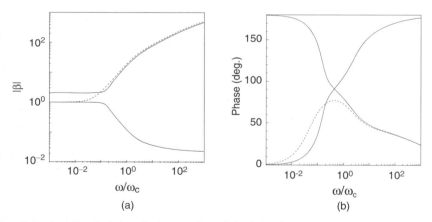

Fig. 9.2. Amplitude (**a**) and phase values (**b**) of the fluid–solid displacement ratios associated with modes 1 and 2 for permeable foam. *Solid curves*: $K_p/K_{f0} = 2$; *dashed curves*: $K_p/K_{f0} = 0$

$\sqrt{2\nu/\omega}$ is large compared to any characteristic pore size. For low frequencies $\sqrt{D}$ can be approximated by

$$\lim_{\omega \to 0} \sqrt{D} \approx i(\omega\theta)^{-1}V_0^2 \left\{ 1 - i\omega\theta V_0^{-4} \left[2V_+^2 V_-^2 - V_0^2(V_+^2 + V_-^2) \right] \right\}, \qquad (9.32)$$

so that it follows from (9.26) that either $c^2 = V_0^2$, or $c^2 = i\omega D_h$, where $D_h = \theta V_+^2 V_-^2/V_0^2$ is the hydraulic diffusivity. In this case one of the longitudinal waves (mode 1) becomes propagatory with real-valued phase velocity V_0. The viscosity of the fluid has caused the fluid motion to lock-on to the solid's $(u_s = u_f)$ with the result that the fast-wave speed is given as an effective modulus divided by the total density. This longitudinal effective modulus was originally derived by Gassmann [16]; this low-frequency limit is often called the Biot-Gassmann result. The other longitudinal mode (mode 2) becomes diffusive for low frequencies since it involves a relative motion of fluid and solid.

9.2.4 High-Frequency Limit

For high frequencies, the two compressional waves become propagatory with speeds determined by the equation

$$c^4 - c^2(V_+^2 + V_-^2) + V_+^2 V_-^2 = 0, \qquad (9.33)$$

so that $c^2 = V_+^2$, or $c^2 = V_-^2$. We call V_+ the fast wave speed and V_- the slow wave speed. We show that the fast wave is always faster than the slow wave. Let us consider in Fig. 9.3 these high-frequency speeds V_+ and V_- as functions of the frame modulus $K_p = K_b + \frac{4}{3}\mu$, keeping K_b/μ (and all other parameters) constant. For example, an unconsolidated (loose) frame has vanishing frame moduli $(K_p = K_b = \mu = 0)$ which corresponds to suspensions and dusty gases. This means that $PR - Q^2 = 0$. From (9.21), we then find that

$$V_+^2(K_p = 0) = \Gamma/(\rho_{11}\rho_{22} - \rho_{12}^2). \qquad (9.34)$$

Substitution of the expressions for P, Q, and R from (9.5), which are given in the unconsolidated frame limit by $\phi_s^2 K_{eff}$, $\phi_s\phi_f K_{eff}$, and $\phi_f^2 K_{eff}$, respectively, where $1/K_{eff} = \phi_f/K_f + \phi_s/K_s$, yields that

$$V_+^2(K_p = 0) = K_{eff} \frac{\phi_f\phi_s\rho_s + \rho_f(\alpha_\infty - 2\phi_f + \phi_f^2)}{\alpha_\infty\rho_f[\phi_s\rho_s + \phi_f\rho_f(1 - \alpha_\infty^{-1})]}. \qquad (9.35)$$

In the case that $\alpha_\infty = 1$, the wave speed reduces to $\sqrt{K_{eff}/\rho_{eff}}$, where $1/\rho_{eff} = \phi_s/\rho_s + \phi_f/\rho_f$. In the limiting case of large tortuosity, corresponding to a locking together of fluid and solid motion, the wave speed reduces to Wood's velocity formula:

$$\lim_{\alpha_\infty \to \infty} V_+^2(K_p = 0) = K_{eff}/\rho, \qquad (9.36)$$

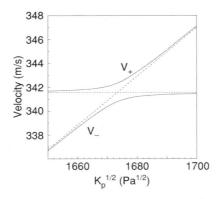

Fig. 9.3. Fast (+) and slow (−) wave velocities versus frame stiffness of permeable foam. The horizontal asymptote $\sqrt{K_f/\rho_f}$ and the asymptote $\sqrt{K_p/(\phi_s\rho_s)}$ are indicated as *dotted lines*

with $\rho = \phi_s\rho_s + \phi_f\rho_f$ the total density. In a similar vein, it is possible to show that in the limiting case for vanishing frame moduli, the slow wave speed goes to zero according to

$$V_-^2(K_p \to 0) = \frac{\phi_f^2 K_p K_f}{\phi_{\text{eff}}\Gamma} = \frac{\phi_f K_p}{\phi_s\phi_f\rho_s + \rho_f(\alpha_\infty - 2\phi_f + \phi_f^2)}. \tag{9.37}$$

At the opposite end of the stiffness spectrum, we find that in the limiting case of large frame moduli the compressional wave speeds are given by

$$V_+^2(K_p \to \infty) = \frac{K_p}{\phi_s\rho_s + \phi_f\rho_f(1 - \alpha_\infty^{-1})}, \tag{9.38}$$

$$\lim_{K_p \to \infty} V_-^2 = \frac{K_f}{\alpha_\infty\rho_f}. \tag{9.39}$$

For large frame moduli, the fast wave corresponds to an oscillation of the skeletal frame and the fluid; the slow wave corresponds to an oscillation of the fluid only, where the speed is decreased from the speed of sound in the fluid because of the tortuous structure of the pore space. We notice that the slow wave speed is always less than that of the fluid. Interestingly, it is noticed from (9.37) and (9.38) that for a light gas in the pores ($\rho_f/\rho_s \ll 1$), the loose frame limit of the slow wave speed is equal to the stiff frame limit of the fast wave speed. This is clearly visible in Fig. 9.3. Moreover, it is seen from (9.35) and (9.39) that for incompressible grains ($1/K_s = 0$) and light gases, the loose frame limit of the fast wave speed is equal to the stiff frame limit of the slow wave speed (see Fig. 9.3).

9.2.5 Loose Frame Limit

As noticed previously, K_p can be set zero for foams without any mechanical strength. Gibson and Ashby [15] argued that foam material in general behaves elastically for small compressive stresses. In this regime the stress–strain proportionality constant that applies for plane wave propagation, is the constrained modulus K_p. At some critical stress the foam collapses and will behave as if it had no strength at all. We have seen that in this case $PR - Q^2 = 0$ and that only the fast wave V_+ remains, for high frequencies. We now investigate the entire frequency domain. Relation (9.24) becomes

$$\left[\frac{\partial^2}{\partial t^2} \left(\frac{\partial^2}{\partial t^2} - V_+^2 \frac{\partial^2}{\partial x^2} \right) + \theta^{-1} \frac{\partial}{\partial t} \left(\frac{\partial^2}{\partial t^2} - V_0^2 \frac{\partial^2}{\partial x^2} \right) \right] u_\mathrm{s} = 0, \qquad (9.40)$$

where V_+ is defined in (9.35) and $V_0 = \sqrt{H/\rho} = \sqrt{K_\mathrm{eff}/\rho}$. Multiplication of (9.40) by θ and integration over time[1] yields that

$$\left[\theta \frac{\partial}{\partial t} \left(\frac{\partial^2}{\partial t^2} - V_+^2 \frac{\partial^2}{\partial x^2} \right) + \left(\frac{\partial^2}{\partial t^2} - V_0^2 \frac{\partial^2}{\partial x^2} \right) \right] u_\mathrm{s} = 0. \qquad (9.41)$$

Writing the solution in the form $\exp \mathrm{i}(\omega t - kx)$, we find that

$$c^2 = V_0^2 \left(1 + \Delta \frac{\omega^2 \theta^2}{F^2 + \omega^2 \theta^2} + \mathrm{i}\Delta \frac{\omega \theta F}{F^2 + \omega^2 \theta^2} \right), \qquad (9.42)$$

where $\Delta = (V_+^2 - V_0^2)/V_0^2$. Note that we have replaced the low-frequency friction factor b_0 by the more realistic $F(\omega)b_0$. The phase velocities and attenuation coefficients are plotted in Fig. 9.1a, b, respectively. For low frequencies, it is clear that $c = V_0$. This velocity is determined by the gas compressibility and the total density because there is no frame stiffness to take into account. According to our definition, it is mode 1 that we discuss here. In the high-frequency limit, $c^2 = V_0^2(1 + \Delta) = V_+^2$, as expected. The fluid–solid displacement ratios are plotted in Fig. 9.2a, b. We note that there is no transition from in-phase to out-of-phase behavior. For this wave the fluid and solid displacements are always more or less in-phase with a maximum phase difference of some $75°$ (see Fig. 9.2b). For low frequencies, the movement of the suspension particles is locked on to that of the fluid ($\beta = 1$). For higher frequencies, the suspension particles cannot catch up to the fluid's motion anymore and their oscillation amplitudes becomes smaller ($|\beta|$ increases, see Fig. 9.2a). If the solid–fluid amplitude ratio becomes very small ($|\beta|$ becomes large), the fluid effectively oscillates in a tortuous path formed by stationary suspension particles. This means that the velocity is now determined by the compressibility and the density of the gas alone. This can be understood from the displacement ratio β. Substitution of (9.35) into (9.31) yields that

$$\lim_{\omega \to \infty} \beta \left(K_\mathrm{p} = 0; 1/K_\mathrm{s} = 0; \alpha_\infty = 1 \right) = \frac{\rho_\mathrm{s}}{\rho_\mathrm{f}}. \qquad (9.43)$$

This ratio is 1,000 in our case, so that the solid particles have low-amplitude displacements indeed. We will next examine two limiting cases where only mode 2 remains.

---

[1] The integration constant is zero.

9.2.6 Stiff Frame Limit

For rigid porous solids, only a fluid-borne wave remains. This is a realistic situation in many shock tube experiments (see e.g., [1, 17, 18]), where relatively weak shocks in air impact upon air-saturated porous samples. Due to the large compressibility contrast between the gas and the solid matrix, the two Biot waves are effectively decoupled. The first wave cannot be detected in the gas anymore because it hardly generates any pressure, and only the second wave remains. Sometimes this wave is denoted the "compaction" wave, which is rather confusing because it has no relation with the compaction of the solid frame or solid constituent. Following Baer ([19], and references therein), we use the term "compaction wave" for the wave that causes non-linear deformation of the frame, usually caused by ramp-wave piston-impact conditions. For stiff frames, the high-frequency velocity of the second wave is given by $\sqrt{K_f/(\alpha_\infty \rho_f)}$ as we have seen in Sect. 9.2.4. In Fig. 9.4 we investigate this second wave over the entire frequency domain. It is possible to define an equivalent-fluid model, where the wave velocity of the equivalent fluid is given by

$$c^2 = K_f(\omega)/(\alpha(\omega)\rho_f), \qquad (9.44)$$

with $\alpha(\omega) = \alpha_\infty(1 - iF(\omega)\omega_c/\omega)$ the dynamic inertial drag factor (dynamic tortuosity) and $K_f(\omega) = K_a/\chi(\omega)$ the dynamic bulk modulus. The viscous correction factor $F(\omega)$ is defined in (9.28), and the thermal correction factor $\chi(\omega)$ with respect to the adiabatic bulk modulus K_a is defined in (9.13). The viscous and inertial effects are now incorporated in an effective density $\alpha(\omega)\rho_f$, and the thermal effects are described by the dynamic bulk modulus $K_f(\omega)$. It can be shown that this wave displays diffusive behavior at low frequencies, which is characteristic for the second wave (see Fig. 9.4).

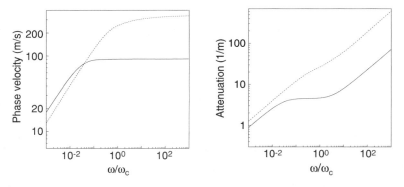

Fig. 9.4. Phase velocities (**a**) and attenuation coefficients (**b**) versus reduced frequency of modes 2 in the stiff frame limit (*dashed lines*) and in the incompressible fluid limit

9.2.7 Incompressible Fluid Limit

Another form of the second wave occurs in the incompressible fluid limit. In this case the compressibility of the system is entirely due to the frame. This is a realistic condition in biomechanical engineering, e.g., for the description of distensible lung tissue. From (9.6) and (9.7) we obtain that ($\alpha_\infty = 1$):

$$\phi_f \phi_s \rho_s \frac{\partial^2 u_s}{\partial t^2} = -\phi_f \phi_s \frac{\partial p}{\partial x} + \phi_f \frac{\partial \sigma}{\partial x} + \phi_f b_0 \frac{\partial}{\partial t}(u_f - u_s), \tag{9.45}$$

$$\phi_f \phi_s \rho_f \frac{\partial^2 u_f}{\partial t^2} = -\phi_f \phi_s \frac{\partial p}{\partial x} - \phi_s b_0 \frac{\partial}{\partial t}(u_f - u_s). \tag{9.46}$$

Subtraction of (9.46) from (9.45) yields that

$$\phi_s \left(\rho_s \frac{\partial^2 u_s}{\partial t^2} - \rho_f \frac{\partial^2 u_f}{\partial t^2} \right) = \frac{\partial \sigma}{\partial x} + \frac{b_0}{\phi_f} \frac{\partial}{\partial t}(u_f - u_s). \tag{9.47}$$

For an incompressible fluid $K_f \to \infty$. No thermal effects are thus taken into account. In the case that also $K_s \to \infty$, we obtain from (9.4) that

$$e_f = -\frac{\phi_s}{\phi_f} e_s, \tag{9.48}$$

where $e_l = \partial u_l / \partial x, l = s, f$. The only compressibility in the system is due to the frame tissue. Substitution of (9.48) into (9.47) yields that

$$\phi_s \rho^* \frac{\partial^2 e_s}{\partial t^2} - \frac{\partial^2 \sigma}{\partial x^2} + \frac{b_0}{\phi_f^2} \frac{\partial e_s}{\partial t} = 0, \tag{9.49}$$

where $\rho^* = \rho_s + \rho_f \phi_s / \phi_f$. Finally substituting the expression for the intergranular stress $\sigma = K_p e_s$ that was found by combining (9.1)–(9.4), yields

$$\frac{\partial^2 e_s}{\partial t^2} - \frac{K_p}{\phi_s \rho^*} \frac{\partial^2 e_s}{\partial x^2} + \frac{b_0}{\phi_s \phi_f^2 \rho^*} \frac{\partial e_s}{\partial t} = 0. \tag{9.50}$$

We again seek solutions in the form $\exp \mathrm{i}(\omega t - kx)$. Substitution into (9.50) yields that

$$k = \frac{\omega}{V^*} \sqrt{1 - \mathrm{i} F \frac{\omega^*}{\omega}}, \tag{9.51}$$

where $V^* = \sqrt{K_p / (\phi_s \rho^*)}$, and $\omega^* = b_0 / (\phi_s \phi_f^2 \rho^*)$ is a characteristic frequency for this configuration. Also here we have replaced the low-frequency friction b_0 by the more realistic friction $F(\omega) b_0$. In the high-frequency limit, $c = V^*$. In Fig. 9.4 the resulting phase velocities and attenuation coefficients are plotted versus the reduced frequency. The characteristic transition from diffusive behavior at low frequencies to propagatory behavior at high frequencies is clearly visible in the plots.

9.3 Wave Interactions and Wave Reflections

The linear theory not only provides the dispersion relation and the fluid–solid displacement ratios β_j, as we have seen, but also yields the relations between the different complex amplitudes. Defining the solid and fluid velocities v_s and v_f, the following results can be obtained:

$$\hat{\sigma}_j = S_j \hat{p}_j, \qquad S_j = \frac{\phi_f(\lambda_j + 2\mu)}{Q + R\beta_j}, \tag{9.52}$$

$$\lambda_j = A - Q\frac{\phi_s}{\phi_f} + \beta_j\left(Q - \frac{\phi_s}{\phi_f}R\right),$$

$$\hat{v}_{sj} = \pm\frac{c_j}{\lambda_j + 2\mu}\hat{\sigma}_j. \tag{9.53}$$

The $\pm$ signs refer to waves traveling in the $\pm$ directions. The factor $\lambda_j + 2\mu$ can be considered as the constrained modulus of the porous medium associated with the wave mode $j = 1, 2$. Indeed, for incompressible grains we find back that $\lambda_j + 2\mu = K_p$, identically. The factor $Q + R\beta_j$ is associated with the fluid compressibility. We are now able to consider the problem of the interaction of harmonic waves with foam interfaces. It is convenient to interpret the complex amplitudes of the two pressure wave modes as the components of a vector $\mathbf{p} = (\hat{p}_1, \hat{p}_2)$. Let a wave with amplitude $\hat{p}_g^+$ and wave speed c_g be incident onto a foam occupying the space $0 < x < L$. At $x = L$, the foam is bounded by a rigid end wall. The gas velocity is $\hat{v}_g^+ = \hat{p}_g^+/Z_g$, with Z_g the acoustic impedance $\rho_g c_g$. From continuity of pressure and volume flux at the foam interface $x = 0$, and using the condition that at a free interface the compressive stress vanishes, we obtain with (9.52) and (9.53):

$$\mathbf{p}_0 = (2/B)(S_2, -S_1)\hat{p}_g^+, \tag{9.54}$$

where

$$B = (E_1 + 1)S_2 - (E_2 + 1)S_1,$$
$$E_j = \phi_f c_j Z_g / K_f. \tag{9.55}$$

It follows directly from (9.52) and (9.53) and the conditions $\hat{v}_{fj}^+ = \hat{v}_{fj}^-$, $\hat{v}_{sj}^+ = \hat{v}_{sj}^-$, that at the wall $x = L$ the pressure vector satisfies

$$\mathbf{p}^- = \mathbf{p}^+. \tag{9.56}$$

Let us now consider the waves that propagate in the negative direction through the foam and reflect from the interface at $x = 0$. The boundary conditions which apply for this interface were discussed earlier. Now each wave mode generates waves of both kinds upon reflection, so a reflection matrix has to be introduced

$$\mathbf{p}^+ = \mathbf{R}\mathbf{p}^-. \tag{9.57}$$

The elements of $\mathbf{R}$ follow from the boundary conditions and (9.52) and (9.53):

$$
\begin{aligned}
R_{11} &= [S_2(E_1 - 1) + S_1(E_2 + 1)]/B, \\
R_{12} &= 2S_2E_2/B, \\
R_{21} &= -2S_1E_1/B, \\
R_{22} &= -[S_2(E_1 + 1) + S_1(E_2 - 1)]/B.
\end{aligned}
\tag{9.58}
$$

The previous results can now be used to solve the problem. At an arbitrary point x in the foam material, waves arrive from both directions. The waves traveling in the positive direction arrive either directly after having entered the foam with amplitude $\mathbf{p}_0^+$, or after having experienced one or more reflections from the solid wall and foam–air interface. The waves traveling in the negative direction must have been reflected at least once from the solid wall. When traveling through the foam over a distance x the pressure vector undergoes a change of phase and amplitude. To describe this we introduce a propagation matrix $\mathbf{T}_x = [\exp(-\mathrm{i}k_1 x), \exp(-\mathrm{i}k_2)]\mathbf{I}$, with $\mathbf{I}$ the unit matrix. The result can then be written as follows:

$$\mathbf{p} = (\mathbf{T}_x + \mathbf{T}_{xw})(1 + \mathbf{R}\mathbf{T}_{2L}(1 + \mathbf{R}\mathbf{T}_{2L}(1 + \cdots)))\mathbf{p}_0^+. \tag{9.59}$$

The subscript xw refers to the position $2L - x$. This result is called Kennett's reflectivity scheme [20]. We finally replace the series expansion in (9.59) to obtain the result

$$\mathbf{p} = (\mathbf{T}_x + \mathbf{T}_{xw})(1 - \mathbf{R}\mathbf{T}_{2L})^{-1}\mathbf{p}_0^+. \tag{9.60}$$

The next step is to decompose a pressure ramp wave with a finite rise time into a finite number of Fourier components. Then the theory described earlier is applied to each component. The solution in the temporal domain is obtained by the inverse numerical Fourier transform [21]. In Fig. 9.5a, b pressures and velocities are shown for wave reflections from a foam adjacent to a rigid wall. The foam is assumed to be fully collapsible: $K_\mathrm{p}/K_{\mathrm{f}0} = 0$. The length of the foam is 50 in reduced coordinates, where the reference length $L_\mathrm{ref} = V_0/\omega_\mathrm{c}$, with V_0 the low-frequency limit of the first wave defined in Sect. 9.2.5. Pressure histories are shown at three positions ahead of the foam and at the end wall. The incident wave front propagates with a relatively high speed, somewhat less than the gas velocity c_g. The reflected waves in the gas phase show a distinct plateau after the first-wave reflection. This pressure level equals the value T_r given by

$$T_\mathrm{r} = \lim_{\omega \to 0} \frac{\hat{p}}{\hat{p}_\mathrm{g}^+} = \frac{2}{1 + \sqrt{\gamma \phi_\mathrm{f} \rho_\mathrm{f}/\rho}}. \tag{9.61}$$

This explains why one-phase wave theories such as that of Gel'fand et al. [3] predict these reflected pressure values quite well. The pressure at the wall shows a very smooth variation, and a maximum value much higher than the maximum gas pressure in front of the foam. The maximum wall pressure has,

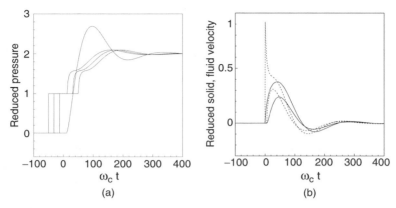

Fig. 9.5. Reflection of a weak shock from a permeable foam adjacent to a solid wall. The thickness of the foam in reduced coordinates is 50. (**a**) Gas pressures at distances of 270, 170, and 70 ahead of the interface, and at the solid wall. (**b**) Velocities of the gas (*dashed lines*) and solid at the interface ($x/L_{\mathrm{ref}} = 0$) and halfway inside the foam ($x/L_{\mathrm{ref}} = 25$). The pressures are normalized by the pressure amplitude of the incident wave; the velocities by the amplitude of the gas velocity induced by the incident wave

in linear theory, an upper bound of twice the value of T_{r}. When looking at the velocities of the gas and solid in Fig. 9.5b, it is clear that in this example the foam and the gas behave as two different interacting mechanical systems. At the interface the gas velocity first jumps to a value somewhat less than unity, because of the high porosity and because there is no inertial interaction, $\alpha_{\infty} = 1$. Then, due to friction, momentum is transferred from the gas to the solid, and reflected pressure waves propagate upstream. Both the solid and gas velocities show oscillatory behavior damped to zero on the longer time scale.

9.4 Nonlinear Models

We now discuss nonlinear extensions of the Biot theory. A theoretical framework was proposed by Baer and Nunziato [22], who discussed a continuum mixture model for the deflagration-to-detonation transition in reactive granular materials. Later, this model was applied successfully to shock-induced flow in low-density foam [23]. For simplicity, the balance equations are given for one-dimensional two-phase flow. Each phase is denoted with a subscript a to identify the solid (a = s) and fluid (a = f) phase. Associated with each phase are the intrinsic densities ρ_{a}, volume fractions ϕ_{a}, particle velocities v_{a}, pressures p_{a}, temperatures T_{a}, and energies E_{a}. The balance laws for mass, momentum, and energy are given as [19]:

$$\frac{\partial}{\partial t}\phi_{\mathrm{a}}\rho_{\mathrm{a}} + \frac{\partial}{\partial x}\phi_{\mathrm{a}}\rho_{\mathrm{a}}v_{\mathrm{a}} = 0, \tag{9.62}$$

$$\frac{\partial}{\partial t}\phi_a\rho_a v_a + \frac{\partial}{\partial x}(\phi_a p_a + \phi_a\rho_a v_a^2) = f_a, \tag{9.63}$$

$$\frac{\partial}{\partial t}(\phi_a\rho_a E_a) + \frac{\partial}{\partial x}((\phi_a\rho_a E_a + \phi_a p_a)v_a) = e_a. \tag{9.64}$$

The energy E_a can be written as the sum of internal and kinetic energy

$$E_a = C_a T_a + \frac{1}{2}v_a^2, \tag{9.65}$$

where C_a is the specific heat. The interaction term m_a is specified as

$$m_f = -m_s = p_f\frac{\partial\phi_f}{\partial x} + C_D(v_s - v_f) + C_F(v_s - v_f)|v_s - v_f|. \tag{9.66}$$

The second and third term represent Darcy and Forchheimer friction, respectively. The coefficient C_D is related to the permeability k_0 and the fluid viscosity η: $C_D = \eta\phi/k_0$. The coefficient C_F depends on the structure of the foam and on the Reynolds number. It is noted that the tortuosity term is neglected in this approach and that linearization of the continuum mixture model results in the original Biot model with $\alpha_\infty = 1$. The energy exchange e_a is represented by

$$e_f = -e_s = m_f v_s + \alpha_T(T_s - T_f). \tag{9.67}$$

Here α_T is the interphase heat transfer coefficient depending on local flow conditions based on Reynolds number scaling. The mechanism of heat transfer from the gas to the solid material is extremely efficient. The specific heat of a foam material is a factor 2 larger than the specific heat of air [15]. The heat capacity, $\phi_s\rho_s C_s$, of a typical foam is therefore about 50 times larger than the heat capacity of the gas. It is therefore reasonable to expect that the propagation of waves in flexible foams will be approximately isothermal. The shock wave propagation in the gas prior to impact upon the foam is obviously not isothermal. The volume fractions are constrained by $\phi_s + \phi_f = 1$. The compressive stress in the solid phase, p_s, consists of two components: $p_s = p_f + \sigma/\phi_s$. The effective stress, σ, is the force per unit total area of the porous material, in excess to the pore pressure p_f, acting inside the solid material. The effective stress is divided by ϕ_s to obtain its contribution to the total normal stress in the solid material (force per unit solid area).

To describe the rate of change of the solid volume fraction Baer [19] proposed a compaction model from thermodynamic considerations

$$\frac{\partial\phi_s}{\partial t} + v_s\frac{\partial\phi_s}{\partial x} = \frac{\phi_s\phi_f}{\eta_c}(p_s - p_g - \sigma/\phi_s). \tag{9.68}$$

The compaction viscosity η_c dictates the rate at which the volume fraction adjusts to the equilibrium pressure state $p_s = p_g + \sigma/\phi_s$. The case for $\eta_c \to \infty$ is in agreement with the conventional Biot approach, where instantaneous adjustment is assumed. The case for $\eta_c \to 0$ corresponds to the "frozen" limit. It is clear that the compaction model essentially allows for an extra degree of freedom with respect to the Biot model. The effect of the compaction is also seen in the energy exchange e_a, which is modified to become

$$e_f = -e_s = m_f v_s + \alpha_T(T_s - T_f) + (p_s - \sigma/\phi_s)\frac{\phi_s\phi_f}{\eta_c}(p_s - p_g - \sigma/\phi_s). \quad (9.69)$$

Model closure was obtained using a nonporous solid thermoelastic equation of state to fit shock Hugoniot data. Additionally, the ideal gas law was used for the gas phase equation of state.

9.4.1 Head-on Collision of Shock Waves with Rigid Porous Media

Unsteady air filtration through rigid porous media induced by shock wave impact was measured and modelled many times (see e.g., [1, 17, 18, 24, 25]). Based on rigorous averaging of the microscopic Navier–Stokes equations over the representative elementary volume (REV), Levy et al. [18] developed a one-dimensional computer code for thermoelastic porous media that undergo small deformations only. Introducing the velocities of the solid and the fluid $\mathbf{v}_s$ and $\mathbf{v}_f$, the macroscopic mass balance equations for the solid and the fluid phases were expressed as

$$\frac{\partial}{\partial t}\phi_s\rho_s + \frac{\partial}{\partial x_j}\phi_s\rho_s v_{sj} = 0, \quad (9.70)$$

$$\frac{\partial}{\partial t}\phi_f\rho_f + \frac{\partial}{\partial x_j}\phi_f\rho_f v_{fj} = 0. \quad (9.71)$$

Denoting $\mathbf{g}$ the acceleration due to gravity, the macroscopic momentum balances for the solid and the fluid were given by

$$\frac{\partial}{\partial t}\phi_s\rho_s v_{si} + \frac{\partial}{\partial x_j}\phi_s\rho_s v_{si}v_{sj} = -\frac{\partial\sigma_{ji}}{\partial x_j} - \phi_s\left(\frac{\partial p}{\partial x_j} + \rho_s g_j\right)T_{ji}^* + f_i, \quad (9.72)$$

$$\frac{\partial}{\partial t}\phi_f\rho_f v_{fi} + \frac{\partial}{\partial x_j}\phi_f\rho_f v_{fi}v_{fj} = -\phi_f\left(\frac{\partial p}{\partial x_j} + \rho_f g_j\right)T_{ji}^* - f_i. \quad (9.73)$$

The interaction term is specified in a nonlinear form:

$$f_i = \phi_f\rho_f w_k w_j F_{kji}, \quad (9.74)$$

where the relative fluid velocity with respect to the solid phase is indicated $\mathbf{w} = \mathbf{v}_f - \mathbf{v}_s$. On the basis of dimensional analysis, Levy et al. [26] argued that these equations need not necessarily contain Darcy-like friction terms. $\mathbf{F}$ and $\mathbf{T}^*$ are the Forchheimer tensor for an isotropic solid matrix and the tensor associated with the directional cosines at the solid–fluid interface. These parameters are defined as simple geometrical quantities, where $\mathbf{T}^*$ follows from integration over the fluid–fluid boundary of the REV, and the Forchheimer tensor follows from integration over the fluid–solid interface area within the REV. Comparing with the linear Biot equations (9.6) and (9.7), convective terms are added and gravity is included.[2] The effective stress for a thermoelastic solid, which undergoes small deformations only, is

$$\sigma_{ij} = \lambda_{\mathrm{s}} e_{kk} \delta_{ij} + 2\mu_{\mathrm{s}} e_{ij} - \eta_{\mathrm{s}}(T_{\mathrm{s}} - T_0), \tag{9.75}$$

where λ_{s}, μ_{s}, and η_{s} denote the Lamé coefficients of a thermoelastic solid. The solid excess temperature $T_{\mathrm{s}} - T_0$ is with respect to the temperature at rest T_0. The macroscopic energy balances for the solid and the fluid were given by

$$\frac{\partial}{\partial t}\left[\phi_{\mathrm{s}}\rho_{\mathrm{s}}\left(C_{\mathrm{s}}T_{\mathrm{s}} + \frac{1}{2}v_{\mathrm{s}i}v_{\mathrm{s}i}\right)\right] + \frac{\partial}{\partial x_j}\left[\phi_{\mathrm{s}}\rho_{\mathrm{s}}v_{\mathrm{s}j}\left(C_{\mathrm{s}}T_{\mathrm{s}} + \frac{1}{2}v_{\mathrm{s}i}v_{\mathrm{s}i}\right)\right]$$
$$= v_{\mathrm{s}i}\frac{\partial}{\partial x_j}\sigma_{ji} - \phi_{\mathrm{s}}v_{\mathrm{s}i}\left(\frac{\partial p}{\partial x_j} + \rho_{\mathrm{s}}g_j\right)T_{ji} + \phi_{\mathrm{f}}\rho_{\mathrm{f}}v_{\mathrm{s}i}w_kw_jF_{kji}, \tag{9.76}$$

$$\frac{\partial}{\partial t}\left[\phi_{\mathrm{f}}\rho_{\mathrm{f}}\left(C_{\mathrm{f}}T_{\mathrm{f}} + \frac{1}{2}v_{\mathrm{f}i}v_{\mathrm{f}i}\right)\right] + \frac{\partial}{\partial x_j}\left[\phi_{\mathrm{f}}\rho_{\mathrm{f}}v_{\mathrm{f}j}\left(C_{\mathrm{f}}T_{\mathrm{f}} + \frac{1}{2}v_{\mathrm{f}i}v_{\mathrm{f}i}\right)\right]$$
$$= -\phi_{\mathrm{f}}v_{\mathrm{f}i}\left(\frac{\partial p}{\partial x_j} + \rho_{\mathrm{f}}g_j\right)T_{ji} - \phi_{\mathrm{f}}\rho_{\mathrm{f}}v_{\mathrm{s}i}w_kw_jF_{kji}, \tag{9.77}$$

where $e_{\mathrm{s,f}}$ is the internal energy of the solid or the fluid, respectively. The fluid was assumed to be a perfect gas, and the solid to be thermoelastic. The computational results were compared with shock tube experiments by Levy et al. [1] on a rigid 4 cm long SiC sample having 10 pores per inch and an incident shock wave Mach number of 1.378. The experimental configuration is depicted in Fig. 9.6. The side-wall is equipped with pressure sensors at different positions along (5, 6, 7) and in front (1, 2, 3, 4) of the porous sample. Usually, sensors are also installed in the end-wall of the tube (8). Good agreement for transmitted and reflected waves was obtained, as can be seen in Fig. 9.7. Measured pressure variations for a 81 mm long specimen of the same material were discussed in the contribution to this Volume by Skews [27]. The most important findings of the experiments can be summarized as follows [1]:

1. The incident shock wave transmits into the pores of the rigid sample as a pressure wave, with a sharp pressure jump that resembles a typical pressure jump across a shock wave. Because of the large compressibility contrast between the gas and the solid matrix, the two Biot compressional wave modes are effectively decoupled. The first wave cannot be detected in the gas anymore (it carries too low gas pressures), and the second cannot be detected in the solid matrix anymore.

---

[2] If $\mathbf{T}^* = \mathbf{I}$, then the equations can be linearized to yield the Biot equations, but only for $\alpha_\infty = 1$. In the more general case a reduction of the nonlinear theory to Biot theory is not straightforward, if possible at all.

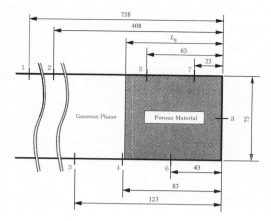

Fig. 9.6. Typical experimental configuration showing part of the test section with different pressure transducers (numbered 1–8) and the porous material installed. The lengths are in millimeters. From [1] with permission

2. The sharp pressure jumps are followed by continuous pressure increases which finally reach asymptotic values.
3. As the pressure wave propagates, the pressure jump across it decreases, but the final asymptotic value remains almost unchanged.
4. The pressure wave velocity has a nearly constant value which is slightly smaller than the speed of sound of the air filling the pores. In the limiting case for weak shocks ($M \approx 1$) and completely rigid solid structures, the pressure wave velocity c_2 is expected to be $c_T/\sqrt{\alpha_\infty} < c_2 < c_S/\sqrt{\alpha_\infty}$, where c_T and c_S are the isothermal and isentropic speeds of sound in the gas, respectively, and α_∞ is the tortuosity.
5. The pressure at the end-wall of the shock tube behind the reflected wave does not remain constant as it should have been had a compaction wave reflected there. Instead, it continuously increases until it levels off at a value slightly higher than the pressure, p_5 that would have been obtained had the incident shock wave collided head-on with the shock tube end-wall without the presence of the porous material.
6. There is no sign of a reflected pressure wave.
7. The pressure jump across the reflected shock was found to increase as it propagated backward.

As a result of these findings it was suggested that the pressure wave broadens as it propagates through the porous material and develops a dispersed structure. The slow down process of the gas which flows through the pores generates compression waves which emerge from the front edge of the porous material and catch up with the reflected shock wave increasing the pressure behind it.

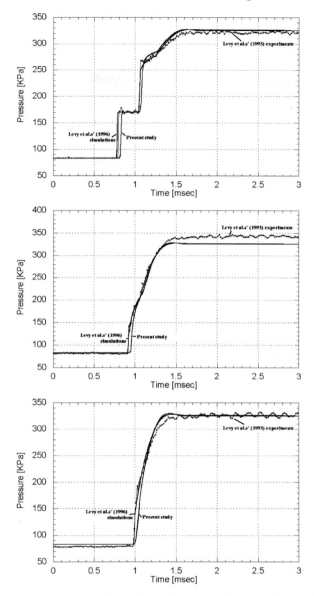

Fig. 9.7. Typical experimental results and numerical simulations with a 40 mm long SiC sample which has 10 pores per inch and an average porosity 0.728±0.016. The incident Mach number was 1.378. The pressure histories 43 mm ahead of the sample (top), 23 mm from the end wall (middle) and at the end wall (bottom) are plotted. From [28] with permission

9.4.2 Head-on Collision of Shock Waves with Flexible Porous Media

Important observations, shadowgraphs and wave diagrams concerning head-on impacts of shock waves with flexible porous media can be found in the contribution to this Volume by Skews [27]. Here, we focus on theoretical aspects and strain-dependent behavior. Levi-Hevroni et al. [28] developed a computer code for a porous matrix that was able to undergo large deformations. Unlike Levy et al. [18], an elastoplastic model was treated to allow for large deformations and the solid was assumed isothermal. Again a perfect gas was used as pore-filler. The effective stress in the flexible porous material was decomposed as $\sigma'_{ij} = -p_s\delta_{ij} + \tau_{ij}$, where it simply holds that $-p_s = K_b e_{kk}$, and $\tau_{ij} = (\lambda - K_b)e_{kk}\delta_{ij} + 2\mu e_{ij}$, with K_b the bulk modulus $\lambda + \frac{2}{3}\mu$. The constitutive equation was [29]:

$$ p_s = \frac{1}{3}\phi(1-\phi)^2 E_s \left[-\ln\left(1 - \frac{\eta'}{\eta_{\max}}\right) - B_s \left(\frac{\eta'}{\eta_{\max}}\right)^n \right], \tag{9.78} $$

where $\eta' = 1 - (1-\phi_0)/(1-\phi)$, and $\eta_{\max} = \phi_0$, with ϕ_0 the initial porosity. E_s is the Young's modulus of the solid material of which the skeleton is made, and B_s and n are adjustable parameters. The code comprised a mixed Eulerian–Lagrangian method to track the interface between the solid and the gaseous phases. The mass and momentum balance equations of the solid matrix were expressed in Lagrangian coordinates along a particle pathline. Computational results were presented for a typical polyurethane foam, having 98% porosity, a Poisson's ratio of 0.45, and a Young's modulus of 45 MPa. The tortuosity and the Forchheimer coefficients were assumed to be 0.78 and $300\,\mathrm{m^{-1}}$, respectively. It has to be remarked that other studies (e.g., [7]) revealed strong dependencies of Forchheimer coefficients and permeabilities on the amount of compression ϕ/ϕ_0. These effects were neglected is this study. The results of this code were also compared with the shock tube experiments on the rigid SiC sample [1] and the one-dimensional numerical results by Levy et al. [18] for an effectively nondeformable structure. Good agreement was obtained between measurements and predictions (see Fig. 9.7). In Sorek et al. [30], some simulations of the Levi-Hevroni code are qualitatively compared with the foam experiments by Skews et al. [31] and Gvozdeva et al. [32]. These comparisons pertain in particular to the observation of the peak effective stress and the deformation of the front surface of the foam.

Other experiments were performed by van Dongen et al. [33]. When a foam is subjected to a mechanical load, it will display a very peculiar mechanical behavior. In Fig. 9.8a this effective stress σ is depicted versus the strain ε for foam Recticel BPS-60, which was used both in the experiments and in the numerical simulations. The strain, which indicates the relative change in length, is related to the porosity by $\varepsilon = (\phi_{f0} - \phi_f)/(1 - \phi_f)$. Here, ϕ_{f0} is the porosity when the foam is uncompressed. Figure 9.8a gives both the measurements with the foam constrained in a tube as well as with the foam

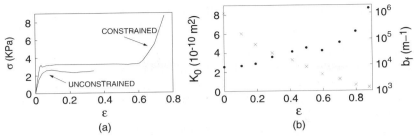

Fig. 9.8. (a) Stress–strain measurements of foam BPS-60. (b) Permeability parameters k_0 ($\times$) and b_f (*filled circle*) as a function of the compressive strain. From [33] with permission

Table 9.2. The physical properties of foam BPS-60. From [33]

ϕ_{f0}	0.983 ± 0.002
K_p (10^5 Pa)	2.12 ± 0.23
E (10^5 Pa)	1.02 ± 0.16
σ_c^* (kPa)	3.53 ± 0.10
σ_u^* (kPa)	2.83 ± 0.15
k_0 (10^{-10} m^2)	8.19 ± 0.08
b_{f0} (10^3 m^{-1})	8.66 ± 0.08
ρ_s (kgm^{-3})	$1,682 \pm 10$

unconstrained. Three regimes are visible: an elastic regime, a collapse regime, and a densification regime. In the elastic regime a linear relation exists between the stress and the strain, characterized by the elasticity modulus E if unconstrained and K_p if constrained. This is followed by a collapse regime in which the stress remains constant at a value σ^*. Then the stress increases rapidly upon further compression. A second important physical property of interest is the permeability, characterized by the parameters k_0 and b_f. Both have been experimentally determined for a wide range of ε values. The results are given in Fig. 9.8b. Permeability is very much affected by the rate of compression. The parameter b_f varies two decades for ε varying between 0.3 and unity. The physical properties are given in Table 9.2. In this table, k_0 and b_{f0} are the permeability and second Forchheimer coefficient in the uncompressed situation. Furthermore, σ_c^* is the constrained value of the collapse stress and σ_u^* the corresponding unconstrained value.

The shock tube experiments were performed in a shock tube whose end wall is depicted in Fig. 9.9. In this figure the positions are given of the pressure transducers and the initial position of the foam material. Transducer T1 is mounted 21.5 cm in front of the end wall, so it measures only the gas pressure. Transducers T2 and T3 are mounted in the end wall of the tube. T2 is in contact with the foam material and T3 measures only the gas pressure [34]. This implies that the difference of T2 and T3 is the effective stress σ. In Fig. 9.10 the time histories of the pressures at the end wall for transducers

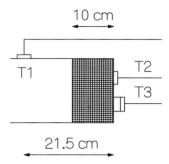

Fig. 9.9. Part of the test section with the different pressure transducers and the initial position of the foam material. From [33] with permission

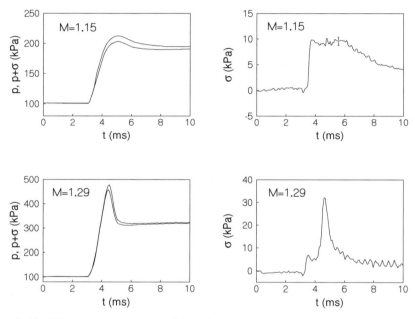

Fig. 9.10. The pressure measured by T3 and the sum of pressure and effective stresses measured by T2 for two different strengths of the incident shock wave at the end wall (left). The difference of the two signals, σ, is depicted on the right. Test gas: air. Initial pressure: 1.0 bar. From [33] with permission

T2 and T3 and the effective stress, σ, are presented. With increasing Mach number the pressure peak increases. After this peak the pressure relaxes toward its value as if no foam were present. Also, it is clearly visible that the wave travels in a dispersive way. For a shock Mach number of 1.15, the stress σ shows a fast increase upon the arrival of the wave front and is then fixed at its dynamic collapse value $\sigma^* = 9.5\pm0.5$ kPa. Apparently the value of the collapse stress is a function of the strain rate of the foam and is therefore different from its measured static value of 3.25±0.10 kPa. After about 2.5 ms σ starts to decrease again: the foam expands. When the Mach number is increased, we notice that the first jump toward the collapse stress is followed by a further increase. Comparison with the static effective-stress measurements shows that the foam must have entered its densification regime. The effects of strain-dependent permeability and Forchheimer coefficient were taken into account by [33]. The governing equations were specified as:

$$\frac{\partial \boldsymbol{U}}{\partial t} + \frac{\partial \boldsymbol{F}}{\partial x} = \boldsymbol{Q}, \tag{9.79}$$

where

$$\boldsymbol{U}^T = [\phi_f \rho_f, \phi_s, \phi_f \rho_f v_f, \phi_s v_s], \tag{9.80}$$

$$\boldsymbol{F}^T = [\phi_f \rho_f v_f, \phi_s v_s, \phi_f \rho_f v_f^2 + \phi_f p_f, \phi_s v_s^2 + \phi_s p_s/\rho_s], \tag{9.81}$$

$$\boldsymbol{Q}^T = [0, 0, m_f, m_s/\rho_s]. \tag{9.82}$$

The term $m_f \ (= -m_s)$ represents the interaction force between the phases:

$$m_f = p_f \frac{\partial \phi_f}{\partial x} + \frac{\phi_f \eta}{k_0}(v_s - v_f) + \phi_f^2 b_f \rho_f(v_s - v_f)|v_s - v_f|. \tag{9.83}$$

In these equations, b_f is the second Forchheimer coefficient. Both the permeability and the Forchheimer coefficient depend on the foam deformation. Due to a large contact surface between the two phases and a very large heat capacity of the porous material it is assumed that the propagation of waves in the foam is an isothermal process. The compressive stress in the solid phase, p_s, consists of two components:

$$p_s = p_f + \sigma/\phi_s, \tag{9.84}$$

where σ represents the effective stress. This is the force per unit (total) area adding up to the pore pressure p_f. For an incompressible solid σ is uniquely related to the porosity ϕ_f. Closure of the model was obtained by the ideal gas equation.

The conservation laws (9.79) are solved for the foam and the gas in front of the foam as if one material were present with varying properties in space (e.g., $\phi_s = 0$ for gas and $\phi_s > 0$ for foam). Since the energy equation is not accounted for in the present isothermal model, this affects the relation between the density and the pressure in the gas. For a more quantitative analysis, temperature effects have to be taken into account. The method used to solve (9.79) is the operator-splitting technique. The homogeneous part is solved using a finite-volume method employing a second-order Runge–Kutta time integration. The inhomogeneous part is solved using a Newton method. An example of the results of the method is depicted in Fig. 9.11. Here a contour plot of the volume fraction of the solid ϕ_s is shown. For the given shock Mach number, ϕ_s reaches a value at the end wall of 0.094, corresponding to a value of the

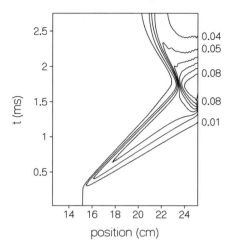

Fig. 9.11. Computed contour plot of the volume fraction of the solid ϕ_s. The shock Mach number was 1.33. Test gas: isothermal. Initial pressure: 1.0 bar. From [33] with permission

strain ε of 0.82. From Fig. 9.8a it becomes clear that the foam has entered its densification regime indeed (as is observed in the experiments, see Fig. 9.10). The contour plot also shows a large foam compression, approximately a factor 5, in agreement with experimental results.

9.5 Conclusions

Important characteristics of shock wave propagation in porous materials can already be found in linear theory. Linear theory predicts the generation of two separate compression waves in the porous sample upon shock wave loading at normal incidence (wave splitting). One of the wave modes has typical in-phase movements of solid and fluid particles, whereas the other typically has out-of-phase movements associated with the propagating wave. Upon impact, the energy distribution over the wave types is determined by the boundary condition, i.e., whether the pore interfaces are open or closed to the surrounding fluid, by the fluid–solid density ratio, and by the fluid–solid compressibility contrast. For relatively stiff frames, the waves are effectively decoupled, and the first wave mode carries too low gas pressures to be detected by pressure gauges. In the sample, a transition from a partly dispersed wave at short distances from the interface to a fully dispersed wave structure farther away from the interface, can be observed. For relatively soft frames, energy is transmitted from the gas phase to the solid material, and large frame deformations may be induced. It is to be expected that for fully liquid-saturated porous media under strong shock wave loading, shock wave splitting phenomena may occur that strongly resemble the wave splitting effects discussed in linear approximation.

References

1. Levy, A., Ben-Dor, G., Skews, B.W., Sorek, S.: Exp. Fluids **15**, 183 (1993)
2. Skews, B.W.: Shock Waves **1**, 205 (1991)
3. Gel'fand, B.E., Gubanov, A.V., Timofeev, E.I.: Izv. Akad. Nauk. SSSR, Mekh. Zhidkostii Gaza **4**, 79 (1983)
4. Bashenova, T.V., Gvozdeva, L.G.: Unsteady interactions of shock waves. In: Grönig H. (ed.) pp. 41–49. Shock Tubes and Waves, VCH Publisher, Weinheim, New York (1988)
5. Baer, M.R.: Shock Waves **2**, 121 (1992)
6. Biot, M.A.: J. Acoust. Soc. Am. **28**, 168 (1956)
7. van Dongen, M.E.H., Smeulders, D.M.J., Kitamura, T., Takayama, K.: Acustica **81**, 63 (1995)
8. Zwikker, C., Kosten, C.W.: Sound Absorbing Materials. Elsevier, New York Amsterdam London Brussels (1949)
9. Frenkel, J.: J. Phys. (USSR) **VIII** (4), 230 (1944)
10. Biot, M.A., Willis, D.G.: J. Appl. Mech. **24**, 594 (1957)
11. Champoux, Y., Allard, J.F.: J. Appl. Phys. **70**, 1975 (1991)
12. Lafarge, D., Lemarinier, P., Allard, J.F., Tarnow, V.: J. Acoust. Soc. Am. **102**, 1995 (1997)
13. Johnson, D.L., Koplik, J., Dashen, R.: J. Fluid Mech. **176**, 379 (1987)
14. Smeulders, D.M.J., Eggels, R.L.G.M., van Dongen, M.E.H.: J. Fluid Mech. **245**, 211 (1992)
15. Gibson, L.J., Ashby, M.: Cellular Solids-Structure and Properties. Pergamon, Oxford (1988)
16. Gassmann, F.: Geophysics **16**, 673 (1951), Vierteljahrsschrift der Naturforschenden Gesellschaft in Zürich **96**, HEFT 1, 1 (1951)
17. van der Grinten, J.G.M., van Dongen, M.E.H., Van der Kogel, H.: J. Appl. Phys. **58**, 2937 (1985)
18. Levy, A., Ben-Dor, G., Sorek, S.: J. Fluid Mech. **324**, 163 (1996)
19. Baer, M.R.: J. Appl. Mech. **55**, 36 (1988)
20. Kennett, B.: Seismic Wave Propagation in Stratified Media. Cambridge University Press, Cambridge (1983)
21. van der Grinten, J.G.M., van Dongen, M.E.H., Van der Kogel, H.: J. Appl. Phys. **62**, 4682 (1987)
22. Baer, M.R., Nunziato, J.W.: Int. J. Multiphase Flow **12**, 861 (1986)
23. Baer, M.R.: A Multiphase model for shock-induced flow in low-density foam. In: Brun, R., Dumitrescu, L.Z. (eds.) Shock Waves @ Marseille, Proceedings of the 19th ISSW vol III. pp. 169–174 Springer-Verlag, Berlin Heidelberg (1995)
24. Britan, A., Ben-Dor, G., Elperin, T., Igra, O., Jiang, J.P.: Int. J. Multiphase Flow **23**, 473 (1997)
25. Gubaidullin, A.A., Britan, A., Dudko, D.N.: Shock Waves **13**, 41 (2003)
26. Levy, A., Sorek, S., Ben-Dor, G., Bear, J.: Transport Porous Media **21**, 241 (1995)
27. Skews, B.W.: Experimental studies of shock wave interactions with porous media. In: van Dongen, M.E.H. (ed.) Shock Waves Science and Technology Library, Vol-1, Shock Waves in Multiphase Media, pp. 272–294, Springer Verlag, Heidelberg (2006)
28. Levi-Hevroni, D., Levy, A., Ben-Dor, G., Sorek, S.: J. Fluid Mech. **462**, 285 (2002)

29. Zeretsky, E., Ben-Dor, G.: J. Engng Mat. Tech. **118**, 493 (1996)
30. Sorek, S., Levi-Hevroni, D., Levy, A., Ben-Dor, G.: Transport in Porous Media **61**, 215 (2005)
31. Skews, B.W., Atkins, M.D., Seitz, M.W.: J. Fluid Mech. **253**, 245 (1993)
32. Gvozdeva, L.G., Farasov, Y.M., Fokeev, V.P.: Sov. Phys. Appl. Math. Technol. Phys. **3**, 111 (1985)
33. van Dongen, M.E.H., Smeets, G.V.R., Hoeijmakers, H.W.M., Smeulders, D.M.J.: Shock-induced wave phenomena in open-cell permeable and flexible foams. In: Sturtevant, B., Shepherd, J.E. Hornung, H.G. (eds.) Proceedings of the 20th ISSW vol II, PP. 1363–1368. World Scientific, Singapore (1996)
34. van Dongen, M.E.H., Smeulders, D.M.J., Kitamura, T., Takayama, K.: On the modelling of wave phenomena in permeable foam. In: Brun, R., Dumitrescu, L.Z. (eds.) Shock Waves @ Marseille, Proceedings of the 19th ISSW vol III, pp. 163–168. Springer, Berlin Heidelberg New York (1995)

10

Shock Waves in Granular Media

Victor Golub and Olga Mirova

10.1 Introduction

A granular medium is characterized by the presence of separate particles or granules (which may be of different size, shape, degree of smoothness) and voids (in what follows, we will refer to them as pores). The granules are loosely bound to one another or not bound at all. The porosity of a granular medium is the ratio of the pore volume V_P to the total volume V_m of a sample i.e., $\phi = V_P/V_m$. It is common practice to describe such a material by the density ratio $\alpha = \rho_s/\rho_m = 1/(1 - \phi)$, where ρ_s is the density of the solid fraction and ρ_m is the mean density of the granular sample.

A shock wave arises and moves in a granular medium as a result of the interaction between the surface of this medium and the front of an air shock or of an explosive wave.

10.2 Shock Adiabat of a Solid

In the general form, the pressure and energy of a solid may be represented as the sum

$$P = P_C + P_T, E = E_C + E_T. \qquad (10.1)$$

Here, P_C and E_C are the elastic or "cold" components of pressure and energy associated solely with the interaction forces acting between particles (atoms or molecules) and independent on temperature; these forces depend only on density.

P_T and E_T are referred to as the thermal components of the same quantities. These components are in turn divided in two parts, namely, the terms responsible for the thermal motion of atoms and the terms responsible for the thermal excitation of electrons. However, the presence of the electron terms becomes perceptible only at temperatures of several tens of thousands degrees; in view of this, the electron components of energy and pressure may be ignored in our treatment of granular media.

Therefore, if the temperature is not too high, the equation of state for a solid may be represented as:

$$P = P_C(V) + P_T(T,V).$$ (10.2)

We use the general thermodynamic identity:

$$\left(\frac{\partial E}{\partial V}\right)_T = T\left(\frac{\partial T}{\partial P}\right)_V - P$$ (10.3)

and, in view of the fact that $E_C = E_C(V)$ and $T = 0$, we derive:

$$P_C = -\frac{\partial E_C}{\partial V} = -\frac{dE_C}{dV}$$ (10.4)

for the elastic component of the pressure.

For the thermal part, we likewise use relation (10.3) and hold that the heat capacity $c_V = 3\,Nk$ is independent on volume (the condition is valid in the case of fairly high temperatures, when T_0 may be ignored) to derive

$$P_T = f(V)T,$$ (10.5)

i.e., the thermal pressure is proportional to temperature.

We rewrite this formula as

$$P_T = \Gamma(V)\frac{c_V T}{V} = \Gamma(V)\frac{E_T}{V}.$$ (10.6)

The introduced coefficient of proportionality between the thermal pressure and thermal energy of the lattice $\Gamma(V)$ is referred to as the Grüneisen coefficient. The Grüneisen coefficient at standard volume, Γ_0, is related to the other parameters of matter [1] as

$$\Gamma_0 = \Gamma(V_0) = \frac{V_0\iota}{c_V\varpi_0} = \frac{\rho_0}{c_V\rho_0\varpi_0} = \frac{\iota c_0^2}{c_V}.$$ (10.7)

Here, ι is the coefficient of volume thermal expansion, ρ_0 is the standard density, and ϖ is the isothermal compressibility of matter under standard conditions.

The approximate equation of state for matter has the following form:

$$P = -\frac{dE_C}{dV} + \Gamma_0\frac{E_T}{V}.$$ (10.8)

We will use the equations of mass and momentum conservation at the shock wave front in order to determine the dynamic compressibility,

$$\rho_0 D = \rho(D - u),\ P - P_0 = \rho_0 u D.$$ (10.9)

Here, ρ_0 is the initial density of matter, ρ is the density of matter behind the shock wave front, D is the shock wave velocity, u is the velocity of particles behind the shock wave front, P_0 is the pressure before the front, and P is the pressure behind the front.

For strong shock waves, the value of P_0 may be ignored compared to P. The variation of the internal energy of matter at the shock wave front is

$$E - E_0 = \frac{P + P_0}{2}(V_0 - V), \tag{10.10}$$

where E_0 is the initial specific internal energy of matter, and E is the specific internal energy behind the shock wave front.

Equations (10.9) and (10.10) yield

$$E - E_0 = \frac{P(V_0 - V)}{2}. \tag{10.11}$$

We use the approximate equation of state (10.8) and relation (10.11) and solve the resultant equation for P [1, 2, 5] to derive

$$P = \frac{(h - 1)p_C(V) - 2E_C(V)/V}{h - V_0/V}, \tag{10.12}$$

where

$$E_C = \int\limits_V^{V_{0k}} p_C(V)dV, \quad h = \frac{2}{\Gamma_0} + 1,$$

and Γ_0 is the Grüneisen coefficient at standard volume.

10.3 Shock Adiabat in Granular Material

The process of shock compression of granular material exhibits a number of special features. Zel'dovich and Raizer [1], Trunin et al. [3] treated the problem of the derivation of the shock adiabat of granular matter and its anomalous behavior.

We will now consider the shock compression of granular material. We will assume that matter in the final state behind the shock wave front is continuous and homogeneous. The point corresponding to standard volume V_s and zero pressure $P = 0$ lies on the shock adiabat (Fig. 10.1). The internal energy acquired by matter in the shock wave $E = \frac{1}{2}P(V_m - V)$ is proportional to the area of the horizontally hatched triangle. Its elastic part is proportional to the area of the curvilinear triangle bounded by the $P_C(V)$ curve and densely hatched in Fig. 10.1. The larger the initial volume V_m, i.e., the higher the porosity of matter, the greater the area difference corresponding to the thermal part of energy during compression of matter to one and the same final volume (the elastic energy at preassigned volume does not vary, and the total energy increases). However, the higher the thermal energy, the higher the thermal

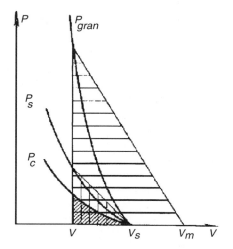

Fig. 10.1. *P-V* diagram of shock compression of granular material: P_{gran}, shock adiabat of a body consisting of granules; P_s, shock adiabat of a continuous body; and P_C, curve of cold compression of a continuous body. (After [1])

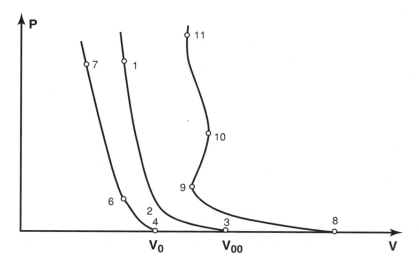

Fig. 10.2. The shock compressibility of granular material. (After [2]). For explanation, see text

pressure as well. Therefore, the higher the porosity, the steeper the shock adiabat. In particular, the shock adiabat of granular matter extends higher than the shock adiabat of continuous matter. Figure 10.2 shows the shock adiabat of continuous matter (4-6-7) and of granular material (1-2-3) whose porosity is relatively low.

Higher pressures are required in order to compress porous matter to the same volume; in so doing, these pressures are the higher, the higher the degree of porosity.

High heating during shock compression of granular materials may result in sharp anomalies in the behavior of the shock adiabat. In the case of compression of highly porous matter to a preassigned pressure, the relative importance of the thermal pressure turns out to be so high that the density in the final state at high pressure is lower than standard, $V > V_s$. In so doing, the volume does not decrease with increasing pressure, as it usually does, but increases, and the shock adiabat exhibits the anomalous behavior shown in Fig. 10.2 (curve 8-9-10-11).

In order to explain the origin of this effect, we will use the equation of the shock adiabat for solid matter (10.12) derived on the assumption that the electron pressure and energy are low, the Grüneisen coefficient is constant, and the initial energy of matter may be ignored.

In deriving (10.12), it was specified here that matter in the initial state was continuous; therefore, this equation may be modified by substitution of V_m for V_s to obtain an equation which describes the family of shock adiabats corresponding to different initial volumes V_m, i.e., to different degrees of porosity which may be characterized by the coefficient $\alpha = V_m/V_s > 1$,

$$P(V, V_m) = \frac{(h-1)p_X(V) - 2E_X(V)/V}{h - V_m/V}. \tag{10.13}$$

Here, Γ_0 is the Grüneisen coefficient of matter under standard conditions, V_0 is the initial volume of solid, and p_X and E_X denote the elastic or cold components of pressure and internal energy, respectively.

At $\alpha = 1$ and $V_m = V_s$, we have a shock adiabat of continuous matter. Point $V = V_0, P = 0$, and the cold compression curve satisfies (10.13) for any initial volume V_m (because $p_X(V_s) = 0$ and $E_X(V_s) = 0$), so that the family of adiabats is a bundle of curves issuing from this point. According to (10.13), at $V_m/V_s \to h$ we have $p_X \to \infty$, i.e., the limiting volume is $V_{lim} = V_m/h$. If this quantity is less than V_s, as is the case for low porosity $\alpha < h$, the shock adiabats exhibit a normal behavior and pass the higher, the larger the initial volume. If $V_{lim} > V_s$ (which is the case for high porosity, when $\alpha > h$), the behavior of the curves is abnormal: the final volume increases with pressure. The family of shock adiabats corresponding to different coefficients of porosity is shown in Fig. 10.3.

The experiment performed by Trunin et al. [3–5] revealed that the process of compaction of granular material of relatively low porosity (Fig. 10.1, adiabatic curve 1-2-3) occurs in two stages. An abrupt increase in density occurs in the segment 3-2 due to the pore collapse at relatively low pressure. Then a relatively slow increase in the density of matter occurs in the segment 2-1 of the adiabatic curve. At a higher value of porosity, the density variation may be divided into three stages. The "packing" of particles at low pressure occurs in the first stage (segment 8-9 of the shock adiabat, Fig. 10.2); in so doing, the

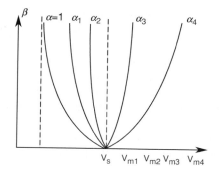

Fig. 10.3. Shock adiabats for different values of density ratios: $\alpha_4 > \alpha_3 > h, h > \alpha_2 > \alpha_1 > 1$. (After [1])

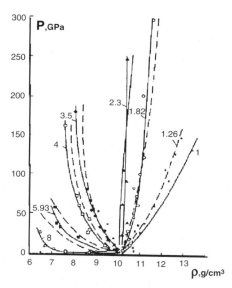

Fig. 10.4. The correlation between P and ρ for molybdenum for different density ratios; the points and solid lines indicate experiment; the *dashed lines* indicate calculation. (After [2])

density does not reach the value of ρ_S (density of the particle material). In the second stage (segment 9-10 of the adiabatic curve), an increase in pressure is accompanied by a decrease in density because of the strong heating of the material. This stage involves the compression of particles, their friction against one another, reorientation, and destruction; these processes cause intensive heating of the material being compacted. In the third stage, the density of matter increases owing to high pressure.

Figure 10.4 gives experimentally obtained shock adiabats of solid and porous molybdenum with the coefficient of porosity $\alpha = \rho_s/\rho_m$ of 1, 1.26,

1.82, 2.3, 3.5, 4, 5.93, and 8. Under shock compression of molybdenum with $\alpha < 2.3$, the density varies in two stages. At $\alpha > 2.3$, the shock adiabat assumes an anomalous direction. With such values of porosity, first the "packing" of particles occurs at low pressure to a density lower than ρ_S, and then the density decreases due to strong heating of porous material with increasing pressure; this decrease in density is the greater, the higher the porosity m.

For example, in the case of compression of molybdenum of porosity $\alpha = 4$, the density $\rho = 7.63\,\mathrm{g\,cm^{-3}}$ is reached even at $P = 1.6\,\mathrm{GPa}$; at $P = 5\,\mathrm{GPa}$, the maximal density is $\rho = 9.86\,\mathrm{g\,cm^{-3}}$; as the pressure continues to increase, the density decreases and, at $P = 158\,\mathrm{GPa}$, $\rho = 8.1\,\mathrm{g\,cm^{-3}}$. In the case of compression of molybdenum of porosity $\alpha = 8$, the density $\rho = 8.9\,\mathrm{g\,cm^{-3}}$ is attained at $P = 2.7\,\mathrm{GPa}$; when the pressure increases to $P = 26.1\,\mathrm{GPa}$, the density is $\rho = 6.44\,\mathrm{g\,cm^{-3}}$.

10.4 Processes of Deformation Under Shock-Wave Loading

When a shock wave moves in a granular medium, the material is compacted irreversibly within the wave front. The process of compaction is accompanied by reorientation, plastic deformation, and destruction of particles as a result of their friction against one another. These processes cause intensive heating of the material being compacted and dissipation of the shock wave energy. The shock-wave front spreads out, and the shock wave transforms to a wave of plastic flow with elastic precursor, whose velocity and amplitude depend on the material density, and then to an elastic stress wave [6].

This complex process cannot be described using the Hugoniot adiabat alone, because the field of thermodynamic quantities behind the shock wavefront is extremely nonuniform. In addition, an increase in the shockwave intensity, which is accompanied by a reduction of its front width, results inevitably in that the wave scale coincides with the characteristic spatial dimension of the medium. This makes the behavior of dissipative processes in the front sensitive to the structure of the material, i.e., a strong shock wave "develops" special features of the structure of the material, and the determining processes are those whose scale coincides with the size of granules, i.e., the events which occur on the particle surfaces such as local heating, melting, chemical reactions, surface deformation, and viscoplastic jet flows. And these events, in turn, are defined by the morphology of particles, their size, strength, roughness, and so on.

10.4.1 Models Used in Calculations

The description of the micromechanics of the deformation of granular media calls for the development of a model of a unit cell. It must accurately convey the similarity with a real medium and must be suitable for treating only the processes required for the purposes of concrete investigations rather than all of the processes occurring during deformation, i.e., this model is defined by the material and the processes whose phenomenology must be identified. Torre [7] suggested a model of hollow sphere of rigid material for describing the correlation between pressure and density in the process of compaction of powders. This model was used to derive the correlation between the current value of porosity and the static pressure:

$$P = \frac{2}{3} Y_0 \ln \frac{1}{1 - \frac{1}{\alpha}}. \tag{10.14}$$

Here, Y_0 is the yield point of the material of the sphere. The pore size does not appear in the foregoing relation. This model for elastoplastic [8] and viscoplastic [10] behavior of the material of the sphere was used to describe the shock-wave processes in porous materials. Based on this model, a correlation was obtained between pressure, porosity, and time derivatives of porosity; this correlation, together with the condition of a stationary shock wave, made it possible to investigate the structure of its front [11]. Dunin and Surkov [12] studied steady-state structures of shock wave fronts in a formulation similar to that of [10]. It was demonstrated that a minimum shock wave velocity exists, as well as a critical velocity at which a pore collapse occurs. Kim and Sohn [13] used this model to treat the thermodynamics of the process of viscoplastic filling of pores; the thermal conductivity was taken into account as well.

We will dwell in more detail on one of the most universally employed models of dynamic compaction of granular and porous materials, namely the Carrol–Holt model. This is likewise a hollow-sphere model (see Fig. 10.5b). The set of equations defining this model was developed in [8]. The size of the spherical pore is taken to be equal to the characteristic pore size in real material, and the mean density of the sphere is equal to the density of the initial undeformed material. These conditions define the geometric parameters of the sphere as follows:

$$a_0 = R \left(\alpha_0 - 1\right)^{1/3}, \quad b_0 = R\, \alpha_0^{1/3}, \tag{10.15}$$

where α_0 is the initial density ratio of the material.

The material of the sphere is taken to be incompressible and ductile; the values of viscosity η and yield point Y are independent of temperature. The

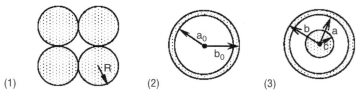

Fig. 10.5. Models of unit cells of granular material (1) real material, (2) Carrol–Holt model, (3) refined Carrol–Holt model. (After [17])

dynamics of pore collapse is described by equations of continuum mechanics for spherically symmetric motion:

$$\rho_S \ddot{r} = \frac{\partial \sigma_{rr}}{\partial r} + \frac{2}{r}(\sigma_{rr} - \sigma_{\Theta\Theta}), \tag{10.16}$$

equation of motion

$$\sigma_{rr} - \sigma_{\Theta\Theta} = Y_0 + 2\eta \left(\frac{\partial v}{\partial r} - \frac{v}{r} \right) = \sigma, \tag{10.17}$$

equation of viscoplastic flow

$$r^3 - r_0^3 = a^3 - a_0^3, \tag{10.18}$$

condition of incompressibility

$$\sigma_{rr}(r = a) = 0; \quad \sigma_{rr}(r = b) = -P(t), \tag{10.19}$$

boundary conditions. The pore pressure is small with respect to the loading pressure, so that we may neglect it. The initial conditions are:

$$a(0) = a_0; \quad \dot{a}(0) = 0. \tag{10.20}$$

The condition of incompressibility makes it possible to derive correlations between the geometric values of the model and density ratio or porosity,

$$\alpha = \left(\frac{b}{b_0} \right)^3 \alpha_0 = \left(\frac{a}{a_0} \right)^3 (\alpha_0 - 1) + 1. \tag{10.21}$$

Here, ρ_s is the density of the material of the sphere; σ_{rr} and $\sigma_{\Theta\Theta} = \sigma_{\phi\phi}$ are components of the stress tensor in the spherically symmetric case; $\sigma = \sigma_{rr} - \sigma_{\Theta\Theta}$ is the shear stress; Y is the yield point; η is the viscosity; and $r, a, b, \alpha, r_0, a_0, b_0$, and α_0 denote the current and initial values of the cell radius, initial radius of a pore, external radius of the sphere, and porosity, respectively. The set of foregoing equations with constant values of Y and η represents the Carrol–Holt model.

The Carrol–Holt model turned out to be remarkably effective from the standpoint of describing the density dependence of pressure in the case of a static deformation of a wide range of materials [14]. In addition, this model plays an important part in analyzing dynamic processes and in understanding the kinetics of compaction of powders and granular materials. Unfortunately, a number of features render this model inadequate to describe the real process of dynamic deformation of granular materials and, thereby, limit its applicability in this field.

A detailed analysis of the advantages and disadvantages of the Carrol–Holt model was made by Nesterenko [14]. This model cannot be used for the simulation of the process of shock-wave compaction of granular materials for the following reasons:

1. In the case of bulk media, the concept of a "characteristic pore size" loses its meaning, although the particle size is defined rigorously.

2. Because of their high porosity ($\alpha_0 \approx 2$), granular media are heated to temperatures comparable to the melting temperature even in relatively weak shock waves with a pressure $P \leq 10Y$. This increase in temperature has a significant effect on η and Y.

3. The plastic filling of pores for these media differs strongly from spherically symmetric motion because of the influence of neighboring cells and is, in fact, a collective process. The hollow sphere model implies that all of the material is plastically deformed during the filling of pores. The results of experiment performed by Staver [15] for the case of shock deformation demonstrated that this is not so. In the case of shock deformation, the monolith density may be attained only owing to plastic deformation of external layers of the material of the granules. The results of experiments performed by Kusubov et al. [16] with quenched copper granules demonstrate clearly that the compaction for homogeneous particles occurs largely owing to plastic flow of their peripheral layer.

4. Even in shock waves with a pressure of $\approx 1\,\text{GPa}(P \approx 2Y)$ at the front, the width of the shock wave front is equal to approximately three times the particle size and approaches the particle size as the pressure increases to 10–13 GPa. Therefore, the deformation of granules at this pressure will occur under conditions of a high gradient in the particle size.

5. The Carrol–Holt model implies that the entire process of compaction occurs within the framework of a spherically symmetric collapse of the pores. However, analysis of the structure of compacts obtained as a result of shock compaction of powders of bulk density [16] reveals that significant viscoplastic deformation may be further observed after compaction of powder to a monolith state.

6. The spherically symmetric pattern of the model is responsible for the emergence of nonphysical features when the radius of the sphere tends to zero.

In view of the foregoing, a refinement of the Carrol–Holt model was suggested in [14, 17]; this refinement consists in introducing an incompressible kernel into the Carrol–Holt sphere (see Fig. 10.5c). This made it possible, while staying within spherical symmetry, to eliminate a number of disadvantages of the Carrol–Holt model and to make it physically adequate to describe the process of shock compaction of granular media. Therefore, the geometry of the model is determined with due regard to the following conditions: the mass of a cell is equal to the mass of real particle (which is more adequate from the physical standpoint for media of bulk density); the specific volume of the model is equal to the specific volume of the real material,

$$a = R; \quad b = R\alpha_0^{1/3}; \quad c = R\left(2 - \alpha_0\right)^{1/3}; \quad \alpha = \left(\frac{a}{a_0}\right)^3 + (\alpha_0 - 1), \quad (10.22)$$

where α_0 is the initial density ratio of the material; in so doing, $\alpha_0 \leq \alpha^* < 2$ in the case of close initial packing of particles, and α^* corresponds to the minimal porosity of the initial material which may be attained only by repacking the particles.

In case the initial α_0 is higher than α^*, one can assume that the compaction of material was accomplished by repacking the particles without their viscoplastic deformation. Therefore, when the density ratio varies from α_0 to α^*, it is appropriate to assume that $Y = Y_0$ and $\eta = 0$. In this case, we have the initial dimensions of the refined model as:

$$a = R\left(\alpha_0 - \alpha^* + 1\right)^{1/3}; \; b = R\alpha_0^{1/3}; \; c = R\left(2 - \alpha^*\right)^{1/3}. \tag{10.23}$$

At $\alpha = \alpha_0$,

$$a^* = R; \; b = R(\alpha^*)^{1/3}. \tag{10.24}$$

In addition, the following temperature dependences of the yield point and viscosity of the envelope material were assumed in the model:

$$Y = Y_1\left(1 - \frac{T}{T_{\mathrm{m}}}\right) \quad \text{for} \quad T \leq T_{\mathrm{m}}, \quad Y = 0 \quad \text{for} \quad T > T_{\mathrm{m}}, \tag{10.25}$$

where Y_1 is the characteristic value of yield at low temperatures, and T_{m} is the melting temperature;

$$\eta = \eta_{\mathrm{m}} \exp B \left(\frac{1}{T} - \frac{1}{T_{\mathrm{m}}}\right), \tag{10.26}$$

where η_{m} is the viscosity of melt, and $B = 5765.26\mathrm{K}^{-1}$.

The introduction of an incompressible kernel into a spherical cell made it possible to divide the process of compaction of granular material into two phases, namely, the stage of pore collapse and the processes occurring after that. Therefore, the parameters attained by the instant of the "termination" of the spherically symmetric model become the initial parameters for the phase of the viscoplastic flow of the material.

The results of calculations performed in [17] using the refined Carrol–Holt model have demonstrated that this model describes adequately the kinematics of compaction of granular materials in a wide range of shock compression and agrees with the known experimental data. The calculation produced the total kinetic energy E_{tkin}, the internal energy E_{int}, and the microkinetic energy E_{mkin} for each element of the computational field.

10.4.2 Phenomenology of the Process of Shock Compaction of Granular Materials

Kusubov et al. [16] performed an experiment on explosive compaction of quenched copper granules. This experiment revealed that a part of samples that survived after loading exhibited special features of the structures which were not observed under static loading. These structures represented regions within which the deformation of particles exhibited a pattern of splashing-out of their surfaces. Figure 10.6a presents powder samples after the densificating into the static mode. It differs from the sample (Fig. 10.6b) subjected to shock loading. Behind the shock front (the direction of shock wave motion is indicated by an arrow) interfaces of the particles are crooked and distorted. Nesterenko [14] and Benson et al. [17] suggested referring to this mode as *dynamic* (Fig. 10.6b). In spite of the shock pattern of loading, the structure of the compacts shown in Fig. 10.6b hardly differs from that in the case of static compaction; as a result, this mode was given the name of *quasistatic*.

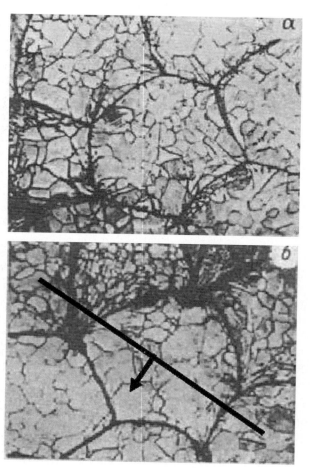

Fig. 10.6. (a) The quasistatic mode of deformation; (b) the dynamic mode of deformation (the arrow indicates the direction of propagation of the shock wave front). (After [14])

The quasistatic mode of deformation may be distinguished from the dynamic one by the form of contacts between particles. In the quasistatic mode of deformation, they remain straight (Fig. 10.6a). In the dynamic mode, the forms of contacts deviate from straight, the particle surfaces become curved, and jets and melts arise on these surfaces (Fig. 10.6b). The temperature of contacts between granules and the emergence of bonds between them depend on the pattern of their deformation. The quasistatic deformation with the contact geometry typical of this deformation is especially unfavorable for the formation of bonds between particles.

The latter fact adds importance to the search for criteria of transition between these two modes. In [14, 17] a transition criterion was suggested, based on the ratio of minimal plastic work required for complete closing of pores to total energy of incident shock wave. This criterion may be written as

$$\frac{E_d}{E} \leq K, \tag{10.27}$$

where E is the total internal energy per unit mass, supplied to the material during shock densification, E_d is the energy dissipates when the pores are closed in the case of static compression of granulated material, and K is a constant.

Benson et al. [17] introduced the term *microkinetic* energy. This quantity may be formally expressed as:

$$E_{mkin} = E - E_d. \tag{10.28}$$

The physical meaning of the suggested criterion is as follows: if the energy which dissipates in the dynamic mode of deformation at the last stage of pore collapse is low compared to the total energy, its remainder in the form of microkinetic energy distributed over the particle surfaces, will cause characteristic changes in the geometry of the particles and in the structure of the material. It will also develop conditions for the emergence of strong bonds between particles in the form of compacts. The remaining amount of microkinetic energy E_{mkin} must be approximately the same as that spent for viscoplastic deformation of the particles until this instant, i.e., it must be sufficient for causing marked changes of the shape of the particles. Benson et al. [17] performed a mathematical simulation of shock compaction of granular material in different modes and obtained the maximum value of the microkinetic energy E_{mkin} at the instant of complete closing of pores for different values of velocity of the incident shock wave, the porosity of the material, and the size of the granules.

The expression for the total internal energy per unit mass, E, supplied to the material during shock densification, can be obtained if the final density is close to the theoretical solid density ρ_S:

$$E = \frac{P}{2\rho_S}(\alpha_0 - 1). \tag{10.29}$$

It follows from the shock adiabat equation:

$$E - E_0 = \frac{1}{2} P \left(\frac{1}{\rho_m} - \frac{1}{\rho_s} \right). \tag{10.30}$$

In which the initial specific energy E_0 of the material may be neglected. Therefore, the criterion of transition from the quasistatic to the dynamic mode may be rewritten as:

$$P \geq P_c = \frac{2\rho_s E_d}{K(\alpha_0 - 1)}, \tag{10.31}$$

where P_c is the pressure at which the transition from quasistatic to dynamic mode occurs [17].

The refined Carrol–Holt model may be used to estimate E_d [14, 17] disregarding the processes of viscous dissipation,

$$E_d = \frac{2}{3} \frac{\overline{Y}}{\rho_s} \left[H(x_0) - H(x_1) \right], \tag{10.32}$$

$$H(x) = x \ln x - (x - 1) \ln x - 1, \tag{10.33}$$

$$x_0 = \frac{\alpha_0}{\alpha_0 - 1}, \quad x_1 = \frac{\alpha_1}{\alpha_0 - 1}, \tag{10.34}$$

where $\overline{Y}$ is the process average value of yield point.

Note that the expression for E_d in the Carrol–Holt model is similar to that given later; however, the values of x_0 and x_1 in this case must be equal to α_1 and α_0, respectively. For the case of practical importance of $\alpha_0 \approx 2$ and $\alpha_1 \approx 1$, the values of E_d obtained for these two models are close,

$$E_d \approx \frac{2\overline{Y} \ln 2}{3\rho_s} \ln \alpha_0 \approx \frac{4}{3\rho_0} \overline{Y} \ln 2. \tag{10.35}$$

In so doing, the criterion $E_d/E \leq K$ takes the form $P \geq 8\overline{Y} \ln 2/3K$.

The critical pressure of transition in this case is independent of the fraction and density ρ_s of the powder material. The criterion must hold well for high-strength materials which exhibit a weak dependence of the strength characteristics on the rate of deformation, for example, for powders of quenched steel. The foregoing inequality indicates that the transition pressure P for materials of close initial porosities such as granules of an approximately spherical shape is uniquely related to their yield point or microhardness,

$$P > HV = P_c. \tag{10.36}$$

10.4.3 Key Features of the Thermodynamics of Shock Compaction of Granular Media

We will now consider in greater detail the thermodynamic features of micromechanics of shock-wave compression of granular material. It was already mentioned that the fields of thermodynamic quantities in this process are extremely nonuniform and depend on the structure of the material. In view of this, it is very important to measure directly the characteristics associated with temperature. In addition, in the case of porous and granular materials, a situation is possible in which the spatial scales of reaching a close-to-equilibrium final state behind the front may be significantly different for different parameters, for example, density and temperature.

One of the key features of granular materials under shock compression is their nonequilibrium heating. The nonequilibrium pattern of heat release in the process of shock compression of powders was first revealed by Blackburn and Seely [18] using optical methods.

In a number of studies [15], the metallographic investigation of compacts resulted in revealing zones of locally elevated temperatures located along the boundaries of particles at pore filling sites. Figure 10.7 shows the structure of VTZ-1 titanium alloy prepared by explosive compaction of granules at standard initial temperature and vacuum of $6.55 \cdot 10^{-3}$ Pa [14]. One can clearly see white zones of local heat release caused by the slip of particles relative to one another; one can further see that the contact temperatures of particles differ significantly depending on the pattern of their deformation, i.e., on the absence or presence of intensive shear deformation of the particles.

Nesterenko [14] used the thermal emf procedure to find the dependence of the nonequilibrium contact temperature on the fraction and pressure of the shock wave. It has been demonstrated that, given the same initial density and conditions of loading, the contact temperature is higher for particles of the larger fraction.

It is demonstrated in [14] that the energy released on the periphery of particles in a copper powder with a fraction size of 0.1–0.5 mm and $\alpha = 2$ at a

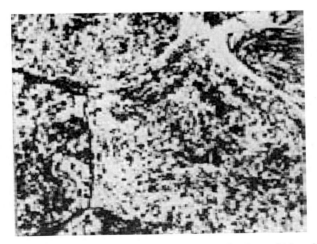

Fig. 10.7. Powder sample after shock-wave loading. (After [14])

shock wave pressure $P = 3$ GPa (where the maximal degree of disequilibrium is attained) is comparable by the order of magnitude to the total internal energy of shock compression E. This indicates that the final equilibrium thermodynamic state behind the shock wave front is reached as a result of friction on the boundaries of powder particles and their viscoplastic flow.

The difference between the contact temperatures of powder particles of different fractions is explained by the dependence of the specific surface of particles on this temperature. Indeed, one and the same energy of shock compression distributed over a smaller surface area of particles leads to a higher temperature. Because, for spherical particles, $S = 6/\rho_s a$ ($a = 2R$ is the particle diameter), a higher value of the fraction leads to a higher contact temperature. Nesterenko [14] observed that the correlation between the contact temperature and the size of fraction is nonlinear, which is defined by the dependence of the zone heated in the process of deformation on the particle size. The difference between temperatures T_1 and T_2 for the large and small fractions decreases with increasing pressure ($P > 3$ GPa). Therefore, the depth of the zone on the periphery of particles, where heat is released as a result of friction and viscoplastic deformation, increases with pressure in this region, as well as the fraction of energy dissipated in the bulk of particles. For the investigated powders at pressures $P > 10$ GPa, the difference between the contact temperatures for different fractions is small. The adiabatic shift may serve as one of the ways of dissipation of internal energy in the bulk of particles. Such processes lead to the emergence of local zones of heat release both over the surface of particles and at sites of pore collapse and in the bulk of particles.

Given the experimentally obtained time dependence of contact temperature rise in the wave front, one can find [14] the minimal estimate of energy E_{mkin} concentrated on contacts. It is assumed that the heat source active on the boundary of particles has zero width. It is further assumed that $(\chi\tau)^{1/2}$ is much less than a, where χ is the thermal diffusivity, τ is the wave width, and a is the average size of particle. This inequality is valid for copper powder with $a \approx 0.3$ mm, $\tau \approx 3 \cdot 10^{-7}$ s, and $\chi = 1$ cm^2 s^{-1}, and the more so for nickel powder. As a result, a one-dimensional approximation may be employed for making energy estimates.

The source intensity $F(t)$ may be found from the time dependence of the contact temperature T_c,

$$F(t) = \frac{k}{\pi^{1/2}\chi^{1/2}} \cdot \frac{d}{dt} \int_0^t \frac{T_k(\xi)\,d\xi}{\sqrt{t-\xi}}, \qquad (10.37)$$

where k and χ are the coefficients of thermal conductivity and thermal diffusivity.

The temperature in the shock wave front may be approximated with adequate accuracy by the following linear dependence:

$$T_k(t) = \frac{T_{k\,max}}{\tau}t, \qquad 0 < t < \tau. \qquad (10.38)$$

Here, S_c is the characteristic specific contact area of particles. Its real values vary in the course of compaction and reach the value of $S/2$ at a predicted density. The estimation of E_{mkin} was made assuming that $S_k \approx S/2$. It will be more correct to assume that $S_c \approx S/2$, which is also a somewhat overstated value and corresponds to the main heat release proceeding in a compacted material.

The values of E_{mkin} calculated by the experimental results using the foregoing integral formula apparently underestimate the energy released on the periphery of particles because of the assumption of zero size of the heat source. On the other hand, the values of energy E_{mkin} for a heat source of zero size obtained by the formula are somewhat overestimated because of the spreading of the front width caused by natural irregularities of the contact interface between two granular media.

The high rate of cooling of contacts may likewise be assigned to typical features of shock compression of granular media. It was shown in [14, 19] that, at a pressure of 6.5 GPa, the time of relaxation of contact temperature in powder with a fraction of $0.1 < a < 0.5$ mm is $\sim 5\,\mu s$, which corresponds to a rate of cooling of the contact zones of $\sim 10^8\,\mathrm{K\,s^{-1}}$ exceeding, for example, the cooling rates that may be reached using the conventional quenching methods.

A feature of fundamental importance in constructing models of heterogeneous heating of powders under shock compression is the asymmetry of the time characteristics of the process of heating and cooling the contacts, which was also observed by Schwarz et al. [20]. This feature consists in that the time of heating is much shorter than the time of temperature decrease, which is clearly indicative of the finite width of the heat source on the boundaries of particles.

The fact that the time of equalization of temperature in powders of large fraction is much longer than the temperature rise time poses the problem of the stationarity and structure of a shock wave front in powder.

Experiments reveal [14] that, for copper powder with a fraction of 0.1–0.5 mm at high pressures, the front width with respect to mechanical variables is $\tau_M \sim a/D \approx 10^{-7}\,s$, and with respect to the contact temperature $-\ \tau_T \sim 1 - 5\,\mu s$, i.e., $\tau_M \ll \tau_T$.

The agreement between the mechanical parameters of powders of different fractions (mass velocity, pressure, shock wave velocity) indicates that the thermodynamic disequilibrium and resultant nonstationarity of motion affect only slightly the mechanics of the process as a whole. Therefore, the process of relaxation may be treated at constant total pressure, and its physical content will reside in the variation of contribution to total pressure between the thermal and "cold" components at each point of compacted material. Note that this process may cause additional shear deformations and defect formation in compacted material. A basically important feature of powder heating, which was revealed in experiments involving thermal emf [14], is the fact that the maximum temperature is reached during time $\tau \approx a/D \approx \tau_M$.

10.4.4 Analysis of Thermodynamic Models of Heterogeneous Heating of Granular Medium Under Conditions of Shock-Wave Deformation

As was discussed previously, the structure of granular material has a cardinal effect on the micromechanics of the behavior of this material under shock-wave compression.

The first model, which enables one to estimate the nonequilibrium heating of contacts, was suggested by Belyakov et al. [21]. The following assumptions formed the basis of this model:

1. The thermal energy is released solely on the surface of particles in a layer of zero thickness.
2. The zone of heating is fully defined by the size of the region covered by the process of heat conduction.
3. The heat source active on the surface of particles has a constant intensity F_0 which provides for the energy release during the time of shock compression τ equal to the thermal part of internal energy. Therefore, $F_0 = E/\tau\, S_c$, $S_c = 3/\rho_0\, a$, where S_c is the specific contact area of particles, ρ_0 is the density, and a is the diameter of particles.
4. The time τ was taken to be $\tau = a/D$, where D is the shock wave velocity. Note that this estimate of the heating time agrees with the wave front width measured by the method of thermal emf at high pressures [14].
5. For short times of shock compression and fairly large particles, the heated zone thickness is much less than the particle radius. This enables one to use a one-dimensional approximation for solving the heat equation.

The solution of the heat conduction problem using this model for a constant heat source which is active in a plane leads to the following expression for the maximal temperature T_c:

$$T_{k\,\max} = \frac{E}{3c}\left(\frac{aD}{\chi\pi}\right)^{1/2},\tag{10.39}$$

where χ is the thermal diffusivity, and c is the heat capacity; these two parameters were taken to be independent of temperature.

The inverse problem was solved by Reybould [22] who made assumptions close to those given earlier and found the correlation between the heating time t, the increase in temperature T_c, and the variation of equilibrium temperature ΔT,

$$t = \frac{1}{12\chi}\left(\frac{a}{2}\frac{\Delta T}{T_k}\right)^2.\tag{10.40}$$

This formula was used by Reybould [22] to find the time t_m during which the melting temperature is reached on the contacts.

The main advantages of the approach employed in [21, 22] are as follows (the model was analyzed in [[14]]):

1. It is possible to make upper estimates of the time of relaxation to thermal equilibrium in a granular sample. Indeed, the inclusion of all omitted factors such as heat release in the bulk of the particles and the finite size of the heat source on contacts will result in a decrease in the time of relaxation to thermal equilibrium.
2. It is possible to reach the maximal attainable surface temperature of particles.
3. The suggested model of thermal effects is a closed model.

At the same time, the approach employed in [21,22] suffers from a number of disadvantages:

1. The dependence of the contact temperature on the particle size is such that T_c is proportional to $a^{1/2}$. This dependence must hold at any increase in pressure. However, experiments reveal a decrease in the difference between the values of T_c for particles of different sizes with rising pressure. This is equivalent to an increase in the area over which the heat release occurs or to an increase in the width of the region of heat release along the boundary with pressure. Therefore, this approach calls for amplitude restrictions on its validity.
2. The need to take into account the real specific surface is not obvious, because the initial irregularities will deform in the front, which will cause a variation of the surface area. In addition, it is hardly advisable to take into account the real surface if the height of irregularities is much less than the size of the region covered by the process of heat conduction under shock compression.
3. A zero thickness of the heat source active on the particle boundaries must result in an approximately symmetric curve describing the rise and fall of the contact temperature, which contradicts the experimental data.
4. The hypothesis of a constant power of the effective heat source on the boundary of granules is the simplest one from the standpoint of obtaining numerical estimates; however, its validity has not been proven. Moreover, the assumption of constant power of flow leads to the shape of the front of mass velocity that is not typical for powders.

Indeed, if we assume that the wave front is steady and the specific internal energy of the material at any point of the front is equal to $0.5u^2$ (where u is the mass velocity at this point), we can write the following equality at $F = F_0$:

$$\frac{1}{2}u^2 = F_0 s_k t \rightarrow u = \sqrt{\frac{t}{\tau}}. \tag{10.41}$$

The instant t is reckoned from the beginning of the front.

It was shown in [14] that this shape of wave profile does not fit that obtained experimentally, i.e., the use of the assumption $F = F_0$ may lead to significant distortion of the front profile compared to profiles typical of powders of bulk density. In view of this, in this approach it is desirable to use, in addition to information about the total width of the front of mass velocity, information about the front shape. The effect of the form of $F(t)$ on the end result (for example, on the maximum value of the contact temperature which defines, in the opinion of Nesterenko [14], the critical modes of compaction) may be studied in the following formulation.

For simplicity, we will treat a plane case in which $T_c(t)$ is determined in terms of $F(t)$ as follows:

$$T_c\left(t\right) = \frac{\chi^{1/2}}{k\pi^{1/2}} \int_0^t F\left(t - \xi\right) \frac{\mathrm{d}\xi}{\xi^{1/2}} \tag{10.42}$$

where k is the thermal conductivity. The use of linear, quadratic, and cubic dependencies for $F(t)$ with the same total energy and τ causes an increase in $T_c(\tau)$ by a factor of 1.33, 1.6, and 1.83, respectively. This emphasizes the importance of taking into account the concrete shape of the profile $F(t)$.

Dunin and Surkov [23] used the Carrol–Holt model to analyze the thermodynamics of shock compression of powder.

The main disadvantage of this approach consists in that the process of pore collapse at a fairly high pressure of compaction may produce any high temperatures. In so doing, the high-temperature zones are concentrated in the vicinity of pores rather than on the particle boundaries. This disadvantage is eliminated partly in the refined model. This model may be used to make upper estimates of temperatures attained in powder in the process of pore collapse but fails to provide an adequate description of transition of microkinetic energy of motion to thermal energy after pore collapse.

Kusubov et al. [16] made an attempt to use the refined Carrol–Holt model and numerical calculations to explain the relaxation to thermal equilibrium behind the front in large-fraction powder at high pressures ($P > 10\,\mathrm{GPa}$), where the front width $\tau \approx 10^{-7}\,\mathrm{s}$ is much less than the time of relaxation to thermal equilibrium owing to heat conduction. It was suggested that the heating of internal layers, which causes a decrease in their strength and viscosity, must lead to more intensive deformation and, consequently, to the heating of external layers, thereby ensuring a uniform heat release over the entire volume. However, the calculations revealed that even the refined version of unit pore model is incapable of explaining the rapid relaxation to thermal equilibrium in the case of a large fraction. At a pressure of $P > 10\,\mathrm{GPa}$, a temperature equal to the melting temperature by the instant of pore collapse was reached in only 20% of the volume of the cell material, and the remaining part exhibited a temperature of 250–300°C. This is fully defined by the spherical geometry of the model; therefore, the model is incapable of describing this feature of thermodynamics of shock-wave deformation of powders of bulk density at high pressures. It is clear from the calculation results that a significant part of thermal energy of the material at $P > 10\,\mathrm{GPa}$ will be released at the stage when the pores collapse and microflows exist in the "monolithic" material.

10.4.5 Conditions Required for Developing High-Strength Compacts

The processes of deformation of granular material under shock-wave loading are of fundamental importance from the standpoint of using them for developing and treating materials of a new class, namely, crystalline and amorphous quenched metal alloys. In addition, explosive compaction may be used to produce metal–ceramic compounds, solid articles of powders of amorphous materials, structural elements of superconducting ceramics.

The importance of local processes of heat release in the case of shock-wave treatment of materials is generally recognized at present. These processes, in turn, are defined by the fraction size of particles, density, thermodynamic characteristics of materials, and loading conditions.

The key hypothesis of the search for criteria of seizure of powder particles is the assumption that the emergence of a strong bond and its further growth are defined by the qualitative change of kinematic modes of deformation of granule contacts.

Nesterenko [14] formulated the assumption that the conditions

$$E > E_{\min}; \; P > P_{\min}; \; E_k > E_{k\,\min}; \; t > t_{\min} \qquad (10.43)$$

must be formulated for the proper choice of the optimum mode of explosive compaction. The physical meaning of the first two conditions consists in that they provide for the minimum pressure $P_{\min}$, which is required for the process of plastic flow that leads to the filling of pores, and the minimal value of total specific energy E depending on the initial specific volume in addition to pressure. The requirement of $E > E_{\min}$ provides for the transition to the dynamic mode of deformation. The value of $E_{k\,\min}$ depends on the initial temperature, porosity, fraction, and so on.

The foregoing three conditions are independent to a certain degree, because the variation of initial porosity makes it possible to gain different values of E for the same values of P. On the other hand, E_k depends to a large degree on the size of granules for the same values of P and E, as was already mentioned. The need for the requirement of $E_k > E_{k\min}$ is demonstrated clearly in Fig. 10.7, where one can see the formation of bonds (white zones) between particles (the particles experienced mutual slip which resulted in the required concentration of surface energy) and the absence of bonds on contacts, where simple deformation of particles was observed without marked slip. This figure demonstrates that reaching a close-to-theoretical density of the material does not necessarily provide the bond over the entire surface of contacting particles.

The requirement of $t > t_{\min}$ provides for a sufficient duration of compression pulse for a complete pore collapse. Different authors suggest that the following parameters should be regarded as $t_{\min}$ [14]:

1. The time required for complete closing of pores.

2. The time required to reach the melting temperature on contacts between particles.
3. The time required for the crystallization of melt on the boundaries of particles.

Useful estimates of the pressure of transition to dynamic mode were made using the results of experiments with copper granules performed by Kusubov et al. [16]. The estimates demonstrated that the detonation pressure for copper granules of low strength compared to steel and titanium ones and with a fraction of 600–850 µs amounts to 6 GPa. These estimates indicate that the explosive methods should not be used for welding high-strength materials at the initial normal temperature.

Benson et al. [17] give the criterion of preparing high-strength compacts, based on the results of comparison between the specific energy dissipated in the neighborhood of a pore at the stage of its collapse and the specific energy of shock compression. The difference between them represents the microkinetic energy required for cleaning the surface contact layers and for developing the desired degree of plastic deformation. For quenched granules, this criterion may be written as

$$P > 2HV, \tag{10.44}$$

where P is the pressure on the front of incident shock wave, and HV denotes the microhardness of the material being compacted. In view of the fact that the transition from the quasistatic mode of deformation lies in the pressure range from $P = HV$ to $P = 2HV$, one can say that close contacts may form only in the case of developed dynamic mode. The technological problems of explosive pressing and welding of granular materials are well covered in [24].

10.5 Attenuation of Explosive Wave as a Result of Deterioration of Granular Material

The deterioration of loosely bound granular material leads to the attenuation of incident explosive waves and shock waves. The attenuation mechanism consists in that the larger part of the explosive wave energy is spent to destroy the obstacle (breakage of bonds between granules) and scatter the particles of the material. The *explosive wave attenuation factor* serves as the measure of the effectiveness of a protective device. It is defined as the ratio of excess pressure during propagation of an explosive wave in open space to the same quantity in the case of an explosion within the attenuation device. Gelfand and others [25] used a basket with sand-filled elastic double walls as the attenuation device. The elastic material of the walls served the function of bonds between particles of sand. The measured values of the attenuation factor were in the range from 2.27 to 3.79. The attenuation of an explosive wave using a cylinder of loosely bound mixture of sand and cement was investigated in [26]. The cylinder was pulverized as a result of explosion. The particles of the cylinder material were scattered by the explosion to a distance of 3 m or less from the epicenter. The experiment revealed that the attenuation factor decreases away from the epicenter; the maximum value of the explosive wave attenuation factor was 4.3.

The process of propagation of a plane shock wave in a granular medium was simulated numerically in [26] using the SPH (Smooth Particle Hydrodynamics) method [27] (see Fig. 10.8). Voids on the figure imitating "pores" (shown as white squares) are introduced into the bulk of sand. The shock wave transmission is accompanied by a pressure rise within the pores, which results in the breaking of bonds between particles and in scattering of the latter. The bulk of energy of the explosive wave transforms to kinetic energy of the accelerating granules. On Fig. 10.8 one can see different moments of process evolution. At zero instant of time (Fig. 10.8a) the air shock wave interacts with the left-hand boundary of the sand screen and the shock wave begin to propagate into sand sample (the position of its front shown by arrow) (Fig. 10.8b). The "pores" of the first row start to deform. On Fig. 10.8c the "pores" of the first row are almost closed, the "pores" of the second row start to deform, the right-hand boundary of the sample comes in motion. In Fig. 10.8d and Fig. 10.8e the final stage of process are presented. The "pores" are closed completely, the sand is compacted and moves to the right.

Figure 10.9 shows qualitatively the process of deterioration of a sand cylinder and scattering of its parts. The simulation was performed using the SPH method. In this code, 2D axisymmetrical hydrodynamic equations are solved by means of Lagrangian smoothed particles, being adapted to local peculiarities of the flow field. The code algorithm was improved by means of adaptive splitting and sticking of SPH particles. In the computations, the detonation period was passed over by substitution of condensed explosive in the initial cells by detonation products at a certain pressure.

$$P_H = \frac{\rho_H H^2}{2\,(\gamma + 1)},\qquad (10.45)$$

where $\gamma = 3, \rho_H = 1.2\,\mathrm{cm}^3,\ H = 6,500\,\mathrm{m\,s}^{-1}$.

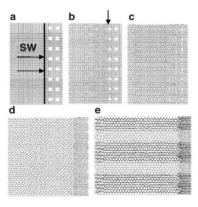

Fig. 10.8. The interaction between an air shock wave with a sand screen (1D case)

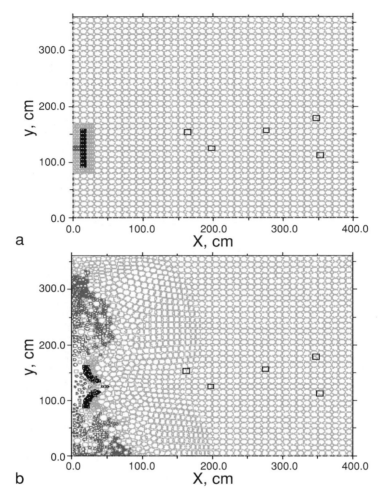

Fig. 10.9. Explosion within a sand cylinder. Instants of time of 0.0, 2.6, 3.5, and 6.5 ms, respectively. The positions of sensors are marked by squares. The material of the cylinder is shown by black circles, the products of detonation – by dark gray circles, and air – by light gray circles. (After [26])

Therefore, an instantaneous detonation was assumed. The equation of state for the detonation products was taken as for a perfect gas:

$$P = (\gamma - 1)\, e. \tag{10.46}$$

An important point in the simulation of an explosion in a sand/cement cylinder is the choice of an equation of state for the heterogeneous medium,

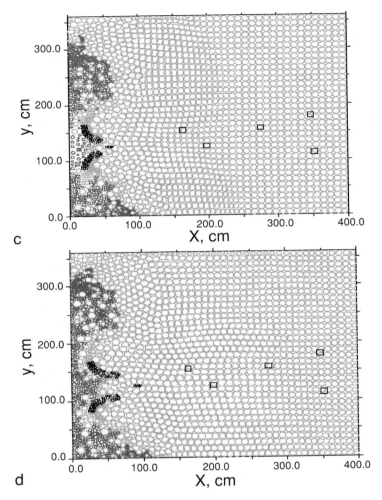

Fig. 10.9. Continued

i.e., sand/air, and the strength properties of the initial phase of cylinder failure. In this simulation we take into account only the first factor. Following [28], we adopted an equation of state that combines properties of liquid and gas:

$$P = C_B^2 \left(\rho - \rho' \right) \delta + \left(\gamma - 1 \right) \rho e, \qquad (10.47)$$

where $C_B = 1,280 \, \mathrm{m \, s^{-1}}, \rho' = 2,200 \, \mathrm{kg \, m^{-3}}, \gamma = 3, \delta = 1$ for $\rho \leq \rho'$ and $\delta = 0$ for $\rho > \rho'$, respectively. The temporal responses of the gauge in the numerical simulation and the experiment were quite similar.

References

1. Zel'dovich, Ya. B., Raizer, Yu. P.: Physics of Shock Waves and High-Temperature Hydrodynamic Phenomena. Academic, New York (1966–1967)
2. Orlenko, L.P. (ed.): Fizika vzryva. Sbornik (Physics of Explosion: A Collection of Papers), Fizmatlit. Moscow. 2 (2002)
3. Trunin, R.F., Krupnikov, K.K., Simakov, G.V., Funtikov, A.I.: High Pressure Shock Compression of Solids VII. Shock Waves and Extreme States of Matter, Springer, Berlin Heidelberg New York, Ch. 5. (2004)
4. Trunin, R.F., Simakov, G.V., Sutulov, Yu.N., et al.: In: Trunin, R.F. (ed.) Svoistva kondensirovannykh veshchestv pri vysokikh davleniyakh i temperaturakh. Sbornik statei (The Properties of Condensed Substances at High Pressures and Temperatures: A Collection of Papers). pp. 123 Sarov: VNIIEF (All-Russia Research Inst. of Experimental Physics) (1992)
5. Trunin, R.F., Medvedev, A.B., Funtikov, A.I., et al.: In: Trunin, R.F. (ed.) Svoistva kondensirovannykh veshchestv pri vysokikh davleniyakh i temperaturakh. Sbornik statei (The Properties of Condensed Substances at High Pressures and Temperatures: A Collection of Papers). pp. 143 Sarov: VNIIEF (All-Russia Research Inst. of Experimental Physics) (1992)
6. Kanel', G.I., Razorenov, S.V., Utkin, A.V., Fortov, V.E.: Udarno-volnovye yavleniya v kondensirovannykh sredakh (Shock-Wave Phenomena in Condensed Media). Yanus-K, Moscow (1996)
7. Torre, C.: Theorie und zusammengeprebteer Pulver. Berg. Hüttenmänn. Monatsh. **93**, 62 (1948)
8. Carrol, M.M., Holt, A.C.: J. Appl. Phys. **43**, 759 (1972)
9. Melosh, H.J.: Impact Cratering. A Geologic Process. Oxford, New York (1989)
10. Holt, A.C., Carrol, M.M., Butcher, B.M.: Application of a new theory for pressure-induced collapse for pores in ductile materials. In: Modry, S. (ed.) Pore Structure and Properties of Materials. Akademia, Prague 5, D63 (1974)
11. Butcher, B.M., Carrol, M.M., Holt, A.C.: J. Appl. Phys. 3864 (1974)
12. Dunin, S.Z., Surkov, V.V.: Prikl. Mekh. Tekh. Fiz. **5**, 106 (1979)
13. Kim, K., Sohn, C.H.: Modelling of Reaction Build-up Process in Shocked Porous Explosives. In: *Proceedings of the 8th International Symposium on Detonation.* **2**, 641 (1984)
14. Nesterenko, V.F.: Impul'snoe nagruzhenie geterogennykh materialov. Nauka, Novosibirsk (1992)
15. Staver, A.M.: Metallurgical effects under shock compression of powder material. In: Mayer, M.A., Murr, L.E. (eds.) Shock waves and high-strain-rate phenomena in metals concepts and applications. pp. 861 Plenum Press, New York (1961)
16. Kusubov, A.S., Nesterenko, V.F., Wilkins, M.L., et al.: Dynamic deformation of powdered materials as a function of particle size, Proceedings of International Seminar on High Energy Working of Rapidly Solidified and High-Temperature Superconducting Materials, p. 139, Novosibirsk (1989)
17. Benson, D.J., Nesterenko, V.F., Jonsdottir, F., Meyers, M.A.: J. Mech. Phys. Solids, **45** (11/12) 1955 (1997)
18. Blackburn, J.H., Seely, L.B.: Nature, **194**, 370 (1962)
19. Nesterenko, V.F., Lazaridi, A.N.: Regimes of shock wave compaction of granular materials. In: *Proceedings of the 12th AIRAPT and 27th EMPRG International Conference on High Pressure Science and Technology.* pp. 835 (1990)

20. Schwarz, R.B., Kasiraj, P., Vreeland, T.: Temperature kinetics during shock wave consolidation of metallic powders. In: Mayer, M.A., Staudhammer, K.S., Murr, L.E. (eds.) Shock Waves and High-Strain-Rate Phenomena in Metals. pp. 313 Marcel Dekker, New York (1986)
21. Belyakov, G.V., Rodionov, V.N., Samosadnyi, V.P.: Fiz. Goreniya Vzryva. **13**(4) 614 (1977)
22. Reybould, D.: Int. J. Powder Metall. Powder Technol. **16**(1) 1 (1980)
23. Dunin, S.Z., Surkov, V.V.: Prikl. Mekh. Tekh. Fiz. **1**, 131 (1982)
24. Prummer, R.: The Explosive Compression of Powdery Substances: Basic Principles, Procedures and Results, Werkstoff-Forschung und –Technik, Band 7, Springer, Berlin Heidelberg New York, ISBN 3-540-17029-4 (1987)
25. Gelfand, B.E., Silnikov, M.A., Mikhailin, A.I., Orlov, A.V.: The attenuation of blast overpressure from liquid in an elastic confinement. In: Proceedings of the 23h, ISSW, pp. 124 (2001)
26. Golub, V.V., Lu, S.A., Medin, S.A., Mirova, O.A., Parshikov, A.N., Petukhov, V.A., Volodin, V.V.: Blast wave attenuation by lightly destructible granular materials. In: Procceedings of the 24t, ISSW, ISBN 7-89494-429-7 (2004)
27. Parshikov, A.N., Medin, S.A.: J. Comput. Phys. **180**, 358 (2002)
28. Schatz, J.F.: Models of inelastic volume deformation for porous geologic materials. In: Cowin, S.C., Carrol, M.M. (eds.) The Effect of Voids on Material Deformations. New York pp. **141** (1976)

Index

Shock Wave Science and Technology Reference Library, Volume 1
Multiphase Flows I

About the Authors

Chapter 1

Leen van Wijngaarden

University of Twente
Physics of Fluids
Enschede
The Netherlands
l.vanwijngaarden
@tnw.utwente.nl

Leen van Wijngaarden is an Emeritus Professor at the University of Twente, the Netherlands. He has a PhD degree from the Technical University of Delft. He was at the Maritime Research Institute of the Netherlands from 1962–1966, where he did research in gravity waves and cavitation. From 1966–1997 he taught fluid dynamics in Twente, both in the Dept. of Mechanical Engineering and in the Dept. of Applied Physics. His, still continuing, research interests are multiphase flow, acoustics and waves of various kinds. He is a member of the Dutch Academy of Arts and Sciences. He was Associate Editor of the Journal of Fluid Mechanics from 1988–2000. He served as Treasurer, President and Vice-President of IUTAM (International Union of Theoretical and Applied Mechanics), and was involved with the Euromech Society in many ways.

Chapter 2

Yukio Tomita

Hokkaido University of Education
Faculty of Education
Hakodate
Japan
tomita@cc.hokkyodai.ac.jp

Dr. Yukio Tomita is a Professor of the Faculty of Education at the Hokkaido University of Education, Hakodate. He obtained his PhD degree in Engineering from Tohoku University in 1986. He was a Royal Society-Japan Society for Promotion of Science Visiting Research Fellow at the University of Birmingham in 1996/1997 and a Visiting Research Fellow at Imperial College London in 2003 through a Short-Term Research Experience Fellowship (MEXT). His research interests include cavitation, especially bubble dynamics, shock wave, jet and drop impact problems in connection with material erosion, medical application and environmental science.

Shock Wave Science and Technology Reference Library, Volume 1
Multiphase Flows I

Chapter 3

Valery K. Kedrinskii

Lavrentyev Institute of
Hydrodynamics
of the Russian Academy
of Science
Novosibirsk
Russian Federation
kedr@hydro.nsc.ru

Professor Valery Kedrinskii is head of the Department of Physical Hydrodynamics and Vice-Director of the Lavrentyev Institute of Hydrodynmaics and lecturer for four-semesters general courses of physics at Novosibirsk State University. He obtained his PhD degree in 1968 from the Joint Scientific Board at the Presidium of the Siberian Branch of the USSR Academy of Sciences, Novosibirsk, and in 1978 his Doctor of Sciences for Physics and Mathematics. He is a State Prize Winner of the USSR (Gold Medal, 1983) for fundamental contributions to multi-phase media mechanics.

Chapter 4

Masahide Murakami

University of Tsukuba
Graduate School of
Systems and Information
Engineering
Tsukuba, Japan
murakami@kz.tsukuba.ac.jp

Dr. Murakami is a Professor at the Graduate School of Systems and Information Engineering at the University of Tsukuba. He obtained his PhD degree in Aeronautics from the University of Tokyo in 1974. He has been working in the fields of superfluid thermo-fluid dynamics, space cryogenics and sports engineering, and received the Russell B. Scott Memorial Award (Best research paper at Cryogenic Engineering Conference) in 2003. He is an international advisory editor of the journal Cryogenics.

Shock Wave Science and Technology Reference Library, Volume 1 Multiphase Flows I

Chapter 5

Abhijit Guha

Aerospace Engineering
Department
University of Bristol
Bristol BS8 1TR
United Kingdom
a.guha@bristol.ac.uk

Dr. Abhijit Guha's research interests are in the thermo-fluid-dynamics of multiphase flow, computational fluid dynamics, gas turbine, and solar energy. He obtained his PhD in Engineering from Trinity College, University of Cambridge, as the prestigious Prince of Wales Scholar. He later became a Senior Rouse Ball Scholar at Trinity College, and then a Fellow of Gonville & Caius College, Cambridge. While at Cambridge, his research was based at the Whittle Laboratory. In 1995 he joined the University of Bristol. He has published many, key and comprehensive, mostly single-authored, fundamental as well as applied research papers in top-ranking journals, books and conferences on a wide range of interdisciplinary topics. He has presented many keynote lectures, short courses and invited seminars, at international conferences and reputed institutions worldwide. In 1995 he delivered the renowned VKI Lecture Series (von Karman Institute, Belgium) on Two-phase Flows with Phase Transition. In 2000 he was elected to the Editorial Board of Journal of Aerosol Science. Dr. Guha taught Two-phase Heat Transfer at University of Cambridge, and now teaches Fluid Mechanics, Thermodynamics, and Aircraft Propulsion at Bristol. He received the University of Bristol Teaching Excellence Award in the very year of its inception.

Chapter 6

Can F. Delale

Faculty of Aeronautics
and Astronautics
Istanbul Technical
University
34469 Maslak, Istanbul,
Turkey
delale@itu.edu.tr

Dr. Can F. Delale is Professor of Applied Mathematics and Aerospace Engineering at Istanbul Technical University. He received his PhD from Brown University in 1983, majoring in Fluid Mechanics and Thermodynamics. His research interests include the kinetic theory of gases, gas/liquid transitions, gas dynamics with condensation, wave phenomena and hydrodynamic cavitation.

Marinus E.H. van Dongen

Applied Physics
Department
Eindhoven University of
Technology
PO Box 513
5600 MB Eindhoven,
Netherlands
m.e.h.v.Dongen@tue.nl

Professor Marinus (Rini) van Dongen is a physicist who has been active in research and education in fluid dynamics, physical gas dynamics, physical transport phenomena, waves in porous media, waves with phase transition, nucleation and condensation in real gases and in bio-fluid dynamics. He is a member of the J.M. Burgerscentrum, Research School for Fluid Mechanics. He has been affiliated with Eindhoven University of Technology and part-time with Twente University, Department of Mechanical Engineering.

Günter H. Schnerr

Chair of Fluid Mechanics
Department for
Gasdynamics
Technical University
Munich
D-85747 Garching,
Germany
schnerr@flm.mw.tu-
muenchen.de

Dr.-Ing.habil. Günter H. Schnerr is Professor for Fluid Mechanics and Gasdynamics at the Technical University of Munich. He received his Dr.-Ing. and Dr.-Ing.habil. degrees from the University of Karlsruhe (TH) in 1977 and 1986, with major scopes in Fluid Mechanics and Nonequilibrium Gasdynamics. From 2000–2003 has been affiliated with the J.M. Burgers Centrum for Fluid Dynamics at Delft and as part time Professor with the University of Twente, The Netherlands. His research interests include theory, computation and experiments on transonic flows, gasdynamics with phase transition – condensation and compressible liquid flows with cavitation in macro and microscale systems, especially in fuel injection systems, with more than 170 publications dedicated to these subjects.

Chapter 7

Gerd E.A. Meier

Am Menzelberg 6
D-37077 Göttingen,
Germany
geameier@web.de

Professor Meier was the director of the Institute of Aerodynamics and Flow Technology in Göttingen, Germany from 1990 to 2002. This institute of the DLR – the German Aerospace Organisation – is devoted to experimental and theoretical flow studies for aerospace applications. From 1964 until 1990 he headed the Transonic Aerodynamics group of the Max Planck Institute for Fluid Mechanics in Göttingen. Both institutes were founded by Ludwig Prandtl. During his career he has authored more than 400 publications, reports, patents etc. on unsteady and steady transonic flows, multiphase flows with phase transition, supersonic and liquid jets, vortex airfoil interaction, turbulence and experimental methods, especially in flow measurement. He has edited five books on fluid dynamics topics, organized more than 10 international conferences in the field fluid mechanics, and is co-editor of two scientific journals (Experiments in Fluids, Progress in Aerospace Sciences). He was previously a member of the congress committees of IUTAM and EUROMECH, founder of the DLR–SCHOOL–LAB and President of the Ludwig–Prandtl–Gesellschaft.

Chapter 8

Beric W. Skews

Flow Research Unit
University of the
Witwatersrand
PO WITS, 2050
Johannesburg
South Africa
Tel: +27 11 7177324
beric.skews@wits.ac.za

Professor Beric Skews is Director of the Flow Research Unit at the University of the Witwatersrand, and has held visiting appointments in Canada, Japan, and Australia. He is an active researcher in the field of unsteady and compressible fluid flow, with particular emphasis on shock waves and flow visualisation. He serves on a number of international bodies in the field, and is an editor of the journal Shock Waves.

Shock Wave Science and Technology Reference Library, Volume 1
Multiphase Flows I

Chapter 9

Marinus E.H. van Dongen

Applied Physics
Department
Eindhoven University of
Technology
PO Box 513
5600 MB Eindhoven,
Netherlands
m.e.h.v.Dongen@tue.nl

Professor Marinus (Rini) van Dongen is a physicist who has been active in research and education in fluid dynamics, physical gas dynamics, physical transport phenomena, waves in porous media, waves with phase transition, nucleation and condensation in real gases and in bio-fluid dynamics. He is a member of the J.M. Burgerscentrum, Research School for Fluid Mechanics. He has been affiliated with Eindhoven University of Technology and part-time with Twente University, Department of Mechanical Engineering.

David Smeulders

GeoTechnology
Department
Delft University of
Technology
PO Box 5028
2600GA Delft,
Netherlands
d.m.j.smeulders@tudelft.nl

David Smeulders holds an MSc in Aeronautics from Delft University, Netherlands and a PhD in Physics from Eindhoven University, Netherlands. He has been working in the field of acoustics and porous media since 1988, currently as an associate professor of Petrophysics at the Delft University, Netherlands.

Chapter 10

Victor Golub

Institute for High Energy
Densities
Russian Academy of
Sciences
golub@ihed.ras.ru

Dr. Victor Golub is a head of the Physical Gasdynamics Department at the Institute for High Energy Densities of the Russian Academy of Sciences. He is working in shock and detonation waves investigations in gas and granular media, flow vizualisation and blast wave attenuation. He has been corresponding member of ERCOFTAC (European Community on Flow Turbulence and Combustion) since 1993, member of the Scientific Council on Combustion and Explosion of the Presidium of the Russian Academy of Sciences since 1999, member of the International Advisory Committee of International Symposium on Shock Waves since 2002, and member of FLUCOME since 2004.

Olga Mirova

Institute for High Energy
Densities
Russian Academy of
Sciences
olga.mirova@ras.ru

Mrs. Olga A. Mirova, is a researcher of Physical Gasdynamics Department of the Institute for High Energy Densities, Russian Academy of Science. She attended the Moscow Institute of Physics and Technology, where she obtained her Master of Science in 1996. She is working in shock wave investigation in gas and granular media and blast wave attenuation.

Printing: Krips bv, Meppel
Binding: Stürtz, Würzburg

RETURN TO: PHYSICS-ASTRONOMY LIBRARY
351 LeConte Hall   510-642-3122

LOAN PERIOD 1	2	3
1-MONTH		
4	5	6

ALL BOOKS MAY BE RECALLED AFTER 7 DAYS.
Renewable by telephone.

DUE AS STAMPED BELOW.

NOV 1 8 2007		

FORM NO. DD 22
2M 7-07

UNIVERSITY OF CALIFORNIA, BERKELEY
Berkeley, California 94720–6000